LONDON MATHEMATICAL SOCIETY LECTURE NOTE SERIES

Managing Editor: Professor N.J. Hitchin, Mathematical Institute,
University of Oxford, 24–29 St Giles, Oxford OX1 3LB, United Kingdom

The titles below are available from booksellers, or from Cambridge University Press at www.cambridge.org/mathematics.

London Mathematical Society Lecture Note Series. 346

Surveys in Combinatorics 2007

Edited by

Anthony Hilton
University of Reading
and
Queen Mary, University of London

John Talbot
University College London

CAMBRIDGE
UNIVERSITY PRESS

CAMBRIDGE UNIVERSITY PRESS
Cambridge, New York, Melbourne, Madrid, Cape Town, Singapore, São Paulo

Cambridge University Press
The Edinburgh Building, Cambridge CB2 8RU, UK

Published in the United States of America by Cambridge University Press, New York

www.cambridge.org
Information on this title: www.cambridge.org/9780521698238

First published 2007

Printed in the United Kingdom at the University Press, Cambridge

A catalogue record for this book is available from the British Library

ISBN   978-0-521-69823-8 paperback

Contents

Preface

The Twenty-first Biennial British Combinatorial Conference was held at Reading in July 2007. The British Combinatorial Committee had invited ten distinguished combinatorial mathematicians to give survey lectures in areas of their expertise, and this volume contains the survey articles on which these lectures were based.

In compiling this volume we are indebted to the authors for preparing their articles so accurately and in such a timely manner and to the referees for their attention to detail while commenting on the articles. We would also like to thank Roger Astley at Cambridge University Press for his advice and help.

The British Combinatorial Committee gratefully acknowledges the financial support provided by the London Mathematical Society, the Institue for Combinatorics and its Applications, and the EPSRC.

Anthony Hilton
Queen Mary, University of London
a.j.w.hilton@reading.ac.uk

John Talbot
University College London
talbot@math.ucl.ac.uk

January 2007

Hereditary and Monotone Properties
of Combinatorial Structures

Béla Bollobás[1]

Abstract

A *hereditary* property of graphs is a collection of (isomorphism classes of)
graphs which is closed under taking induced graphs, and contains arbitrarily
large structures. Given a family $\mathcal{F}$ of graphs, the family $\mathcal{P}(\mathcal{F})$ of graphs containing no member of $\mathcal{F}$ as an induced subgraph is a hereditary property, and
every hereditary property of graphs arises in this way. A hereditary property of
other combinatorial structures is defined analogously. A property is *monotone*
if it is closed under taking (not necessarily induced) substructures.

Given a property $\mathcal{P}$, we write $\mathcal{P}^n$ for the number of distinct structures with
vertices labelled $1, \ldots, n$, and call the function $n \mapsto |\mathcal{P}^n|$ the *labelled speed* of
$\mathcal{P}$. Similarly, the unlabelled speed is $n \mapsto |\mathcal{P}_n|$, where $\mathcal{P}_n$ is the set of distinct
structures with n unlabelled vertices. The study of hereditary properties is on
the borderline of extremal, enumerative, and probabilistic combinatorics. Thus,
for a family $\mathcal{F}$ of graphs, the problem of determining the speed of $\mathcal{P}(\mathcal{F})$ is a
natural extension of the basic question in extremal graph theory concerned with
the maximal number of edges in a graph of order n containing no member of $\mathcal{F}$
as a subgraph.

For many a combinatorial structure (graphs, posets, partitions, words, etc.),
there is a surprising phase transition: the speed *jumps* from one range to a much
higher one. Thus the speed of a property is either not much larger than a certain
function $f(n)$ or is at least as large as a function F which is *much larger* than
f. Although the jumps may look fairly similar for a variety of combinatorial
structures, much of the time their proofs need new ideas, and give deep insights
into the structures.

In the past few decades, much research has been done on hereditary and
monotone properties of a number of combinatorial structures: the aim of this
paper is to review some of these results, with special emphasis on the most
recent results.

1 Introduction

The roots of the theory we are about to discuss go back to the basic problem of
extremal graph theory: given a graph F, what is $\mathrm{ex}(n; F)$, the maximal number of
edges in a graph of order n not containing F as a subgraph? The traditional starting
point of extremal graph theory, Turán's theorem [118], proved in 1941, answers this
question when F is a complete graph. A few years later, Erdős and Stone [69] proved
that slightly more edges than are needed to guarantee a complete $(r + 1)$-graph as
a subgraph also guarantee a complete $(r + 1)$-partite graph with s vertices in each
class, provided the number of vertices is greater than a certain function of r and s.
(For considerable extensions of this theorem, see [38], [40], [41], [57], [80].)

With this result, Erdős and Stone had given very good bounds on $\mathrm{ex}(n; F)$ for
every F *decades before* the function $\mathrm{ex}(n; F)$ was even defined and the problem of

[1]Research supported in part by NSF grants CCR-0225610, DMS-0505550 and W911NF-06-1-
0076.

determining it was posed. Indeed, the Erdős–Stone theorem is often viewed as the fundamental theorem of *extremal graph theory* (see [36]).

In the 1960s, Erdős and Simonovits [61, 62, 67, 68, 114] launched the study of the more general function $ex(n; \mathcal{F})$, the maximal number of edges in a graph of order n without a subgraph belonging to a certain family $\mathcal{F}$ of graphs. Putting it slightly differently, let $\mathcal{P}_{\mathcal{F}}$ be the class ('property') of graphs containing no element of $\mathcal{F}$ as a subgraph. (For a good reason, in §2 we shall use a different notation.) Thus, for $\mathcal{F} = \{K_{r+1}\}$, $\mathcal{P}_{\mathcal{F}}$ is the class of K_{r+1}-free graphs, i.e., the *property* of not containing a complete graph of order $r + 1$. Then $ex(n; \mathcal{F})$ is the maximal size of a graph of order n belonging to $\mathcal{P}_{\mathcal{F}}$.

Erdős, Kleitman and Rothschild [64] studied a very different problem concerning $\mathcal{P}_{\mathcal{F}}$ in the case when $\mathcal{F}$ consists of a single complete graph, K_{r+1}, namely the number of graphs on $[n] = \{1, \ldots, n\}$ that belong to $\mathcal{P}_{\mathcal{F}}$, with $\mathcal{F} = \{K_{r+1}\}$. Ten years or so later, this result was extended in two directions. First, Erdős, Frankl and Rödl [63] extended this 'asymptotic enumeration' result to that of $\mathcal{P}_{\mathcal{F}}$ for $\mathcal{F} = \{F\}$, where F is any fixed graph. Second, the 'asymptotic structure' of K_{r+1}-free graphs was determined by Kolaitis, Prömel and Rothschild [90].

In addition, in a series of papers, Prömel and Steger [102, 103, 104, 105, 106] followed up a suggestion of Erdős, and investigated the number and asymptotic structure of graphs that contain no *induced* quadrilaterals and other *induced* subgraphs. Not surprisingly, the problems involving induced subgraphs turned out to be much harder than those concerning subgraphs. The reason is that the property of not containing a certain graph is *monotone*: if G has this property $\mathcal{P}$ then *every* subgraph of G has $\mathcal{P}$. On the other hand, the property $\mathcal{P}^*$ of not containing a certain graph as an *induced* subgraph is only *hereditary*: if G has $\mathcal{P}^*$ then every *induced* subgraph of G has $\mathcal{P}^*$, but a non-induced subgraph need not have this property. We shall see in §2 and the rest of the paper that this distinction is rather important.

In 1994, Scheinerman and Zito [112] gave a new direction to the study of (general) hereditary properties, when they showed that the growth of the number of graphs as the function of the order of the graphs with a hereditary property is very restricted: no matter what property we take, only a handful of ranges of growth are possible. (Later, we shall state this more precisely.) Thus the innovation of Scheinerman and Zito was that rather than attempting the asymptotic enumeration of graphs not containing certain induced subgraphs (like quadrilaterals), we try to get information about the growth rate of *every* hereditary property, no matter how it is defined.

Since the appearance of this seminal paper of Scheinerman and Zito, much research has been done on the 'growth' and structure of hereditary and monotone properties. In this paper, we shall give a brief account of these developments, and shall sketch some related results concerning other combinatorial structures. It is regrettable that, for lack of space, we can give only a fraction of the results we should like to give. Also, we do not always give the results in chronological order, and fail to give as much emphasis to a number of contributions as they deserve.

The rest of the paper is organized as follows. First, we define general monotone and hereditary properties of graphs, and then we introduce ways of measuring their sizes. In §4 we prove an isoperimetric inequality concerning projections of bodies, and apply it to fast-growing hereditary properties of graphs and hypergraphs. In the next two sections, forming the heart of this paper, we discuss the possible ranges of growth of hereditary properties of graphs: the highest range in §5, and the other ranges in §6. The next section is devoted to monotone properties: we shall give some results that have been proved only for monotone properties, although they may well be true for hereditary ones as well.

In §8 we discuss the unlabelled speed of a hereditary property, and in §9 we describe some of the delicate structural results that enable us to obtain good bounds on the $\mathcal{P}$-chromatic number of a random graph for a hereditary property $\mathcal{P}$. The natural question as to how far a graph can be from a hereditary property is examined in §10. In §11 we turn to other combinatorial structures: we shall touch on properties of posets, permutations, ordered hypergraphs and partitions. Finally, §12 is about properties of words.

I should like to emphasize that the choice of topics and results has been strongly influenced by my own preferences: similar papers could be written emphasizing a rather different set of topics. Needless to say, this short review cannot do justice to the wealth of results in the active area of hereditary properties of combinatorial structures.

2 Hereditary and Monotone Properties

We shall study properties of various combinatorial structures, with emphasis on (finite) graphs. Much of the time, it is easy to extend the notions from graphs to other structures, *mutatis mutandis*. Our terminology and notation are standard, see, e.g., [37]. Thus, $V(G)$ denotes the vertex set of a graph G and $E(G)$ its edge set; the number of vertices, $|G| = |V(G)|$, is the *order* of G, and the number of edges, $e(G) = |E(G)|$, is its *size*. For a graph G and vertex $x \in V(G)$, we write $G - x$ for the graph obtained from G by deleting x and the edges incident with it. Also, C_k denotes a k-cycle, and K_r is a complete graph of order r.

A *property of graphs* is an infinite collection $\mathcal{P}$ of graphs closed under isomorphism. Without any restriction, a property is too general to be of much interest, but even mild restrictions lead to interesting and difficult problems. Our main aim in this paper is to study hereditary and monotone properties.

A property $\mathcal{P}$ of graphs is *hereditary* if it is closed under taking *induced* subgraphs. Equivalently, $\mathcal{P}$ is hereditary if whenever $G \in \mathcal{P}$ and $x \in V(G)$, the graph $G - x$ also belongs to $\mathcal{P}$. Also, a property $\mathcal{P}$ is *monotone decreasing* or, simply, *monotone*, if it is closed under taking subgraphs: if $G \in \mathcal{P}$ and $H \subset G$, i.e., H is a (not necessarily induced) subgraph of G, then $H \in \mathcal{P}$. The complement $\overline{\mathcal{P}}$ of a monotone decreasing property $\mathcal{P}$ is a *monotone increasing* property: if $G \in \overline{\mathcal{P}}$ and $G \subset H$ then $H \in \overline{\mathcal{P}}$. (The complement $\overline{\mathcal{P}}$ is defined as usual: every graph belongs to either $\mathcal{P}$ or $\overline{\mathcal{P}}$, but not both.)

Clearly, a monotone (decreasing) property is hereditary, but the converse does not hold. For example, the class of planar graphs is monotone, as is the class of triangle-free graphs, but the class of perfect graphs is hereditary but not monotone.

A useful class of hereditary properties is obtained by taking an infinite graph G and letting $\mathcal{P}(G)$ be the collection of finite induced subgraphs of G. Clearly, $\mathcal{P}(G)$ is a hereditary property. For example, for the infinite complete graph K_∞ and the infinite star $K_{1,\infty}$ we have $\mathcal{P}(K_\infty) = \{K_n : n \in \mathbb{N}\}$ and $\mathcal{P}(K_{1,\infty}) = \{H : H$ is a star or an empty graph$\}$. Similarly, we may take $\mathcal{P}(G_1, G_2, \dots) = \bigcup_i \mathcal{P}(G_i)$

for any finite or infinite set of graphs $\{G_i\}$.

We shall always assume that our properties are non-trivial: not only are there infinitely many graphs with the property, but there are also infinitely many graphs without the property. Occasionally we shall emphasize that we make this assumption.

Every property $\mathcal{P}$ is defined by its complement; however, hereditary and monotone properties are also characterized by much smaller classes. Thus, given a set $\mathcal{F}$ of (finite) graphs, let $\mathcal{P} = \mathrm{Her}(\mathcal{F})$ be the class of graphs containing *no* member of $\mathcal{F}$ as an induced subgraph. We call $\mathcal{F}$ the set of forbidden induced subgraphs for $\mathcal{P}$. Clearly, $\mathcal{P}$ is a hereditary property, and $\mathcal{F} \subset \overline{\mathcal{P}}$. By definition, $\mathcal{P} = \mathrm{Her}(\overline{\mathcal{P}})$ for every hereditary property $\mathcal{P}$, but to study $\mathcal{P}$ it is better to choose a small set $\mathcal{F} \subset \overline{\mathcal{P}}$ with $\mathcal{P} = \mathrm{Her}(\mathcal{F})$. The smallest such set $\mathcal{F}$ consists of the minimal elements of $\overline{\mathcal{P}}$, i.e., of the graphs $F \in \overline{\mathcal{P}}$ whose every proper induced subgraph belongs to $\mathcal{P}$. For example, $\mathcal{F}_0 = \{C_4, C_6, C_8, \dots\}$ may be such a set: if $\mathrm{Her}(\mathcal{F}) = \mathrm{Her}(\mathcal{F}_0)$ then $\mathcal{F} \supset \mathcal{F}_0$.

Similarly, given a family $\mathcal{F}$ of graphs, let $\mathcal{P} = \mathrm{Mon}(\mathcal{F})$ be the class of graphs containing no member of $\mathcal{F}$ as a (not necessarily induced) subgraph; clearly, $\mathcal{P}$ is a monotone property. As before, $\mathcal{F}$ is the set of *forbidden* graphs defining the monotone property $\mathcal{P}$; the minimal elements of $\overline{\mathcal{P}}$ form the unique minimal set of forbidden graphs for $\mathcal{P}$. A (monotone or hereditary) property defined by a single forbidden graph is said to be *principal*; not surprisingly, the first properties to be studied were principal.

By definition, $\mathrm{Mon}(\mathcal{F}) \subset \mathrm{Her}(\mathcal{F})$ for every family $\mathcal{F}$. For example, if $\mathcal{F} = \{K_4\}$ then $\mathrm{Mon}(\mathcal{F}) = \mathrm{Her}(\mathcal{F})$ is the class of graphs containing no complete graph on four vertices, i.e., the set of K_4-free graphs, but if $\mathcal{F} = \{C_6\}$ then $\mathrm{Mon}(\mathcal{F})$ is the class of C_6-free graphs, while $\mathrm{Her}(\mathcal{F})$ contains, e.g., all connected graphs whose blocks are complete graphs. To describe the principal monotone property $\mathrm{Mon}(C_6)$ as a hereditary property, i.e., by forbidding induced subgraphs, we have to take the family $\mathcal{F}'$ of all (non-isomorphic) Hamiltonian graphs of order 6: for this family $\mathcal{F}'$ we do have $\mathrm{Mon}(C_6) = \mathrm{Her}(\mathcal{F}')$. (Strictly speaking, we should have written $\mathrm{Mon}(\{C_6\})$ for the property, but to reduce the clutter, here and elsewhere we omit the braces.) In fact, a principal monotone property is also a principal hereditary property if and only if a single complete graph is forbidden.

The notions of hereditary and monotone graphs have natural extensions to other combinatorial structures; here we shall note only some of these. First of all, everything above carries over verbatim to multigraphs, directed graphs, oriented graphs, hypergraphs, directed hypergraphs, and so on. A less trivial extension is obtained if, rather than considering the class of all graphs on finite sets of vertices, we consider subgraphs of certain 'ground graphs'. Thus, writing Q^n for the graph of the n-dimensional cube on $[2]^n$ with 2^n vertices and $n2^{n-1}$ edges (with each vertex having degree n), we may consider subgraphs of these cubes. In fact, we shall consider only spanning subgraphs, subgraphs whose vertex set is the entire set $V(Q^n) = [2]^n$. The definition of a monotone property is as before; however, some care is needed to define a hereditary property. Let $\mathcal{P}$ be a set of subgraphs of the cubes Q^0, Q^1, Q^2, $\dots$ with each $G \in \mathcal{P}$ having vertex set $V(Q^n)$ for some n. Then $\mathcal{P}$ (as a set of subgraphs of the cubes) is *hereditary* if whenever $G \subset Q^n$ (with $V(G) = V(Q^n)$) belongs to $\mathcal{P}$, every face (subcube) of Q^n induces a graph that also belongs to $\mathcal{P}$. (Needless to say, it suffices to demand that the restriction of G to each of the $2n$

one-codimensional faces belongs to $\mathcal{P}$.) For example, the set of spanning subgraphs of cubes containing no 3-dimensional subcube minus an edge is a hereditary (but not monotone) property.

Instead of the cubes Q^n, we may take the grids P_ℓ^n; here P_ℓ^n has vertex set $[\ell]^n$, and the edge set is as usual (so that $[\ell]$ is the path on $[\ell] = \{1, \ldots, \ell\}$). In defining a hereditary property, from a grid P_ℓ^n we pass on to any of its subgrids (isomorphic to P_ℓ^k for some k) or just to any of its one-codimensional subgrids.

Another natural example of a collection of graphs with restricted ground sets is the set of subgraphs of the symmetric groups S_n. The vertex set of S_n as a graph is the set S_n of $n!$ permutations of $[n]$, and two permutations, $\pi, \rho \in S_n$, are adjacent if one is obtained from the other by multiplying it (on the right, say) by a transposition. (Thus our graph is just the Cayley graph of the group S_n, with the transpositions as the generators.) Although S_n is used for three different objects (the symmetric group, the set of $n!$ elements of this group, and the graph with this vertex set), this multiple usage is unlikely to lead to any confusion. Note that the graph S_n has $n!$ vertices, and each vertex has degree $\binom{n}{2}$.

To define a hereditary property of subgraphs of the symmetric groups (with each $G \subset S_n$ spanning the vertex set of S_n), we represent a permutation π of $[n]$ as a sequence $x_1 x_2 \ldots x_n$ with $x_i = \pi(i)$. Given a position i and a value k, the set $S_{i,k;n}$ of sequences $x = x_1 x_2 \ldots x_n$ with $x_i = k$ is naturally identified with the set S_{n-1}; a sequence $x \in S_{i,k;n}$ is mapped into S_{n-1} as follows: we delete k (from position i), decrease by one each term that is greater than k, and decrease by one the position of each term after the ith position. (Turning this around, this gives an embedding of S_{n-1} into S_n.) For example, for $i = 4$ and $k = 3$ the sequence $526341 \in S_{4,3;6}$ is mapped into $42531 \in S_5$. In the definition of a hereditary property we demand that the restriction of a graph $G \subset S_n$ in $\mathcal{P}$ to such a set belongs to $\mathcal{P}$ for all choices of i and k.

Rather than subgraphs of the symmetric groups, we may consider *sets of permutations*, with a suitable (and natural definition) of a hereditary property. These properties have been studied extensively, and we shall review some of the results in this paper. We shall also consider posets, tournaments, oriented graphs, and ordered graphs; as we shall see, the problems and results concerning them are intimately connected.

For lack of space, we shall concentrate on properties of graphs: much of the time it will be clear how the notions to be defined can be carried over to other combinatorial structures.

3 Measures of Properties

There are several natural ways of measuring the size of a property of graphs. Each of these measures is a function $f(n)$ of the order n of the set of graphs in the property, and we are interested in the growth of this function as $n \to \infty$. For some pleasant properties we may even be able to determine the exact value of $f(n)$ – however, this is the exception rather than the rule.

Every graph property $\mathcal{P}$ is the disjoint union of its *levels*: $\mathcal{P} = \cup_{n=1}^{\infty} \mathcal{P}_n$, where $\mathcal{P}_n$ is the set of (isomorphism classes of) graphs in $\mathcal{P}$ with n vertices. We call the function $n \to |\mathcal{P}_n|$ the *unlabelled speed* of $\mathcal{P}$. Similarly, the *labelled speed* or, simply, the *speed* of $\mathcal{P}$ is the function $n \to |\mathcal{P}^n|$, where $\mathcal{P}^n$ is the set of graphs with vertex set $[n]$ that are in $\mathcal{P}$.

Clearly, no labelled speed can be more than $2^{\binom{n}{2}}$, the speed of the trivial property of all graphs. Similarly, the unlabelled speed of a property is at most that of the trivial property, which is about $2^{\binom{n}{2}}/n!$, since almost every graph has a trivial automorphism group.

As an example of a non-trivial property, if $\mathcal{P} = \mathcal{P}(K_{1,\infty})$, i.e., $\mathcal{P}$ consists of the empty graph and all the stars (i.e., each $G \in \mathcal{P}$ is either empty or a star) then

$$|\mathcal{P}^n| = n + 1 \quad \text{and} \quad |\mathcal{P}_n| = 2.$$

Also, let $\mathcal{R}$ be the set of graphs consisting of a (possibly trivial) star and isolated vertices. Then

$$|\mathcal{R}^n| = \sum_{i=3}^{n} i \binom{n}{i} + \binom{n}{2} + 1 \ \sim \ n \, 2^{n-1},$$

and

$$|\mathcal{R}_n| = n.$$

Given a property $\mathcal{P}$, if $G \in \mathcal{P}^n$ then $\mathcal{P}^n$ contains all graphs on $[n]$ that are isomorphic to G, so the labelled speed $|\mathcal{P}^n|$ is at least as large as the number of non-isomorphic labellings of G. Also, trivially,

$$|\mathcal{P}^n|/n! \leq |\mathcal{P}_n| \leq |\mathcal{P}^n|, \tag{3.1}$$

so for large speeds the logarithms of the labelled and unlabelled speeds are about the same.

For large speeds it is customary to take the *logarithmic density* c_n of the set of graphs of order n in a property $\mathcal{P}$: this is defined by

$$|\mathcal{P}^n| = 2^{c_n \binom{n}{2}}.$$

As we shall see in §4, Theorem 4.4, if $\mathcal{P}$ is hereditary then the sequence (c_n) is monotone decreasing and so tends to a limit $c \geq 0$, the *asymptotic logarithmic density* or *entropy* of the property $\mathcal{P}$, so that

$$|\mathcal{P}^n| = 2^{(c+o(1))\binom{n}{2}}.$$

Clearly, this formula is informative for $c > 0$ but too crude for $c = 0$, in which case we have to use finer measures to distinguish the speeds.

For a monotone property there is another natural measure, its size. The *size* $e(\mathcal{P}^n)$ *of a monotone property* $\mathcal{P}$ *at level* n is the maximum of the size of a graph in $\mathcal{P}^n$, and its *normalized size* d_n is the density of of this graph:

$$e(\mathcal{P}^n) = d_n \binom{n}{2} = \max\{|E(G)| : \ G \in \mathcal{P}^n\}.$$

The *size of* $\mathcal{P}$ is the sequence $(e(\mathcal{P}^n))$, and its *normalized size* is (d_n).

The size and the speed of a monotone property $\mathcal{P}$ are intimately related: if one is large, so is the other. First,

$$|\mathcal{P}^n| \geq 2^{e(\mathcal{P}^n)}, \tag{3.2}$$

since if $G \in \mathcal{P}^n$ then $\mathcal{P}^n$ contains all $2^{e(G)}$ subgraphs of G. This inequality states precisely that $c_n \geq d_n$. Second,

$$|\mathcal{P}^n| \leq \sum_{m=0}^{e(\mathcal{P}^n)} \binom{\binom{n}{2}}{m}, \tag{3.3}$$

since the right-hand side above is the total number of graphs on $[n]$ with at most $e(\mathcal{P}^n)$ edges.

The problem of determining the size of a monotone property predates by several decades the study of the speed (labelled or unlabelled) of a hereditary property: if $\mathcal{P} = \mathrm{Mon}(\mathcal{F})$ then $e_{\mathcal{P}}(n)$ is precisely the *extremal function* $\mathrm{ex}(n; \mathcal{F})$ for the family $\mathcal{F}$ of forbidden graphs, i.e., $e(\mathcal{P}^n) = \mathrm{ex}(n; \mathcal{F})$. Indeed, Mantel [96] showed over a hundred years ago that $\mathrm{ex}(n; K_3) = \lfloor n^2/4 \rfloor$, and the extension of this easy result from triangles to complete graphs, proved by Turán [118] in 1941, was the starting point of extremal graph theory:

$$\mathrm{ex}(n; K_{r+1}) = e(T_r(n)) = \left(1 - \frac{1}{r}\right)\binom{n}{2} + O(n). \tag{3.4}$$

Here $T_r(n)$ is the *r-partite Turán graph of order n*, the complete r-partite graph with n vertices and as equal classes as possible, i.e., the unique (up to isomorphism, as always) r-colourable graph of order n with maximal size. Thus, if $n = rk + s$, $0 \leq s < r$, then the size $t_r(n) = e(T_r(n))$ of the Turán graph $T_r(n)$ is as follows:

$$t_r(n) = \binom{n}{2} - s\binom{k+1}{2} - (r - s)\binom{k}{2} = \binom{n}{2} - r\binom{k}{2} - s.$$

In particular,

$$\left(1 - \frac{1}{r}\right)\binom{n}{2} < t_r(n) \leq \left(1 - \frac{1}{r}\right)\frac{n^2}{2}.$$

In 1946, Erdős and Stone [69] proved a considerable extension of Turán's theorem. Their result, which can be considered to be the Fundamental Theorem of extremal graph theory, states, roughly, that a graph of order n with slightly more than $t_r(n)$ edges contains not only a complete $(r + 1)$-graph with one vertex in each class (i.e., K_{r+1}), as guaranteed by Turán's theorem, but one with many vertices in each class.

Theorem 3.1 *Given $\varepsilon > 0$ and $r, s \in \mathbb{N}$, there is an $n_0 = n_0(\varepsilon, r, s)$ such that if $n \geq n_0$ then*

$$\left(1 - \frac{1}{r}\right)\binom{n}{2} < t_r(n) \leq \mathrm{ex}(n, K_{r+1}) < \left(1 - \frac{1}{r} + \varepsilon\right)\binom{n}{2},$$

where $K_{r+1}(s)$ is the complete $(r + 1)$-graph with s vertices in each class. □

In fact, the Erdős-Stone theorem has a formulation which *seems* to say more than the original form, Theorem 3.1. Note that, trivially, if the chromatic number $\chi(F)$ of a graph F is $r + 1$ then $F \not\subset T_r(n)$ for every n, and $F \subset K_{r+1}(s)$ if s is large enough, say, $s \geq |F|$. Hence, if $r + 1 = \min_{F \in \mathcal{F}} \chi(F)$ then

$$t_r(n) \leq \mathrm{ex}(n; \mathcal{F}) \leq \mathrm{ex}(n; K_r(s)),$$

provided s is large enough. Hence, Theorem 3.1 has the following reformulation, first noted by Erdős and Simonovits [67].

Theorem 3.1′. *Let $\mathcal{F}$ be a family of graphs with $r + 1 = \min_{F \in \mathcal{F}} \chi(F) \geq 2$. Then*

$$\left(1 - \frac{1}{r}\right)\binom{n}{2} \leq \mathrm{ex}(n; \mathcal{F}) = \left(1 - \frac{1}{r}\right)\binom{n}{2} + o(n^2).$$

Equivalently, if (d_n) is the normalized size of the monotone property $\mathcal{P} = \mathrm{Mon}(\mathcal{F})$ then

$$\lim_{n \to \infty} d_n = 1 - 1/r. \qquad \square$$

For $r = 1$ it is easy to improve the very weak bound in Theorem 3.1′. Indeed, if $\min_{F \in \mathcal{F}} \chi(F) = 2$ so that $F_0 \subset K_2(s)$ for some $F_0 \in \mathcal{F}$ and $s \geq 1$, then

$$\mathrm{ex}(n; \mathcal{F}) \leq \mathrm{ex}(n; F_0) \leq \mathrm{ex}(n; K_2(s)) = O(n^{2-1/s}).$$

For a hereditary property the above naive definition of the size is clearly inadequate since the *very* small property of being complete (and so having speed 1) would have maximal size, $\binom{n}{2}$. In order to define the size in such a way that inequality (3.2) holds for hereditary properties as well (in fact, for all properties), we need a little preparation. First, a *pregraph* $\widetilde{G}$ is a triple $(V; \widetilde{E}, \widetilde{N})$, where V is a finite set, the set of *vertices*, and $\widetilde{E}$ and $\widetilde{N}$ are disjoint subsets of $V^{(2)}$, the set of unordered pairs of vertices; $\widetilde{E}$ is the set of *edges* and $\widetilde{N}$ is the set of *non-edges* of $\widetilde{G}$. A graph $G = (V, E)$ is said to *extend* $\widetilde{G}$ if G contains every edge of $\widetilde{G}$, but no non-edge:

$$\widetilde{E} \subset E \subset V^{(2)} \setminus \widetilde{N}.$$

We say that $\widetilde{G}$ *belongs* to $\mathcal{P}^n$ if every graph extending $\widetilde{G}$ belongs to $\mathcal{P}^n$.

The *size* $e(\widetilde{G})$ of a pregraph $\widetilde{G}$ is $|V^{(2)} \setminus (\widetilde{E} \cup \widetilde{N})|$, so that there are $2^{e(\widetilde{G})}$ graphs extending it. Finally, the *size* $e(\mathcal{P}^n)$ of the nth level $\mathcal{P}^n$ of a property $\mathcal{P}$ and the *normalized size* d_n of $\mathcal{P}^n$ are given by

$$e(\mathcal{P}^n) = d_n\binom{n}{2} = \max\{e(\widetilde{G}): \ \widetilde{G} \in \mathcal{P}^n\}.$$

The *size* of a property $\mathcal{P}$ is the sequence $(e(\mathcal{P}^n))$, and its *normalized size* is (d_n).

A graph $G = (V, E)$ is naturally identified with the pregraph $\widetilde{G} = (V; \emptyset, V^{(2)} \setminus E)$, so that the extensions of $\widetilde{G}$ are precisely the subgraphs of G; with this identification we find that $e(G) = e(\widetilde{G})$. Hence, for a monotone property $\mathcal{P}$, the two definitions do give the same value for $e(\mathcal{P}^n)$. Furthermore, as 2^m graphs extend a pregraph of size m, (3.2) holds for *every* property $\mathcal{P}$; equivalently,

$$c_n \geq d_n$$

for every property $\mathcal{P}$. We shall see later that for a hereditary property the sequences (c_n) and (d_n) converge to the same limit. There is no obvious analogue of inequality (3.3) for hereditary properties (let alone general properties).

4   The Cover Inequality, the Box Theorem, and Large Hereditary Properties

The main aim of this section is to show that every hereditary property of hypergraphs has an asymptotic logarithmic density, i.e., the sequence (c_n) tends to a limit. (The existence of a limit is more or less content-free if our property is so small that, trivially, $c_n \to 0$, so in this section we shall be interested in 'large' properties.) As we shall see, this result is an easy consequence of an inequality concerning projections of sets in $\mathbb{R}^n$; we shall start with this inequality.

We shall call a compact convex subset of $\mathbb{R}^n$ which is the closure of its interior a *body* in $\mathbb{R}^n$. Let $(v_1, \ldots, v_n)$ be the standard basis of $\mathbb{R}^n$, so that $\mathbb{R}^n$ is the linear span of these vectors: $\mathbb{R}^n = \lin\{v_1, \ldots, v_n\}$. For a subset A of $[n] = \{1, \ldots, n\}$, write k_A for the orthogonal projection of a body K onto $\lin\{v_j : j \in A\}$, and denote by $|K_A|$ the $|A|$-*dimensional volume* of K_A. In particular, $|K| = |K_{[n]}|$ is the volume of K. With $\beta(K) = (|K_A| : A \subset [n]) = (|K_A|)_{A \subset [n]} \in \mathbb{R}^{\mathcal{P}(n)} = \mathbb{R}^{2^n}$, the map $K \to \beta(K)$ can be considered to measure the size of the boundary of K. (As usual, $\mathcal{P}(n)$ is the collection of all 2^n subsets of $[n]$.)

By a *cover* of $[n]$ we mean a multiset $\mathcal{C}$ of subsets of $[n]$ such that each element $i \in [n]$ is in at least one of the members of $\mathcal{C}$. A *k-cover* is a cover in which each element of $[n]$ is in exactly k of the members of $\mathcal{C}$. Note that the sets in a cover *need not be distinct*; for example, $\{\{1, 2\}, \{1, 2\}, \{3\}, \{3\}\}$ is a 2-cover of [3]. A *uniform cover* of $[n]$ is a k-cover for some $k \geq 1$. We call the 1-cover $\{[n]\}$ of [n] *trivial* and all other covers non-trivial.

A uniform cover of $[n]$ which is not the disjoint union of two uniform covers of $[n]$ is said to be *irreducible*. A simple compactness argument tells us that there are only a finite number of irreducible uniform covers of $[n]$. In fact, writing $D(n)$ for the number of irreducible covers of $[n]$, Huckeman, Jurkat and Shapley proved (see Graver [77]) that $D(n) \leq (n+1)^{(n+1)/2}$ for all n. (For related results see Alon and Berman [7] and Füredi [74].)

The following *Cover Inequality* and its proof are from Bollobás and Thomason [42]; the result also follows from Shearer's inequality (see [58]) concerning entropy. This inequality is a considerable extension of the classical *Loomis–Whitney* inequality [92] (see also [53, page 95] and [79, page 162]; the inequality was rediscovered by Allan [6], who gave a more streamlined proof), which claims it for the $(n-1)$-uniform cover $\{[n] \setminus \{i\}\}_{i=1}^n$ of $[n]$, or for the $\binom{n-1}{k-1}$-cover of $[n]$ by *all* k-subsets.

Theorem 4.1 *Let K be a body in $\mathbb{R}^n$, and let $\mathcal{C}$ be a k-cover of $[n]$. Then*

$$|K|^k \leq \prod_{A \in \mathcal{C}} |K_A|.$$

Proof We apply induction on n. As the case $n = 1$ is trivial, we turn to the induction step. For each $x \in \mathbb{R}$ let $K(x)$ be the section of K consisting of points with nth coordinate equal to x, so that $|K| = \int |K(x)|\, dx$. Let us split $\mathcal{C}$ as follows: $\mathcal{C}' = \{A \in \mathcal{C}; \ n \in A\}$, so that $|\mathcal{C}'| = k$, and $\mathcal{C}'' = \mathcal{C} \setminus \mathcal{C}'$. Then $\{A \setminus \{n\}; \ A \in \mathcal{C}'\} \cup \mathcal{C}''$ is a k-cover of $[n-1]$, so, by the induction hypothesis,

$$\prod_{A \in \mathcal{C}'} |K(x)_{A \setminus \{n\}}| \prod_{A \in \mathcal{C}''} |K(x)_A| \geq |K(x)|^k.$$

Moreover, we have

$$|K_A| = \int |K(x)_{A \setminus \{n\}}| \, dx \quad \text{for} \quad A \in \mathcal{C}'$$

and

$$|K_A| \geq |K(x)_A| \quad \text{for} \quad A \in \mathcal{C}''.$$

Consequently, by Hölder's inequality,

$$
\begin{aligned}
|K| &= \int |K(x)| \, dx \leq \int \left[\prod_{A \in \mathcal{C}'} |K(x)_{A \setminus \{n\}}| \prod_{A \in \mathcal{C}''} |K(x)_A| \right]^{1/k} dx \\
&\leq \left[\prod_{A \in \mathcal{C}''} |K_A| \right]^{1/k} \int \prod_{A \in \mathcal{C}'} |K(x)_{A \setminus \{n\}}|^{1/k} \, dx \\
&\leq \left[\prod_{A \in \mathcal{C}''} |K_A| \right]^{1/k} \prod_{A \in \mathcal{C}'} \left[\int |K(x)_{A \setminus \{n\}}| \, dx \right]^{1/k} \\
&= \left[\prod_{A \in \mathcal{C}} |K_A| \right]^{1/k},
\end{aligned}
$$

as claimed. □

Now, let us turn to the boundary function $\beta(K) = (|K_A|)_{A \subset [n]} \in \mathbb{R}^{2^n}$. What is the solution of the isoperimetric problem for this 'boundary': what body $K \subset \mathbb{R}^n$ of volume 1 should we choose to minimize *all* boundary volumes $|K_A|$? In other words, if K is a body of volume 1 such that no volume $|K_A|$ can be decreased without increasing some other boundary volume $|K_B|$, what is K? The answer is surprisingly simple: a *box*, a rectangular parallelepiped whose sides are parallel to the coordinate axes. This result, the *Box Theorem* of Bollobás and Thomason [42], should surely have been discovered by the 19th century geometers, but they missed it.

Theorem 4.2 *Let K be a body in $\mathbb{R}^n$. Then there is a box B in $\mathbb{R}^n$, with $|B| = |K|$ and $|B_A| \leq |K_A|$ for every $A \subset [n]$.*

Proof If $k \geq 1$ and $\mathcal{C}$ is an irreducible k-cover of $[n]$, by Theorem 4.1 we have

$$\prod_{A \in \mathcal{C}} |K_A| \geq |K|^k;$$

also, rather trivially,

$$\prod_{i \in S} |K_{\{i\}}| \geq |K_S|$$

for every $S \subset [n]$. Since there are only finitely many irreducible uniform covers, this gives us a finite set of inequalities involving the numbers $\{|K_A|; \ A \subset [n]\}$.

Let $\{x_A; \ A \subset [n]\}$ be a collection of positive numbers with $x_A \leq |K_A|$ and $x_{[n]} = |K|$, which are minimal subject to satisfying all the above inequalities with

x_A in place of $|K_A|$ for all A. Note that we are applying only a finite number of constraints to the x_A, but that nevertheless $\prod_{A\in\mathcal{C}} x_A \geq |K|^k$ for *every* k-cover $\mathcal{C}$, since every cover is a disjoint union of irreducible uniform covers.

As the numbers $x_{\{i\}}$, $i \in [n]$, are minimal, for each i there is an inequality involving $x_{\{i\}}$ in which equality holds. If the inequality is of the first kind, then we have a k_i-cover $\mathcal{C}_i$ of $[n]$ with $\{i\} \in \mathcal{C}_i$ and $\prod_{A\in\mathcal{C}_i} x_A = |K|^{k_i}$. The same is true if the inequality is of the second kind, namely $\prod_{i\in S} x_{\{i\}} = x_S$ for some $S \subset [n]$, because in this case the minimality of x_S implies $\prod_{A\in\mathcal{C}} x_A = |K|^{k_i}$ for some k_i-cover $\mathcal{C}$ of $[n]$ containing S, and we can take $\mathcal{C}_i = (\mathcal{C} \setminus \{S\}) \cup \{\{i\} : i \in S\}$.

Now, set $\mathcal{C} = \bigcup_{i=1}^n \mathcal{C}_i$ and $k = \sum_{i=1}^n k_i$. Then $\mathcal{C}$ is a k-cover of $[n]$ containing $\{i\}$ for all $i \in [n]$, and $\prod_{A\in\mathcal{C}} x_A = |K|^k$. But $\mathcal{C}' = \mathcal{C} \setminus \{\{i\} : i \in [n]\}$ is a $(k-1)$-cover of $[n]$, so $\prod_{A\in\mathcal{C}'} x_A \geq |K|^{k-1}$, which implies that $\prod_{i=1}^n x_{\{i\}} \leq |K|$. Since $\{\{i\} : i \in [n]\}$ is a 1-cover of $[n]$, the reverse inequality also holds here, and hence, in fact, equality holds.

Finally, note that, for any $A \subset [n]$, the set $\{A\} \cup \{\{i\}; i \notin A\}$ is a 1-cover of $[n]$, so

$$|K| \leq x_A \prod_{i\notin A} x_{\{i\}} \leq \prod_{i\in A} x_{\{i\}} \prod_{i\notin A} x_{\{i\}} = |K|,$$

whence $x_A = \prod_{i\in A} x_{\{i\}}$. Consequently, the box B, of side length $x_{\{i\}}$ in the direction of v_i, satisfies $|B| = |K|$ and $|B_A| = x_A \leq |K_A|$ for all $A \subset [n]$. $\square$

Note that the Cover Inequality (Theorem 4.1) is a trivial consequence of BTBT (Theorem 4.2). In fact, BTBT implies more: if the volume of a box can be bounded in terms of the volumes of a certain collection of projections, then the same bound holds for all bodies.

An easy consequence of the Cover Inequality is that every hereditary property has an asymptotic logarithmic density, as we remarked in §2. This can be seen by representing each k-uniform hypergraph on $[n]$ by a unit cube in $\mathbb{R}^N$, where $N = \binom{n}{k}$ and the coordinates are indexed by the k-subsets of $[n]$. Rather than spelling out this argument, we deduce it from an inequality concerning traces of set systems.

Given a system $\mathcal{F}$ of subsets of $[n]$ and a subset $A \subset [n]$, the *trace of $\mathcal{F}$ on A* is $\mathcal{F}_A = \{F \cap A : F \in \mathcal{F}\}$. In the present context, it is perhaps more natural to call $\mathcal{F}_A$ the *projection of $\mathcal{F}$ onto A*.

Theorem 4.3 *Let $\mathcal{F}$ be a set of subsets of $[n]$, and let $\mathcal{C}$ be a k-cover of $[n]$. If $c \geq 1$ is such that $|\mathcal{F}_A| \leq c^{|A|}$ for all $A \in \mathcal{C}$, then $|\mathcal{F}| \leq c^n$.*

Proof Let us associate to each subset $A \in [n]$ its characteristic vector $v_A = \sum_{i\in A} v_i \in \mathbb{R}^n$. Let I be the unit cube and write $I + v_A$ for the translate of I by v_A. Finally, put $K = \bigcup_{F\in\mathcal{F}}(I + v_A)$. This is the set that corresponds to K: indeed, $|K| = |\mathcal{F}|$, and $|K_A| = |\mathcal{F}_A|$ for all $A \subset [n]$. Applying Theorem 4.1 to the body K and its projections K_A we find that

$$|\mathcal{F}|^k = |K|^k \leq \prod_{A\in\mathcal{C}} |K_A| \leq \prod_{A\in\mathcal{C}} c^{|A|} = c^{\sum_{A\in\mathcal{C}} |A|} = c^{kn},$$

because $\mathcal{C}$ is a k-cover of $[n]$. $\square$

It is just as easy to show that every hereditary property of uniform hypergraphs has an asymptotic logarithmic density or entropy, so this is what we shall show. Let us spell out the notions we need to state this result. A *property* $\mathcal{P}$ of r-uniform hypergraphs is an infinite class of r-uniform hypergraphs which is closed under isomorphism, and $\mathcal{P}$ is *hereditary* if every induced subgraph of every member of $\mathcal{P}$ is also in $\mathcal{P}$. As always, $\mathcal{P}^n$ is the set of hypergraphs in $\mathcal{P}$ with vertex set $[n]$. The *logarithmic density* of $\mathcal{P}$ at level n is $c_n \geq 0$, defined by

$$|\mathcal{P}^n| = 2^{c_n\binom{n}{r}}.$$

Note that eventually $0 < c_n < 1$ unless $\mathcal{P}$ consists of hypergraphs with no edges only, or complete hypergraphs only, or of all hypergraphs. As pointed out in [42], an easy consequence of Theorem 4.3 is that the sequence (c_n) is monotone decreasing; in particular, as shown by Alekseev [3] (see also [4]), it is convergent.

Theorem 4.4 *Let $\mathcal{P}$ be a hereditary property of r-uniform hypergraphs and let $|\mathcal{P}^n| = 2^{c_n\binom{n}{r}}$. Then $c_{n-1} \geq c_n$ for $n \geq 2$. In particular, the asymptotic logarithmic density $c = \lim_{n\to\infty} c_n$ exists.*

Proof Let us identify a hypergraph with the subset of $[n]^{(r)}$ which is its edge set where, as usual, $[n]^{(r)}$ is the set of r-subsets of $[n]$.

This identification turns $\mathcal{P}^n$ into a set system on $[n]^{(r)}$. Let $A(i)$ be the set of r-subsets of $[n] \setminus \{i\}$. Then $\mathcal{P}^n_{A(i)}$, the projection of the set system $\mathcal{P}^n$ onto $A(i)$, is the set of hypergraphs induced by the hypergraphs in $\mathcal{P}^n$ on the vertex set $[n] \setminus \{i\}$. Since $\mathcal{P}$ is hereditary, $|\mathcal{P}^n_{A(i)}| \leq |\mathcal{P}^{n-1}|$ for every such $A(i)$.

Now the sets $A(i)$, $1 \leq i \leq n$, form an $(n-1)$-cover of $[n]^{(r)}$. Also $|A(i)| = \binom{n-1}{r}$ and $|\mathcal{P}^n_{A(i)}| \leq |P^{n-1}| = 2^{c_{n-1}|A(i)|}$. By Theorem 4.3,

$$|\mathcal{P}^n| \leq 2^{c_{n-1}|[n]^{(r)}|} = 2^{c_{n-1}\binom{n}{r}},$$

completing the proof. $\square$

As we shall see in the next section, for $r = 2$, i.e., for graphs, the limit $\lim c_n$ has a simple description in terms of the property and its family of forbidden induced subgraphs; however, for $r \geq 3$ very little is known about the dependence of the limit on the property, and about the set of limit points.

5   Hereditary Properties of Graphs

In this section we return to the study of *graphs* rather than hypergraphs. Our main aim is to describe the limit c in Theorem 4.4 for graphs, and to give a simple characterization of the properties with a given limit. In particular, we shall see that, as in the Erdős–Stone theorem, only countably many values appear as limits: $0, 1/2, 2/3, 3/4, \ldots$ and 1, for the trivial property of all graphs.

To prepare the ground, we have to introduce some notions. First, let G be a graph, $0 \leq s \leq r$ integers, and ψ a map $V(G) \to [r]$. Following Prömel and Steger [104], we call ψ an (r, s)-*colouring* of G if $G[\psi^{-1}(i)]$ is complete for $1 \leq i \leq s$ and is empty otherwise.

For example, the complete graph K_r has no $(r-1, 0)$-colouring, but it is $(1, 1)$-colourable. Also, the quadrilateral C_4 is $(2, 2)$-colourable and $(2, 0)$-colourable, but not $(2, 1)$-colourable.

Note that a graph G is $(r, 0)$-colourable if and only if $\chi(G) \leq r$, since an $(r, 0)$-colouring is just an r-colouring in the usual sense. Note also that a graph is (r, r)-colourable if and only if its complement is r-colourable. Now let

$$\mathcal{C}^n(r, s) = \{G : V(G) = [n] \text{ and } G \text{ is } (r, s)\text{-colourable}\}$$

and set

$$\mathcal{C}(r, s) = \bigcup_{n \geq 1} \mathcal{C}^n(r, s),$$

so that $\mathcal{C}(r, s)$ is the property of being (r, s)-colourable, and $\mathcal{C}^n(r, s)$ is the nth level of this property.

It is easily shown that

$$e(\mathcal{C}^n(r, s)) = t_r(n) \geq (1 - 1/r) \binom{n}{2}$$

and

$$|\mathcal{C}^n(r, 0)| = |\mathcal{C}^n(r, r)| \leq |\mathcal{C}^n(r, s)| = 2^{(1-1/r)n^2/2 + O(n)}$$

for every s; here $\mathcal{C}(r, 0)$ is the property of being r-colourable, and $\mathcal{C}(r, r)$ is the property of having an r-colourable complement.

Then the *colouring number* $r(\mathcal{P})$ of a hereditary property $\mathcal{P}$ is defined by

$$r(\mathcal{P}) = \max\{r : \text{ there exists } 0 \leq s \leq r \text{ such that } \mathcal{P} \supset \mathcal{C}(r, s)\},$$

that is, $r(\mathcal{P})$ is the largest integer r such that, for some s, the property $\mathcal{P}$ contains every (r, s)-colourable graph. By our remarks above,

$$c_n \geq d_n \geq 1 - 1/r(\mathcal{P}) \tag{5.1}$$

for our sequences (c_n), (d_n).

Clearly, if $\mathcal{P} = \mathrm{Her}(\mathcal{F})$ is the hereditary property given by a family $\mathcal{F}$ of forbidden graphs then

$$
\begin{aligned}
r(\mathcal{P}) &= \max\{r : \text{ for some } 0 \leq s \leq r \text{ no } F \in \mathcal{F} \text{ is } (r, s)\text{-colourable}\}\\
&= \max\{r : \mathcal{F} \cap \mathcal{C}(r, s) = \emptyset \text{ for some } 0 \leq s \leq r\}.
\end{aligned}
$$

For example, as C_4 is not $(2, 1)$-colourable, but it is $(3, s)$-colourable for every s, $0 \leq s \leq 3$, the property $\mathrm{Her}(C_4)$ of not containing an induced quadrilateral has colouring number 2.

Since a graph of order r is (r, s)-colourable for every s, $0 \leq s \leq r$, it follows that $r(\mathcal{P})$ is finite if $\mathcal{P}$ is non-trivial. Note also that the only $(1, 0)$-colourable graphs are empty, and the only $(1, 1)$-colourable graphs are complete, so by Ramsey's theorem $r(\mathcal{P}) \geq 1$.

For a monotone property, (r, s)-colourability can be replaced by colourability since every (r, s)-colourable graph contains an r-colourable subgraph, so that if $\mathcal{P} = \mathrm{Mon}(\mathcal{F})$ then

$$r(\mathcal{P}) = \max\{r : \text{ no } F \in \mathcal{F} \text{ is } r\text{-colourable}\} = \min_{f \in \mathcal{F}} \chi(F) - 1.$$

Extending results of Prömel and Steger [104] concerning specific and principal hereditary properties, Bollobás and Thomason [44] proved that, in fact, the trivial inequalities (5.1) are close to best possible.

Theorem 5.1 *Let $\mathcal{P}$ be a non-trivial hereditary property of graphs with colouring number $r = r(\mathcal{P})$. Then $c = d = 1 - 1/r$: the asymptotic logarithmic density and asymptotic normalized size of $\mathcal{P}$ are both $1 - 1/r$. Thus, if $|\mathcal{P}^n| = 2^{c_n \binom{n}{2}}$ and $e(\mathcal{P}^n) = d_n \binom{n}{2}$ then*

$$\lim_{n \to \infty} c_n = \lim_{n \to \infty} d_n = 1 - 1/r(\mathcal{P}).$$

Furthermore, there is an s, $0 \le s \le r$, such that $\mathcal{P} \supset C(r, s)$ and

$$|\mathcal{P}^n| = |\mathcal{C}^n(r, s)| \, 2^{o(n^2)}. \qquad \qquad \square$$

We see from this result that not only are there only countably many well-separated ranges for the speeds: $2^{o(n^2)}$, $2^{(1-1/2+o(1))n^2/2}$, $2^{(1-1/3+o(1))n^2/2}$, and so on, but also in each of these ranges there are two properties of *minimal speed*: in the range $2^{(1-1/r+o(1))n^2/2}$ the property $\mathcal{C}(r, 0)$ of being r-colourable has the smallest speed, and so does the property $\mathcal{C}(r, r)$ consisting of the complementary graphs. Thus from $2^{(1-1/(r-1)+o(1))n^2/2}$ the speed has to jump to at least $|\mathcal{C}^n(r, 0)| = |\mathcal{C}^n(r, r)| = 2^{(1-1/r)n^2/2 + O(n)}$.

The proof of Theorem 5.1 is based on the three fundamental results of combinatorics: Ramsey's theorem [107], the Erdős–Stone theorem [69], and Szemerédi's Regularity Lemma [116]. Only the very simplest case of Ramsey's theorem is needed, that the diagonal graph Ramsey function is finite: $R(k) < \infty$ for every k. On the other hand, one needs a little more than the Erdős–Stone theorem, as formulated in Theorem 3.1: we have to find a $K_{r+1}(s)$ that spans no 'forbidden' edge.

Theorem 5.2 *Given $r \ge 1$, $t \ge 1$ and $\varepsilon > 0$, there exist $\delta = \delta(r, t, \varepsilon)$ and $n_0 = n_0(r, t, \varepsilon)$ such that the following holds. Let F and G be graphs on the same vertex set of order $n \ge n_0$ with $e(F) \le \delta n^2$ and*

$$e(G) \ge \left(1 - \frac{1}{r} + \varepsilon\right) \binom{n}{2}.$$

Then G contains a $K_{r+1}(t)$ subgraph that spans no edge of F. $\qquad \square$

The most important ingredient of the proof of Theorem 5.1 is SRL, *Szemerédi's Regularity Lemma* [116]. Before we can state it, we need a basic definition.

Given a graph $G = (V, E)$, and subsets $A, B, \subset V$, the *density* $d(A, B)$ is defined as

$$d(A, B) = \frac{e(A, B)}{|A||B|},$$

where $e(A, B)$ is the number of A-B edges. A pair (A, B) is (ε, δ)-uniform if

$$|d(A', B') - d(A, B)| \le \varepsilon$$

whenever $A' \subset A$, $B' \subset B$, $|A'| \ge \delta|A|$ and $|B'| \ge \delta|B|$. Here is then SRL; due to its importance, we state it as a theorem.

Theorem 5.3 *For all $\varepsilon, \delta, \eta > 0$ there is an $M = M(\varepsilon, \delta, \eta)$ such that the vertex set of every graph G can be partitioned into M sets $U_1, \ldots, U_M$ of sizes differing by at most 1, such that at least $(1-\eta)M^2$ of the (ordered) pairs (U_i, U_j) are (ε, δ)-uniform.* □

The smaller M can be taken the more powerful SRL is; unfortunately, when $\varepsilon = \delta = \eta$, all we know about it that it is at most a tower of 2s of height proportional to ε^{-5}. It is somewhat disappointing that, in spite of the easy and seemingly wasteful proof of SRL, one can not guarantee a small bound on $M(\varepsilon, \delta, \eta)$: as proved by Gowers [76] it can not be less than of tower type in $1/\delta$, even when ε and η are kept large.

To prove Theorem 5.1, Bollobás and Thomason [44] used the results above to deduce the theorem below stating, very roughly, that every large family of graphs has a member with *many* induced subgraphs. Thus, in order to find every member of $\mathcal{C}^k(r, s)$ as an induced subgraph, we do not have to take induced subgraphs of many members of the family: it suffices to find one which on its own contains every member of $\mathcal{C}^k(r, s)$ as an induced subgraph.

Theorem 5.4 *Let $r, k \in \mathbb{N}$, $r \geq 2$ and $\varepsilon > 0$ be given. Then there exists $n_0 = n_0(r, k, \epsilon)$ such that if $n > n_0$ and $\mathcal{Q}^n$ is a collection of at least $2^{(1-1/r+\varepsilon)\binom{n}{2}}$ labelled graphs with vertex set $[n]$, then $\mathcal{Q}^n$ contains a graph G_0 such that for some s, $0 \leq s \leq r + 1$, every member of $\mathcal{C}^k(r + 1, s)$ is an induced subgraph of G_0.* □

To all intents and purposes, Theorem 5.4 is stronger than Theorem 5.1. Indeed, let $\mathcal{P}$ be a non-trivial hereditary property of graphs with colouring number $r = r(\mathcal{P})$. Our task is to prove that $\limsup_{n \to \infty} c_n \leq 1 - 1/r$. Suppose that this is false, so that there exists $\varepsilon > 0$ such that $|\mathcal{P}^n| > 2^{(1-1/r+\varepsilon)\binom{n}{2}}$ for infinitely many values of n. Theorem 5.4 implies that for each integer k there is an integer s_k, $0 \leq s_k \leq r+1$, such that for some n every graph in $\mathcal{C}^k(r + 1, s_k)$ is an induced subgraph of some $G_0 \in \mathcal{P}^n$. A fortiori, as $\mathcal{P}$ is hereditary, $\mathcal{C}^k(r + 1, s_k) \subset \mathcal{P}^k$. Consequently, for some value of s, we have $\mathcal{C}^k(r + 1, s) \subset \mathcal{P}^k$ for infinitely many k, and hence for all k. But this contradicts the definition of $r(\mathcal{P})$, completing the proof of Theorem 5.1.

To conclude this section, let us note a striking difference between monotone and hereditary properties concerning the (very crude) measure $c = d$. For *monotone* properties, the measure of the intersection of finitely many properties is just the minimum of the measures: if $\mathcal{P}_1$ and $\mathcal{P}_2$ are monotone properties then

$$c(\mathcal{P}_1 \cap \mathcal{P}_2) = \min\{c(\mathcal{P}_1), c(\mathcal{P}_2)\}. \tag{5.2}$$

Indeed, (5.2) holds since for $\mathcal{P}_i = \mathrm{Mon}(\mathcal{F}_i)$, $i = 1, 2$, we have

$$\min\{\chi(F) : F \in \mathcal{F}_1 \cup \mathcal{F}_2\} = \min\{\min\{\chi(F) : F \in \mathcal{F}_1\}, \min\{\chi(F) : F \in \mathcal{F}_2\}\}.$$

However, equality (5.2) does not hold for hereditary properties since the definition of the colouring number depends on the existence of an extra variable, s. For example, let $\mathcal{P}_1 = \mathrm{Mon}(K_4) = \mathrm{Her}(K_4)$ and $\mathcal{P}_2 = \mathrm{Her}(C_7)$, so that $\mathcal{P}_1$ is monotone and $\mathcal{P}_2$ is hereditary. What are the colouring numbers of $\mathcal{P}_1$, $\mathcal{P}_2$ and $\mathcal{P}_1 \cap \mathcal{P}_2$? Since $\chi(K_4) = 4$, we have $r(\mathcal{P}_1) = 3$. Also, C_7 is $(4, s)$-colourable for every s, but not $(3, 3)$-colourable, so $r(\mathcal{P}_2) = 3$ as well. What about the intersection? The graph K_4

is $(3, s)$-colourable for every $s \geq 1$, and C_7 is $(3, 0)$-colourable, so $r(\mathcal{P}_1 \cap \mathcal{P}_2) \leq 2$. Finally, neither K_4 nor C_7 is $(2, 0)$-colourable, so $r(\mathcal{P}_1 \cap \mathcal{P}_2) \geq 2$, implying that $r(\mathcal{P}_1 \cap \mathcal{P}_2) = 2$. By Theorem 5.1, this shows that

$$c(\mathcal{P}_1) = c(\mathcal{P}_2) = 2/3, \quad \text{but} \quad c(P_1 \cap \mathcal{P}_2) = 1/2,$$

i.e., $\mathcal{P}_1$ and $\mathcal{P}_2$ are 'large' properties: $|\mathcal{P}_1^n| \approx 2^{n^2/3}$ and $|\mathcal{P}_2^n| \approx 2^{n^2/3}$, whereas their intersection is much smaller: $|\mathcal{P}^n| \approx 2^{n^2/4}$.

6   Classifying Labelled Speeds

As we have already remarked, the study of *general* hereditary properties of graphs started with the discovery of Scheinerman and Zito [112] that the speed $|\mathcal{P}^n|$ is severely constrained. (Here, and throughout this section, $\mathcal{P}$ stands for a (non-trivial) property of graphs.) Later Balogh, Bollobás and Weinreich [27, 28, 29, 30] considerably sharpened and extended this result by characterizing the possible speeds and the structures giving rise to those speeds. In this brief section we shall give only a fraction of the results and describe only some of the structures.

Note that when talking of possible speeds (e.g., that the speed is polynomial) we *have to* mean that the speed $|\mathcal{P}^n|$ has the stated properties (e.g., it is a polynomial in n) *provided n is large enough*. Indeed, we are always free to add small graphs to the property if we make sure that the property remains hereditary. For example, let $\mathcal{P}$ consist of all complete graphs and all graphs of order at most 100. Then $|\mathcal{P}^n|$ is $2^{\binom{n}{2}}$ for $n \leq 100$, and one for $n > 100$. In view of this, we call two properties, $\mathcal{P}$ and $\mathcal{Q}$, *equivalent* if $|\mathcal{P}^n| = |\mathcal{Q}^n|$ for n large enough, i.e., the symmetric difference $\mathcal{P} \triangle \mathcal{Q}$ consists of finitely many elements. Using this terminology, $|\mathcal{P}^n| = f(n)$ means that there is a property $\mathcal{Q}$ equivalent to $\mathcal{P}$ such that $|\mathcal{Q}^n| = f(n)$ for every n.

In the theorem below we have collected assertions from [112, 44, 27] and [30]. We write $B(n)$ for the nth Bell number, the number of partitions of an n-element set, so that $B(1) = 1$, $B(2) = 2$, $B(3) = 5$, $B(4) = 15$, and $B(n) \sim ((1 + o(1))n/\log n)^n$.

Theorem 6.1 *Let $\mathcal{P}$ be a hereditary property of graphs. Then one of the following cases holds for sufficiently large n.*

   (i) *$|\mathcal{P}^n|$ is identically zero, one or two.*

   (ii) *There is an integer $k > 0$ such that $|\mathcal{P}^n|$ is a polynomial of degree k in n.*

   (iii) *There is an integer $k > 1$ such that $|\mathcal{P}^n|$ has exponential order of the form $\sum_{i=1}^{k} p_i(n)i^n$, where $p_i(n)$ is a polynomial in n, with $p_k(n)$ non-zero.*

   (iv) *There is an integer $r > 1$ such that $|\mathcal{P}^n| = n^{(1-1/r+o(1))}$.*

   (v) *$B(n) \leq |\mathcal{P}^n| = 2^{o(n^2)}$.*

   (vi) *There is an integer $r > 1$ such that $|\mathcal{P}^n| = 2^{(1-1/r+o(1))n^2/2}$.* □

We call a property $\mathcal{P}$ *polynomial* if $|\mathcal{P}^n| = O(n^k)$ for some fixed k, *exponential* if $|\mathcal{P}^n| \neq O(n^k)$ for every k and $|\mathcal{P}^n| = n^{o(n)}$, and *factorial* if $k^n \leq |\mathcal{P}^n| \leq n^{(1-\varepsilon)n}$ for every $k > 0$ and some $\varepsilon > 0$, for all n large enough. By Theorem 6.1, in each of these ranges the growth rate of $|\mathcal{P}^n|$ is much more restricted than the definitions indicate. As we shall see, in the *superfactorial range* (v), when $|\mathcal{P}^n| \geq n^{(1+o(1))n}$, a fair amount of oscillation is possible. On the other hand, by Theorem 4.4, in the *high range* covered by (vi) we have monotonicity after a suitable normalization; in particular, the labelled speed is a well-behaved function.

Note that Theorem 5.1 tells us (vi) and the second half of (v): either $|\mathcal{P}^n| = 2^{o(n^2)}$ or $|\mathcal{P}^n| = 2^{(1-1/r+o(1))n^2/2}$ for some $r > 1$.

Let us emphasize that in several ranges there are 'hard boundaries': minimal and maximal speeds. Starting with the top speeds, if $\mathcal{P}$ is a hereditary property with $|\mathcal{P}^n| = 2^{(1-1/r+o(1))n^2/2}$ for some $r \geq 2$, then

$$|P^n| \geq |\mathcal{C}^n(r,0)| = 2^{(1-1/r+o(1))n^2/2}$$

for every n, where, as before, $\mathcal{C}^n(r,0)$ is the set of r-colourable graphs with vertex set $[n]$. Indeed, by Theorem 5.1,

$$\mathcal{P} \supset \mathcal{C}(r,s)$$

for some s, $0 \leq s \leq 1$, and one can check that if n is large enough,

$$|\mathcal{C}^n(r,s)| > |C^n(r,0)| = |C^n(r,r)|$$

for all s, $1 \leq s \leq r-1$. Thus, either $|\mathcal{P}^n| \leq 2^{(1-1/r+o(1))n^2/2}$ or $|\mathcal{P}^n|$ is at least as large as the number of r-colourable graphs with vertex set $[n]$.

As shown in [30], the jump from speeds of the type $n^{(1-1/k+o(1))n}$ to $n^{(1+o(1))n}$ is similarly clean. This time there are again two (complementary) properties with minimal speeds. Let $\mathcal{S}$ denote the property that consists of all graphs whose components are cliques, and set $\overline{\mathcal{S}} = \{G : \overline{G} \in \mathcal{S}\}$, i.e., let $\overline{\mathcal{S}}$ be the class of complete k-partite graphs for $k \geq 1$. Clearly, $|\mathcal{S}^n| = \overline{\mathcal{S}}^n|$ is the number of partitions of $[n]$ into (any number of) sets, i.e., the nth Bell number $B(n)$. Parts (iv) and (v) of the theorem above tell us that if $|\mathcal{P}^n| \geq n^{(1+o(1))n}$ then $|\mathcal{P}^n| \geq B(n)$.

At the other end of the range, it is not surprising that it is rather easy to prove precise results for small speeds. For example, as noted by Scheinerman and Zito [112], either $|\mathcal{P}^n| \leq 2$ for n large enough (and if $G \in \mathcal{P}$ then G is either empty or complete), or else $|\mathcal{P}^n| \geq n-1$ for every n. Indeed, if $|\mathcal{P}^n| \geq 3$ then there is a graph $G \in \mathcal{P}$ which is neither empty nor complete. Hence it contains a vertex x with $1 \leq d(x) \leq n-2$. Giving x label 1, we see that there are at least $\binom{n-1}{d(x)}$ ways of labelling G, so $|\mathcal{P}^n| \geq n-1$. Furthermore, if $2 \leq d(x) \leq n-3$ then $|\mathcal{P}| \geq \binom{n-1}{d(x)} > n$. A little more analysis tells us that $|\mathcal{P}^n| > n$ unless $\mathcal{P}^n$ consists of the n stars on $[n]$ or their complements. Now, if a hereditary property contains the star $K_{1,n}$ then it also contains the empty graph on n vertices. Consequently, if $\mathcal{P}$ is a hereditary property then either $|\mathcal{P}^n| \leq 2$ for every n, or else $|\mathcal{P}^n| \geq n+1$ for every $n \geq 3$.

With a little work one can show that for linear speeds there is not only a minimum, but also a maximum. In fact, as shown in [27], the entire polynomial range is divided into well-separated subranges with minima and maxima: for every $k \geq 1$, the set of speeds of order $\Theta(n^k)$ has minimal and maximal members. To define the minimal speeds, let $\mathcal{L}_1$ be the property we have just encountered consisting of stars and empty graphs, so that $|\mathcal{L}_1^n| = n+1$. (In fact, this is also the property $\mathcal{R}$ we gave as an example in §3.) Also, let $\mathcal{L}_k$ consist of graphs made up of a clique of order at most k and isolated vertices, so that

$$|\mathcal{L}_k^n| = \binom{n}{k} + \binom{n}{k-1} + \cdots + \binom{n}{2} + 1.$$

The maximal speeds are given by slightly more complicated properties. We say that two vertices of a graph, x and y, are *twins* or *equivalent*, and write $x \sim y$, if they are joined to the same set of vertices outside $\{x, y\}$: if $z \neq x, y$ then z is joined to x if and only if it is joined to y. This relation $\sim$ is an equivalence relation, so our terminology is justified; we call its equivalence classes *homogeneous sets*. Clearly, the subgraph induced by a homogeneous set is either complete or empty. Note that the vertex set V of a graph G with precisely two equivalence classes has a bipartition $V = V_1 \cup V_2$ such that each V_i spans either a complete graph or an empty graph, and G contains either all the edges from V_1 to V_2 or none of them.

For polynomial speeds there must be a very large homogeneous set. Let $\mathcal{U}_k$ be the hereditary property consisting of all graphs in which all but at most k of the vertices are equivalent. Thus $\mathcal{U}_1$ consists of the stars and the empty graphs together with the complements of these graphs, so that $|\mathcal{U}_1^n| = 2n - 2$. It is easy to see that

$$|\mathcal{U}_k^n| \leq \frac{1}{k!} 2^{\binom{k+1}{2}+1} n^k$$

for $k \geq 2$.

Here is then the result from [27] for polynomial speeds.

Theorem 6.2 *Let $\mathcal{P}$ be a hereditary property with $|\mathcal{P}^n| = \Theta(n^k)$ for some $k \geq 1$, and let $\mathcal{L}_k$ and $\mathcal{U}_k$ be the properties defined above. Then*

$$|\mathcal{L}_k^n| \leq |\mathcal{P}^n| \leq |\mathcal{U}_k^n|. \qquad \square$$

As stated in Theorem 6.1 (iii), not only do we have these bounds, but $|\mathcal{P}^n|$ itself is a polynomial. In fact, more is true.

Theorem 6.3 *If $|\mathcal{P}^n| = O(n^k)$ then, for n large enough, $|\mathcal{P}^n|$ is a polynomial. More precisely, there are integers a_j, $0 \leq a_j \leq 2^{\binom{j+1}{2}+1}$, depending on $\mathcal{P}$ such that*

$$|\mathcal{P}^n| = \sum_{j=1}^{k} a_k \binom{n}{j}$$

if n is large enough. Furthermore, for each $k \in \mathbb{N}$, there are only finitely many non-equivalent hereditary properties with polynomial growth $\Theta(n^k)$. $\qquad \square$

For exponential properties one has to work harder to show that their speeds are severely restricted. In this range the rough order of the speed is determined by the numbers and sizes of homogeneous sets. More precisely, given a property $\mathcal{P}$, write $\ell_\mathcal{P}$ for the maximal number of homogeneous sets in a graph $G \in \mathcal{P}$. As shown in [27], $\ell_\mathcal{P} < \infty$ if and only if $|\mathcal{P}^n| = \Theta(k^n)$ for some k. Assuming that $\ell_\mathcal{P} < \infty$, let $k_\mathcal{P}$ be the maximal k such that for every $s \geq 1$ there is a graph $G \in \mathcal{P}$ with k homogeneous classes, each with at least s vertices. Polynomial speeds arise when $k_\mathcal{P} = 1$. If $k_\mathcal{P} \geq 2$ then $|\mathcal{P}^n|$ is exponential, and the following theorem holds.

Theorem 6.4 *Let $\mathcal{P}$ be a hereditary property with parameters $\ell_\mathcal{P}$ and $k_\mathcal{P}$, as above. If $\ell_\mathcal{P} < \infty$ then there exist polynomials $p_1, \ldots, p_k$ with integer coefficients such that $k = k_\mathcal{P}$, p_k is not the zero polynomial, and $|\mathcal{P}^n| = \sum_{i=1}^{k} p_i(n) i^n$ if n is sufficiently large.* $\qquad \square$

In order to prove the results above, we have to study the *structure* of a property, i.e., the structure of the graphs making up the property. Having decided what the structure is like, counting the graphs is a simple matter. Here we shall describe briefly the structure of polynomial and exponential properties. Let $G(A, B)$ denote a graph whose vertex set is partitioned into two classes, A and B: the *anchor* and the *body* of the graph. We call $G(A, B)$ a *template*, as we shall use it to construct a family of graphs. A template is allowed to contain loops, but only at the vertices of B. This notation suggests that when we construct graphs from $G(A, B)$ as a code, the vertices in B will be 'blown up' into many vertices, while those in A will just 'anchor' the new graph.

Let $x_1, \ldots, x_b$ be an enumeration of the vertices in B, so that $|B| = b$. Given a template $G_{A,B}$ and non-negative integers $m_1, \ldots, m_b$, we denote by $G(A; B \cdot (m_i)_1^b)$ the graph obtained from $G_{A,B}$ by replacing each x_i by m_i vertices, and joining two new vertices if the original vertices were joined by an edge or loop. (note that $m_i = 0$ is allowed.) Thus, $H = G(A; B \cdot (m_i)_1^b)$ has $|A| + \sum_{i=1}^b m_i$ vertices; also, two of the m_i vertices replacing x_i are joined by an edge if and only if $G_{A,B}$ has a loop at x_i, and a vertex replacing x_i is joined to a vertex replacing x_j if and only if $x_i x_j$ is an edge of $G_{A,B}$. We say that H has been obtained from $G_{A,B}$ by blowing up the vertices of B, or multiplying each vertex x_i by m_i.

Let us use a template $G(A, B)$ to define a family $\mathcal{P}(G_{A,B})$ of graphs as follows:

$$\mathcal{P}(G(A, B)) = \{G : \ G \cong G(A; B \cdot (m_i)_1^b), \ m_i \geq 1 \text{ for every } i\}.$$

Clearly, $\mathcal{P}(G(A, B))$ is a hereditary property of graphs; with a slight change of notation, we write $\mathcal{P}^n(G(A, B))$ for the set of graphs in $\mathcal{P}(G(A, B))$ with vertex set $[n]$.

Using templates, the structure of a polynomial hereditary property $\mathcal{P}$ is easily described: there is a finite set of templates G_{A_i, B_i}, $i = 1, \ldots, \ell$, with each B_i a single vertex or a vertex with a loop, such that $\mathcal{P}^n = \cup_{i=1}^\ell \mathcal{P}^n(A_i, B_i)$ if n is large enough.

Exponential properties are more complicated: in this case the templates determining the properties are not restricted. However, surprisingly, every exponential property is equivalent to a property determined by *finitely many* templates; this is an important step in the proof of Theorem 6.4.

Theorem 6.5 *Let $\mathcal{P}$ be an exponential hereditary graph property. Then there are finitely many templates, $G_1(A_1, B_1), \ldots, G_s(A_s, B_s)$, such that*

$$\mathcal{P}^n = \bigcup_{i=1}^s \mathcal{P}^n(G_i(A_i, B_i))$$

if n is sufficiently large. □

The results above seem to indicate that the unlabelled speed of a hereditary property is a rather pleasant, well-behaved function. Indeed, we know from Theorem 6.1 that the limits

$$\lim_{n \to \infty} \frac{\log |\mathcal{P}^n|}{n} \quad \text{and} \quad \lim_{n \to \infty} \frac{\log |\mathcal{P}^n|}{n^2}$$

exist for every hereditary property $\mathcal{P}$. However, this is somewhat misleading: in the penultimate range, (part (v) of Theorem 6.1) the speed may 'oscillate' so much that

the limit
$$\lim_{n\to\infty} \frac{\log\log|\mathcal{P}^n|}{\log n}$$
does not exist. In fact, in a large range we can come close to prescribing the exact value of $|\mathcal{P}^n|$ even for a monotone property $\mathcal{P}$.

Theorem 6.6 *Let* $1 < c < c'$ *and* $\varepsilon > 1/c$. *Let* $f(n)$ *be a function such that* $n^{c'n} < f(n) < 2^{n^{2-\varepsilon}}$ *for all* n. *Then there are integer sequences* (r_i) *and* (s_i) *and a monotone property* $\mathcal{P}$ *such that*
   (1) $|\mathcal{P}^n| = n^{(c+o(1))n}$ *whenever* $n = r_i$,
   (2) $|\mathcal{P}^n| > f(n) - n!$ *whenever* $n = s_i$,
   (3) $|\mathcal{P}^n| \leq f(n)$ *for all* n,
   (4) $|\mathcal{P}^n| \geq n^{(c+o(1))n}$.   □

Note that the bounds in the oscillation above, c and ε, depend on each other. It has not been ruled out that this dependence is unnecessary, i.e., the assertions above would hold whenever $c > 1$ and $\varepsilon > 0$. In [28] it was conjectured that this is not the case, and the dependence is necessary: for all $c > 1$ there is an $\varepsilon > 0$ such that if $\mathcal{P}$ is a hereditary property and $|\mathcal{P}^n| \geq 2^{n^{2-\varepsilon}}$ holds infinitely often, then $|\mathcal{P}^n| \geq n^{(c+o(1))n}$, and, conversely, for all $d > 1$ there is a $\delta > 0$ such that if $|\mathcal{P}^n| \leq n^{dn}$ infinitely often then $|\mathcal{P}^n| \leq 2^{n^{2-\delta+o(1)}}$.

7 Monotone Properties

For monotone properties the speeds and structures are even more restricted than for hereditary properties. For example, polynomial monotone properties are easily identified. Indeed, for a monotone property $\mathcal{P}$, let $v^*(\mathcal{P})$ be the supremum of the order of a graph $G \in \mathcal{P}$ without isolated vertices. It is easily seen that if $v^*(\mathcal{P}) = k < \infty$ then
$$\binom{n}{k} \leq |\mathcal{P}^n| \leq k!\binom{n}{k},$$
hence, as pointed out in [29], a monotone property $\mathcal{P}$ has polynomial speed if and only if $v^*(\mathcal{P}) < \infty$. With considerably more work, it is shown in [28] that if the speed of a monotone property is exponential then its size is linear in n: $e(\mathcal{P}) = O(n)$.

Another easy result is that, as shown in [28], in the superfactorial range $B(n) \leq |\mathcal{P}^n| = 2^{o(n^2)}$ (case (v) of Theorem 6.1) the upper bound can be improved considerably.

Theorem 7.1 *Let* $\mathcal{P}$ *be a monotone property with* $|\mathcal{P}^n| = 2^{o(n^2)}$. *Then there is a* $t \geq 1$ *such that* $|\mathcal{P}^n| \leq 2^{n^{2-1/t+o(1)}}$.

Proof Let $\mathcal{P} = \text{Mon}(\mathcal{F})$ be a monotone property with speed $|\mathcal{P}^n| = 2^{o(n^2)}$. If no member of $\mathcal{F}$ were bipartite, then $\mathcal{P}$ would contain the property $\mathcal{B}$ consisting of bipartite graphs, so we should have $2^{\lfloor n^2/4 \rfloor} \leq |\mathcal{B}^n| \leq |\mathcal{P}^n|$. Hence, $F \in \mathcal{F}$ for some bipartite graph F. We may clearly assume that F is a non-empty bipartite graph: let t be the order of a larger set in a bipartition.

We claim that $|\mathcal{P}^n| \leq 2^{n^{2-1/t+o(1)}}$. Suppose for a contradiction that for some $\varepsilon > 0$ the inequality $|\mathcal{P}^n| \geq 2^{n^{2-1/t+\varepsilon}}$ holds for infinitely many values of n. Inequality

(3.3) is easily seen to imply that for infinitely many n we have $e(\mathcal{P}^n) \geq n^{2-1/t}$, i.e., $e(G_n) \geq n^{2-1/t}$ for some $G_n \in \mathcal{P}_n$. By a theorem of Kővári, Sós and Turán [91] (see also [36]), if n is large enough then such a G_n contains the complete bipartite graph $K_{t,t}$. But then $F \subset K_{t,t} \subset G_n$, a contradiction. $\qquad\square$

Balogh, Bollobás and Simonovits [25] considerably sharpened the upper bound in Theorem 5.1 (i.e., Theorem 6.1 (vi)) for *monotone* properties. (As we remarked earlier, for monotone, rather than hereditary, properties this was proved by Erdős, Frankl and Rödl [63].) Recall that for a monotone property $\mathcal{P} = \mathrm{Mon}(\mathcal{F})$ the colouring number $r(\mathcal{P})$ is one less than the minimal chromatic number of a forbidden graph: $r(\mathcal{P}) = \min\{\chi(F) : f \in \mathcal{F}\} - 1$.

Theorem 7.2 *For every monotone property $\mathcal{P}$ there is a constant $\gamma = \gamma(\mathcal{P}) > 0$ such that*

$$|\mathcal{P}^n| \leq 2^{(1-1/r)n^2/2 + O(n^{2-\gamma})}. \qquad\square$$

In fact, Balogh, Bollobás and Simonovits [25] proved more: they determined the exact order of the error term in the exponent above.

The proof of Theorem 7.2 is rather involved: it is based on Szemerédi's Regularity Lemma[116], and the Stability Theorem of Erdős and Simonovits (see [61, 67, 68, 114]). In a subsequent paper, Balogh, Bollobás and Simonovits [26] went considerably further: they proved the stability result that almost all graphs in $\mathcal{P} = \mathrm{Mon}(\mathcal{P})$ are rather close to the Turán graph $T_r(n)$, where $r = r(\mathcal{P})$.

It is likely that, as conjectured in [28], Theorem 7.2 holds for hereditary properties as well. It is almost certain that new methods are needed to prove this conjecture, since the proof of Theorem 7.2 seems to break down irretrievably.

8   Unlabelled Speed

Recall that the *unlabelled speed* of a hereditary property $\mathcal{P}$ is the function $n \mapsto |\mathcal{P}_n|$ where $\mathcal{P}_n$ is the set of (isomorphism classes of) graphs in $\mathcal{P}$ with n vertices. In (3.1) we noted the triviality that the labelled and unlabelled speeds differ by at most a factor $n!$, so we shall study the unlabelled speed of a hereditary property $\mathcal{P}$ with labelled speed not much larger than $n!$. More precisely, we shall be interested in properties covered by cases (i)–(iv) of Theorem 6.1, and the lower end of the range in (v). In particular, we are not interested in the high range covered by (vi).

As shown in [30], the jump from speeds of the type $n^{(1-1/k+o(1))n}$ to $n^{(1+o(1))n}$ is similarly clean. This time there are again two (complementary) properties with minimal speeds. Let $\mathcal{S}$ denote the property that consists of all graphs whose components are cliques, and set $\overline{\mathcal{S}} = \{G : \overline{G} \in \mathcal{S}\}$, i.e., let $\overline{\mathcal{S}}$ be the class of complete k-partite graphs for $k \geq 1$. Clearly, $|\mathcal{S}^n| = \overline{\mathcal{S}}^n|$ is the number of partitions of $[n]$ into (any number of) sets, i.e., the nth Bell number $B(n)$. Parts (iv) and (v) of the theorem above tell us that if $|\mathcal{P}^n| \geq n^{(1+o(1))n}$ then $|\mathcal{P}^n| \geq B(n)$.

Balogh, Bollobás, Saks and Sós [24] proved fairly precise results about the unlabelled speed of a property in cases (i)–(iv) of Theorem 6.1. As in Theorems 6.2–6.4, the unlabelled speed $|\mathcal{P}_n|$ greatly depends on the homogeneous sets found in a graph $G \in \mathcal{P}$. Of special importance are *homogeneous k-partite graphs*, graphs G with

$V(G) = V_1 \cup \cdots \cup V_k$ such that any two vertices belonging to the same part V_i are twins.

We have already encountered the property $\mathcal{S}$ consisting of graphs in which every component is a complete graph, and the complementary property $\overline{\mathcal{S}}$ of complete k-partite graphs for $k = 1, 2, \ldots$. We know that these properties $\mathcal{S}$ and $\overline{\mathcal{S}}$ have the smallest labelled speed in Theorem 6.1 (v), namely $|\mathcal{S}^n| = |\overline{\mathcal{S}}^n| = B(n)$, the nth Bell number. The unlabelled speed of these properties is $|\mathcal{S}_n| = |\overline{\mathcal{S}}_n| = S(n)$, the number of partitions of a set with n indistinguishable elements into nonempty subsets, so that $S(1) = 1$, $S(2) = 2$, $S(3) = 3$, $S(4) = 5$, $S(5) = 7$ and $S(6) = 11$.

Before we state the main result from [24], we introduce two more properties and their complements. Let $\mathcal{T}$ denote the property consisting of all *star* forests, i.e., graphs whose components are stars, and put $\overline{\mathcal{T}} = \{G : \overline{G} \in \mathcal{T}\}$. Also, denote by $\mathcal{F}$ the property consisting of all the *path* forests, i.e., graphs whose components are paths, and set $\overline{\mathcal{F}} = \{G : \overline{G} \in \mathcal{F}\}$. Clearly, each of $\mathcal{T}$, $\overline{\mathcal{T}}$, $\mathcal{F}$ and $\overline{\mathcal{F}}$ has unlabelled speed $S(n)$, and labelled speed greater than $B(n)$.

Theorem 8.1 *For every hereditary graph property $\mathcal{P}$ one of the following assertions holds.*

(i) There are integers ℓ, t and C such that if n is large enough then every graph $G \in \mathcal{P}^n$ is such that for some set V_0 of at most C vertices, the graph $G - V_0$ is the symmetric difference of a homogeneous ℓ-partite graph and a graph in which every component has at most t vertices. The unlabelled speed $\mathcal{P}_n$ is polynomially bounded; even more, there is a positive integer k and a rational number c such that

$$|\mathcal{P}_n| = c \cdot n^k + O(n^{k-1}). \tag{8.1}$$

(ii) If n is large enough, $|\mathcal{P}| \geq S(n)$. Furthermore, equality holds for infinitely many n if and only if for n large enough $\mathcal{P}$ the n unlabelled slice of one of the six hereditary properties $\mathcal{S}, \overline{\mathcal{S}}, \mathcal{T}, \overline{\mathcal{T}}, \mathcal{F}$ and $\overline{\mathcal{F}}$. □

It would be good to prove more precise results. For example, it is very likely that if (8.1) holds then, for large enough n, the unlabelled speed $|\mathcal{P}_n|$ is essentially a polynomial.

9   Colouring Random Graphs with Hereditary Properties

As customary, denote by $G_{n,p}$ a random graph with vertex set $[n]$, whose edges are selected independently, with probability p. The probability space of these graphs is $\mathcal{G}(n, p)$. In particular, $\mathcal{G}(n, 1/2)$ is the space of all $2^{\binom{n}{2}}$ graphs on $[n]$ with the uniform distribution.

What can one say about the chromatic number $\chi(G_{n,p})$ of a random graph $G_{n,p}$ with $p = p(n)$, $0 < p(n) < 1$? This was perhaps the most important question proposed and left open by Erdős and Rényi [65, 66], when around 1960 they founded the theory of random graphs (see [33]). After partial results by Grimmett and McDiarmid [78], Matula [98], Bollobás and Erdős [39], Shamir and Spencer [113], and others , in 1988 it was proved [34] that if $0 < p < 1$ is fixed and $q = 1 - p$ then

$$\chi(G_{n,p}) = (1 + o(1))\frac{n}{2\log_{1/q} n} \tag{9.1}$$

for almost every $G_{n,p}$. Substantial extensions of this result were proved by Łuczak [94, 95], Frieze and Łuczak [73], Alon and Krivelevich [11], Achlioptas and Naor [1]. The proofs of these results are based on martingale inequalities (see [35]).

Here we are concerned with certain generalized colourings introduced by Scheinerman [111]. For a property $\mathcal{P}$, a $\mathcal{P}$-*colouring* of a graph $G = (V, E)$ with k colours is a partition $V = V_1 \cup \cdots \cup V_k$ of the vertex set such that every class V_i induces a $\mathcal{P}$-graph: $G[V_i] \in \mathcal{P}, i = 1, \ldots, k$. The $\mathcal{P}$-*chromatic number* $\chi_\mathcal{P}(G)$ of a graph G is the minimal number of classes in a $\mathcal{P}$-colouring of G. Thus $\chi_{\mathcal{C}(1,0)}(G) = \chi(G)$ and $\chi_{\mathcal{C}(1,1)}(G) = \chi(\overline{G})$, where, as before, $\mathcal{C}(r, s)$ is the property of being (r, s)-colourable. Scheinerman [111] was the first to study the $\mathcal{P}$-chromatic number of random graphs: he noted that if $\mathcal{P}$ is a hereditary property then either $\mathcal{C}(1,0) \subset \mathcal{P}$ or $\mathcal{C}(1,1) \subset \mathcal{P}$ so $\chi_\mathcal{P}(G) \leq \max\{\chi(G), \chi(\overline{G})\}$, which implies that $\chi_\mathcal{P}(G_{n,p}) = O(n \log n)$ for every fixed $0 < p < 1$ and hereditary property $\mathcal{P}$. In fact, it is easily seen that $\chi_\mathcal{P}(G_{n,p}) = \Theta(n \log n)$.

In 1995, Bollobás and Thomason [43] proved an analogue of (9.1) for a general hereditary property in the case $p = 1/2$: *if $\mathcal{P}$ is a non-trivial hereditary property of graphs with colouring number $r = r(\mathcal{P})$, then*

$$\chi_\mathcal{P}(G_{n,1/2}) = \left(\frac{1}{2r} + o(1) \right) \frac{n}{\log_2 n} \tag{9.2}$$

for almost every $G_{n,1/2}$.

To prove (9.2), one shows that $\chi_\mathcal{P}(G_{n,1/2})$ is unlikely to be much smaller than $n/(2r \log_2 n)$ since $|\mathcal{P}^n| = 2^{(1-1/r+o(1))n^2/2}$, and it is unlikely to be much larger than $n/(2r \log_2 n)$ since $\mathcal{C}(r, s) \subset \mathcal{P}$ for some s with $0 \leq s \leq r$.

Unlike other graph parameters, for which results concerning $G_{n,1/2}$ are just about equivalent to those about $G_{n,p}$ for fixed $p \neq 1/2$, the $\mathcal{P}$-chromatic number of $G_{n,p}$ is much easier to determine for $p = 1/2$ than for $p \neq 1/2$. However, it is easy to obtain a good lower bound from the following result, which is a consequence of Theorem 4.2.

Theorem 9.1 *Let $\mathcal{P}$ be a hereditary graph property, let $0 < p < 1$ and let the constants $e_{k,p}(\mathcal{P})$ be defined by $\mathbf{P}(G_{k,p} \in \mathcal{P}) = 2^{-e_{k,p}(\mathcal{P})\binom{k}{2}}$. Then $e_{k,p}(\mathcal{P})$ increases with k. In particular, $e_{k,p}(\mathcal{P})$ tends to a limit $e_p(\mathcal{P})$ as $k \to \infty$. Furthermore, $e_p(\mathcal{P}) > 0$ if $\mathcal{P}$ is non-trivial, i.e., if not every graph has $\mathcal{P}$.* □

Theorem 9.1 implies that, for $\varepsilon > 0$, the expected number of induced subgraphs of order k in a random graph $G_{n,p}$ having property $\mathcal{P}$ is $o(1)$ for $k \geq (2/e_p+\epsilon) \log_2 n$, and tends to infinity for $k \leq (2/e_p - \epsilon) \log_2 n$. From this it follows that

$$\chi_\mathcal{P}(G_{n,p}) \geq (e_p + o(1))n/(2 \log_2 n) \tag{9.3}$$

almost surely.

The proof of (9.2) was based on the fact that for $p = 1/2$ the constant $e_p(\mathcal{P})$ has a simple interpretation in terms of the values (r, s) for which $\mathcal{C}(r, s) \subset \mathcal{P}$. Unfortunately, for $p \neq 1/2$ this is no longer true: $e_p(\mathcal{P})$ cannot be characterized solely in terms of these values (r, s). Nevertheless, Bollobás and Thomason [45] proved that, as conjectured in [43] and claimed by (9.2) for $p = 1/2$, (9.3) holds with equality.

Theorem 9.2 *Let $\mathcal{P}$ be a hereditary graph property, $0 < p < 1$, and $e_p = e_p(\mathcal{P})$ the constant defined in Theorem 9.1. Then*

$$\chi_{\mathcal{P}}(G_{n,p}) = (e_p + o(1))n/(2\log_2 n)$$

almost surely.   □

The proof of Theorem 9.2 is considerably more difficult than that of (9.1): not only does it use martingale inequalities and Szemerédi's Regularity Lemma, but also relies heavily on a careful analysis of the structure of a general hereditary property.

The *product* $\prod_{\gamma \in \Gamma} \mathcal{P}_\gamma$ of hereditary properties $\mathcal{P}_\gamma$, $\gamma \in \Gamma$, is the class of graphs G with vertex sets $\bigcup_{\gamma \in \Gamma} V_\gamma$ such that $G[V_\gamma] \in \mathcal{P}_\gamma$ for every $\gamma \in \Gamma$. A hereditary property is *irreducible* if it is not the product of two other hereditary properties. It is easily shown that every hereditary property is the product of a finite collection of irreducible hereditary properties. Also, if $\mathcal{P} = \prod_{\gamma \in \Gamma} \mathcal{P}_\gamma$ then

$$e_p(\mathcal{P})^{-1} = \sum_{\gamma \in \Gamma} e_p(\mathcal{P}_\gamma)^{-1}.$$

One can show that if Theorem 9.2 holds for the properties $\mathcal{P}_1, \ldots, \mathcal{P}_k$, then it holds for $\prod_{i=1}^{k} \mathcal{P}_i$ as well. Consequently, it suffices to prove Theorem 9.2 for irreducible properties.

In fact, the heart of the proof of Theorem 9.2 is the assertion that it holds for every 'typed' property $\mathcal{P} = \mathcal{P}(\tau)$. A *type* is a labelled graph, with the vertices and the edges coloured black or white. Given a type τ, the property $\mathcal{P}(\tau)$ consists of those graphs G for which $V(G)$ has a partition $\bigcup_{t \in V(\tau)} V_t$ such that $G[V_t]$ is complete or empty according as t is black or white, and moreover, if the edge tu is in τ then $G[V_t, V_u]$ is a complete or empty bipartite graph according as the edge tu is black or white. To prove that Theorem 9.2 holds for typed properties $\mathcal{P}(\tau)$ we have to give a careful analysis of the maximal number of induced edge-disjoint subgraphs of a given order having property $\mathcal{P}$.

10   The Distance from a Hereditary Property

Following Axenovitch, Kézdi and Martin [17] (see also [18]), define the *edit distance* of two graphs, G_1 and G_2, on the same vertex set as

$$\triangle(G_1, G_2) = |E(G_1) \triangle E(G_2)|,$$

i.e., the minimal number of edges whose deletion and addition turns G_1 into G_2. Also, given a graph property $\mathcal{P}$ and a graph G on $[n]$, write $\triangle(G, \mathcal{P})$ for the edit distance of G from $\mathcal{P}$:

$$\triangle(G, \mathcal{P}) = \min\{\triangle(G, H) : H \in \mathcal{P}^n\}.$$

Finally, set

$$\mathrm{ed}(n, \mathcal{P}) = \max\{\triangle(G, \mathcal{P}) : V(G) = [n]\}.$$

Thus $\mathrm{ed}(n, \mathcal{P})$ is the maximal number of edges we may have to alter to make a graph on $[n]$ have property $\mathcal{P}$.

What can we say about $\text{ed}(n, \mathcal{P})$ for various hereditary properties? Axenovitch, Kézdi and Martin [17], who posed this question, proved that for a principal hereditary property $\mathcal{P} = \text{Her}(H)$ with colouring number r we have

$$\left(\frac{1}{2r} + o(1)\right) \leq \text{ed}(n, \mathcal{P}) \leq \frac{1}{r}\binom{n}{2}.$$

In fact, in [17] it was also proved that for some families $\mathcal{P}$ of graphs $\text{ed}(n, \mathcal{P})$ is the lower bound above.

Alon and Stav [14] have gone considerably further: they proved that $\text{ed}(n, \mathcal{P})$ is essentially attained on a random graph $G_{n,p}$.

Theorem 10.1 *Given a hereditary property $\mathcal{P}$, there is a constant $p = p(\mathcal{P})$, $0 < p < 1$, such that if $\varepsilon > 0$ then*

$$\mathbb{P}\big(\text{ed}(n, \mathcal{P}) - \text{ed}(G_{n,p}, \mathcal{P}) \geq \varepsilon n^2\big) \to 0,$$

as $n \to \infty$. $\square$

The result above and its proof are related to *testable* properties, i.e., properties $\mathcal{P}$ for which there is a probabilistic algorithm that samples small portions of a graph, and decides whether the graph belongs to $\mathcal{P}$. Alon, Fischer, Krivelevich and Szegedy [8, 9] proved that every principal hereditary property is testable, and this result was extended to all hereditary properties by Alon and Shapira [12] (see also Lovász and Szegedy [93]). The proof of Theorem 10.1 by Alon and Stav was based on the techniques in [12] and [45]. Concerning *monotone* properties, Alon, Shapira and Sudakov [13] gave a polynomial time algorithm for approximating the edit distance of an input graph from a monotone property.

In [15], Alon and Stav went further: they showed that for a hereditary property $\mathcal{P}$ not in the highest range, the probability $p(\mathcal{P})$ is 0, 1, or 1/2.

Theorem 10.2 *Let $\mathcal{P}$ be a hereditary graph property of speed $|\mathcal{P}^n| = 2^{o(n^2)}$. Then one of the following assertions holds.*
1. *$e(G) = o(n^2)$ for every $G \in \mathcal{P}^n$; $\text{ed}(n, \mathcal{P}) = (1 + o(1))\binom{n}{2}$, and $p(\mathcal{P}) = 1$.*
2. *$e(\overline{G}) = o(n^2)$ for every $G \in \mathcal{P}^n$; $\text{ed}(n, \mathcal{P}) = (1 + o(1))\binom{n}{2}$, and $p(\mathcal{P}) = 0$.*
3. *There is a constant $c > 0$ such that for every n there are graphs $G_1, G_2 \in \mathcal{P}^n$ with $e(G_1) > cn^2$ and $e(\overline{G_1}) > cn^2$; $\text{ed}(n, \mathcal{P}) = (\frac{1}{2} + o(1))\binom{n}{2}$, and $p(\mathcal{P}) = \frac{1}{2}$.* $\square$

Alon and Stav determined the probability $p(\mathcal{P})$ for several frequently used hereditary property. For example, they proved that if $\mathcal{P} = \text{Her}(K_{1,3})$, i.e., $\mathcal{P}$ is the probability of being claw-free, then $p(\mathcal{P}) = \frac{1}{3}$ and $\text{ed}(n, \mathcal{P}) = (\frac{1}{3} + o(1))\frac{n}{2}$.

11 Posets, Permutations, Ordered Hypergraphs, and Partitions

The aim of this brief section is to draw attention to the work on a variety of structures, rather than give an account of the results and proofs.

Let us start with posets. Kleitman and Rothschild [88] were the first to give good bounds on N_n, the number of posets on $[n]$. To get a (pretty good) lower bound on N_n, partition $[n]$ into two classes at random, V_1 and V_2, say, then select edges v_1v_2 from V_1 to V_2 with probability $1/2$, and declare v_1 greater than v_2. From this it is easy to see that $N_n \geq 2^{n^2/4}$. Kleitman and Rothschild [88] proved that this lower bound is not too far from an upper bound; the method of proof in this paper was used by Erdős, Kleitman and Rothschild [64] to prove their result on K_{r+1}-free graphs. As we mentioned in §1, this result was a precursor of the study of general hereditary properties.

A few years later Kleitman and Rothschild [89] improved their own result considerably, and another twenty years later Brightwell, Prömel and Steger [52] proved an astonishingly precise and surprising result: not only did they determine the asymptotic value of N_n with great precision, but also discovered that it depends substantially on the parity of n.

Theorem 11.1 *There is an absolute constant $C > 1$ such that*

$$N_n = \left(1 + O(C^{-n})\right) \sum_{s=0}^{n} \binom{n}{s} 2^{(s+1)(n-s)}.$$

Furthermore,

$$N_n = (1 + O(1/n))\varphi_{i(n)} 2^{(n+1)^2/4} \binom{n}{\lfloor n/2 \rfloor},$$

where $i(n)$ is 1 if n is odd, and 2 if it is even; also $\varphi_1 = \sum_{j=-\infty}^{\infty} 2^{-(j+1/2)^2} = 2.1289312\cdots$ and $\varphi_2 = \sum_{j=-\infty}^{\infty} 2^{-j^2} = 2.1289368\cdots$. □

Brightwell, Grable and Prömel [51] considered the speed of the principal hereditary property $\mathcal{P}(P)$ of posets not containing a fixed poset P. Clearly, if P has height at least 3, i.e., does not contain elements $x < y < z$, then $\mathcal{P}(P)$ contains all posets of height 2. In particular, if $[n] = V_1 \cup V_2$ is a partition of $[n]$ then $\mathcal{P}^n(P)$ contains every poset on $[n]$ in which $x > y$ implies that $x \in V_1$ and $y \in V_2$, so $|\mathcal{P}^n(P)| \geq 2^{n^2/4}$. Brightwell, Grable and Prömel [51] proved that, in fact, $|\mathcal{P}^n(P)| = 2^{(1+o(1))n^2/4}$ whenever P has height at least 3. On the other hand, if $\mathcal{P}$ has height 2 then $|\mathcal{P}^n(P)| = 2^{o(n^2)}$. Even more, they proved that $|\mathcal{P}^n(P)| \leq n!c^n$ for some constant c if and only if P is either an antichain or one of ten small partial orders.

Since posets are in one-to-one correspondence with the closures of oriented graphs, Theorem 11.1 is also a result about certain hereditary properties of oriented graphs. (To spell it out, an acyclic oriented graph $\overrightarrow{G}$ corresponds to the poset on $V(\overrightarrow{G})$ in which $u > v$ if $\overrightarrow{G}$ contains a $u - v$ path oriented from u to v.)

Concerning *general* properties of oriented graphs, Alekseev and Sorochan [5] proved that for the labelled speed there is a large jump in the highest range.

Theorem 11.2 *The labelled speed of a hereditary property of oriented graphs is either $2^{o(n^2)}$ or at least $2^{n^2/4+o(n^2)}$.* □

Turning to permutations, we take S_n to be the set of permutations of $[n]$, and write $\pi \in S_n$ as the sequence $\pi(1)\pi(2)\ldots\pi(n)$. Given a permutation $\pi = x_1x_2\ldots x_n$ of $[n]$, if $\rho = y_1y_2\ldots y_k$ is obtained from π by deleting $n-k$ terms of π and fitting the others into $[k]$, while keeping their order, then we say that π *contains* ρ. Otherwise π is said to *avoid* ρ. More formally, a permutation $\pi \in S_n$ is said to *contain* a permutation $\rho \in S_k$ if there are $1 \leq n_1 < \cdots < n_k$ such that $\pi(n_i) < \pi(n_j)$ if and only if $\rho(i) < \rho(j)$. For example, 36182475 contains 321: keeping 8, 7 and 5, the order of these terms is the same as that of 3, 2 and 1. Also, there is only one permutation $\pi \in S_n$ that avoids $21 \in S_2$, the increasing permutation $\pi = 12 \ldots n$.

A property $\mathcal{P}$ of permutations is taken to be a subset of $\bigcup S_n$. We call $\mathcal{P}$ *hereditary* if it is closed under containment; $\mathcal{P}^n = \mathcal{P} \cap S_n$ is the set of permutations of *length n* in $\mathcal{P}$, and the function $n \mapsto |\mathcal{P}^n|$ is the *speed* of $\mathcal{P}$. Given a permutation $\rho \in S_k$, write $\mathcal{P}(\rho)$ for the hereditary property of permutations avoiding ρ. Note that if π is a non-trivial hereditary property then $\mathcal{P} \subset \mathcal{P}(\rho)$ for some ρ, in fact, for every $\rho \notin \mathcal{P}$. Thus, properties of type $\mathcal{P}(\rho)$ are the largest non-trivial hereditary properties.

What can we say about the range of speeds $n \mapsto |\mathcal{P}^n|$? If $\mathcal{P}$ is the trivial property consisting of all permutations then $|\mathcal{P}^n| = n! = n^{(1+o(1))n}$. What about the speed of a non-trivial hereditary property? R.P. Stanley and H.S. Wilf conjectured that such a speed is at most c^n for some constant c depending on $\mathcal{P}$. (This conjecture was first published by Bóna [46, 47].) Equivalently, for every permutation ρ there is a constant c such that $|\mathcal{P}^n(\rho)| \le c^n$. In fact, Arratia [16] showed that, as expected, $|\mathcal{P}^n(\rho)|^{1/n}$ tends to a limit for every ρ, since $|\mathcal{P}^n(\rho)|$ is submultiplicative; consequently, if the Stanley–Wilf conjecture is true then for every permutation ρ there is a constant $c(\rho) \ge 1$ such that $|\mathcal{P}^n(\rho)| = c(\rho)^{(1+o(1))n}$.

The Stanley–Wilf conjecture was proved by Bóna [46, 47, 48, 49] in some special cases, and Alon and Friedgut [10] came close to proving it in full when they gave an upper bound only slightly larger than c^n. In spite of this progress, a proof of the conjecture itself seemed to be out of reach. Thus it was quite a surprise when the combined efforts of Klazar [84], and Marcus and Tardos [97] brought about a very elegant and rather simple proof of the full Stanley–Wilf conjecture. The starting point was a conjecture made by Hajnal and Füredi [75] 1992 concerning extremal properties of 0–1 matrices. Then, in 2000, Klazar [84] proved that this conjecture implied the Stanley–Wilf conjecture. Finally, in 2004, Marcus and Tardos [97] proved the Hajnal–Füredi [75] conjecture, and so the Stanley–Wilf conjecture as well.

To state these results, we have to extend the notion of containment from permutations to 0–1 matrices. Let $A = (a_{ij})$ be an $n \times n$ matrix and $B = (b_{ij})$ a $k \times k$ matrix with the entries 0s and 1s. We say that A *contains b* if there are $1 \le n_1 < \cdots < n_k \le n$ such that $a_{n_i n_j} = 1$ whenever $b_{ij} = 1$. In particular, if A is the matrix of a permutation π and B is that of ρ then A contains B if and only if π contains ρ. We write $||A||$ for the sum of entries of a 0–1 matrix A, i.e., the number of 1s in A. Here is then the theorem conjectured by Füredi and Hajnal [75] and proved by Marcus and Tardos [97].

Theorem 11.3 *For every k there is a constant c_k such that if A is an $n \times n$ matrix whose entries are 0s and 1s and $||A|| \ge c_k n$ then it contains every $k \times k$ permutation matrix.* $\square$

To prove Theorem 11.3, Marcus and Tardos gave a beautiful combinatorial argument based on the pigeon-hole principle to find a recursive estimate for the maximal number of ones in a 0–1 matrix not containing a given permutation matrix, and deduced from this a bound for c_k. This bound is expected to be very far from the best possible value for c_k.

As we remarked above, before Theorem 11.3 was proved, Klazar [84] had proved that it implied the Stanley–Wilf conjecture.

Theorem 11.4 *For every permutation* $\rho \in S_k$ *there is a constant* c *such that* $|\mathcal{P}^n(\rho)| \leq c^n$ *for every* n. $\qquad\square$

Not much is known about the constant $c(\rho) = |\mathcal{P}^n(\rho)|^{1/n}$ which Theorem 11.4 guarantees to be finite. Arratia conjectured that if $\rho \in S_k$ then $c(\rho) \leq (k-1)^2$, but Albert, Elder, Rechnitzer, Westcott and Zabrocki [2] disproved this conjecture by showing that $c(\rho) > 9.47$ for the permutation $\rho = 4231$. (In fact, the estimated value of $c(4231)$ is between 11 and 12.) Furthermore, Bóna [50] showed that, contrary to expectations, $c(\rho)$ need not be rational; in particular, for $c(12453) = 9 + 4\sqrt{2}$.

In fact, $c(\rho)$ cannot be too small either: improving a result of M. Petkovšek (see [119, Theorem 4]), P. Valtr (see [81]) proved that for every c, $0 < c < e^{-3} = 0.04978\ldots$ there is a $k(c)$ such that if $k > k(c)$ and $\rho \in S_k$ then $c(\rho) > ck^2$.

Concerning general hereditary properties of permutations, Kaiser and Klazar [81] proved that the speed is considerably restricted. To state this result, for a fixed integer k, we write $F_{n,k}$ for the *generalized Fibonacci numbers* defined as follows: $F_{n,k} = 0$ for $n < 0$, $F_{n,1} = 1$, and

$$F_{n,k} = F_{n-1,k} + F_{n-2,k} + \cdots + F_{n-k,k}$$

for $n > 0$. Thus, as a function of n, $F_{n,k}$ grows roughly like α_k^n, where α_k is the largest positive real root of $x^k - x^{k-1} - x^{k-2} - \cdots - 1$. Clearly, for $k = 2$ we get the standard Fibonacci numbers.

Theorem 11.5 *Let* $\mathcal{P}$ *be a non-trivial hereditary property of permutations. Then* $\mathcal{P}$ *is either bounded, or exactly one of the following possibilities holds.*

(i)   *There are integers* $k, \ell \geq 1$ *such that*

$$F_{n,k} \leq |\mathcal{P}^n| \leq n^\ell F_{n,k}$$

for every n.

(ii) $|\mathcal{P}^n| \geq 2^{n-1}$ *for every* n. $\qquad\square$

Let us say a few words about ordered graphs and partitions. An *ordered hypergraph* $\mathcal{H} = (V, E, <)$ is a hypergraph $\mathcal{H} = (V, E)$ with a linear order $<$ on its vertex set V. Thus, V is a finite set of vertices, and E, the set of edges, is a collection of subsets of V. We shall also assume that every edge has at least two vertices. (Equivalently, we could demand that the edge set contains every singleton.) We call $\mathcal{K} = (U, F, <)$ an *induced sub-hypergraph* of $\mathcal{H}$ if $U \subset V$, $F = \{e \cap U : e \in E, |e \cap U| \geq 2\}$, and the ordering on U is the restriction of $<$ to U. Hereditary properties are defined in the obvious way, as are sub-hypergraphs and monotone properties.

Let us remark that a permutation π of $[n]$ is encoded by the graph G_π on $[n]$ in which i is joined to j, $i < j$, if π reverses their order, i.e., $\pi(i) > \pi(j)$. Furthermore, it is easily seen that for a hereditary property $\mathcal{P}$ of permutations the family $\mathcal{G}_{\mathcal{P}} = \bigcup_{n=1}^{\infty} \{G_\pi : \pi \in \mathcal{P}^n\}$ is a hereditary property of ordered graphs. Thus ordered graphs and their hereditary properties generalize permutations and their hereditary properties. In view of these remarks, Balogh, Bollobás and Morris [22] extended Theorem 11.5 when they proved that the assertions hold for hereditary properties of ordered graphs as well.

A *partition* of $[n]$ is an unordered collection of disjoint, non-empty sets $\{A_1, \ldots, A_k\}$ with $\bigcup A_i = [n]$. Note that a partition can be thought of as an ordered graph in which every component is a clique.

Properties of ordered hypergraphs, partitions, and other related structures were studied in detail by Klazar [85, 86] and Balogh, Bollobás and Morris [20]–[23]. In particular, in [21] several results were proved, each of which generalizes the Klazar–Marcus–Tardos theorem, Theorem 11.4.

The theorem below was conjectured by Klazar [85] and proved independently by Balogh, Bollobás and Morris [21], and Klazar and Marcus [87]: it claims that the speed of a hereditary property jumps from the exponential range to the factorial, and there is a unique minimal property in the factorial range.

Theorem 11.6 *Let $\mathcal{P}$ be a hereditary property of partitions. If for every constant $c > 0$ we have $|\mathcal{P}^N| > c^N$ for some N then*

$$|\mathcal{P}^n| \geq \sum_{k=0}^{\lfloor n/2 \rfloor} \binom{n}{2k} k! = n^{n/2 + o(n)}$$

for every n. This lower bound is best possible, and there is a unique hereditary property of partitions with this speed. □

Balogh, Bollobás and Morris [21] also proved that exactly the same bounds hold for *monotone* properties of ordered graphs as well. They also made the much stronger conjecture that precisely the same result is true for *hereditary* properties of ordered hypergraphs. The rather special case of this conjecture concerning *graphs* is also open. However, in [21] the conjecture was proved for hereditary graph properties consisting of graphs of size $o(n^2)$.

Theorem 11.7 *Let $\mathcal{P}$ be a hereditary property of graphs such that for some function $f(n) = o(n^2)$ we have $e(G) \leq f(n)$ for every $G \in \mathcal{P}^n$ and every n. If for every $c > 0$ we have $|\mathcal{P}^N| \geq c^N$ for some N, then*

$$|\mathcal{P}^n| \geq \sum_{k=0}^{\lfloor n/2 \rfloor} \binom{n}{2k} k! = n^{n/2 + o(n)} \tag{11.1}$$

for every n. □

In fact, this theorem is deduced from the result that (11.1) holds for properties not containing the complete bipartite graph $K_{t,t}$ for t large enough.

To conclude this section, let us say a few words about *tournaments*, complete oriented graphs. Balogh, Bollobás and Morris [23] proved that the unlabelled speed jumps from polynomial to exponential. More precisely, define a Fibonacci-type sequence of integers by setting $F_0^* = F_1^* = F_2^* = 1$, and $F_n^* = F_{n-1}^* + F_{n-3}^*$ for $n \geq 3$, so that $F_n^* = c^{(1+o(1))n}$ as $n \to \infty$, where $c = 1.47\ldots$ is the largest real root of the polynomial $x^3 = x^2 + 1$.

Theorem 11.8 *Let $\mathcal{P}$ be a hereditary property of tournaments. Then either*

(i) $|\mathcal{P}_n| = \Theta(n^k)$ *for some $k \in \mathbb{N}$, or*

(ii) $|\mathcal{P}_n| \geq F_n^*$ *for every $4 \neq n \in \mathbb{N}$.*

Moreover, this lower bound is attained on a unique property.   □

To describe the unique property $\mathcal{T}$ of tournaments in the theorem above, for $a_1, \ldots, a_m$, $a_i \in \{1, 3\}$, let $T = T(a_1, \ldots, a_m)$ be the tournament with vertex set $\{x(i, j) : i \in [m], j \in [a_i]\}$, in which $x(i, j)$ sends an edge to $x(k, \ell)$ if either $i < k$, or $i = k$ and $\ell - j \equiv 1 \pmod 3$. (Thus T has $\sum_{i=1}^m a_i$ vertices; if $a_i = 1$ from every i then T is the transitive tournament on m vertices; if $a_i = 3$ for every i then T is obtained from the transitive tournament on m vertices by replacing each vertex by a cyclic triangle.) Let $\mathcal{T}$ be the family of (isomorphism classes) of such tournaments. It is easy to check that $\mathcal{T}$ is a hereditary property of tournaments. As the sequence $(a_1, \ldots, a_m)$ can be reconstructed from a tournament $T = T(a_1, \ldots, a_m)$, it follows that $|\mathcal{T}_n| = F_n^*$ for every n, so in Theorem 11.8 (ii) equality is indeed attained on the property $\mathcal{T}$. However, the proof of the inequality is a different matter. In fact, unlike in the case of unlabelled graphs and ordered graphs, in proving Theorem 11.8 we cannot make use of the classification of labelled speeds of graphs in Theorem 6.1.

12   Words

For a set A and a natural number n, a *word of length n over the alphabet A*, or, simply, an *n-word*, is a sequence $w = w_1 w_2 \ldots w_n = (w_i)_1^n$ with $w_i \in A$ for every i. A *finite word* is an n-word for some n. A *$\mathbb{Z}$-word* over A is a $\mathbb{Z}$-sequence $w = \ldots w_{-2} w_{-1} w_0 w_1 w_2 \ldots$ with $w_i \in A$ for every i, and an *$\mathbb{N}$-word* is an $\mathbb{N}$-sequence $w = w_1 w_2 \ldots$. Thus $A^{\mathbb{Z}}$ is the set of $\mathbb{Z}$-words, and $A^{\mathbb{N}}$ is the set of $\mathbb{N}$-words. An *infinite word* is a $\mathbb{Z}$-word or an $\mathbb{N}$-word. From now on we take $A = \{0, 1\}$: this assumption makes no difference to the results.

An *n-block* of a word $w = (w_i)$ is an n-word of the form $w_{j+1} w_{j+2} \ldots w_{j+n}$ for some j; a *block* is an n-block for some n. Note that a word of length N has $N - n + 1$ n-blocks; in particular, a word of length $n + 1$ has two blocks of length n. A word (w_i) is *p-periodic* or *periodic with period p* if $w_{i+p} = w_i$ whenever w_{i+p} and w_i are letters of w. An $\mathbb{N}$-word (w_i) is *eventually periodic* if $(w_i)_{i \geq k}$ is periodic for some $k \geq 1$.

A set $\mathcal{P}$ of finite words over an alphabet A is said to be *hereditary* if any block of a word in $\mathcal{P}$ is also in $\mathcal{P}$. For example, if W is a set of (finite or infinite) words, then the set $\mathcal{P}(W)$ formed by the blocks of the words in W is clearly hereditary. If W consists of a single word w then we write $\mathcal{P}(w)$ for $\mathcal{P}(W)$. Given a property $\mathcal{P}$, we denote by $\mathcal{P}^n$ the set of n-words in $\mathcal{P}$; thus, $\mathcal{P}^n(w)$ is the set of n-blocks of w. The cardinality $|\mathcal{P}^n|$ as a function of n is called the *speed* or *complexity* of $\mathcal{P}$; similarly, $n \mapsto |\mathcal{P}^n(w)|$ is the *complexity of the word w*. For example, if $w = \ldots 010101 \ldots$, then $\mathcal{P}^3(w) = \{010, 101\}$, and $|\mathcal{P}^n(w)| = 2$ for $n \geq 2$. Also, if $W = \{\ldots 00100 \ldots, \ldots 11011 \ldots, \ldots 000111 \ldots\}$ then $\mathcal{P}^n(W) = 3n - 1$ for $n \geq 3$.

The basic result concerning the complexity of an infinite word is the following theorem of Morse and Hedlund [99].

Theorem 12.1 (i) *Let w be a $\mathbb{Z}$-word such that $|\mathcal{P}^k(w)| \leq k$ for some k. Then there is an n such that w is n-periodic, $|\mathcal{P}^\ell(w)| = n$ for every $\ell \geq n$, and $|\mathcal{P}^\ell(w)| \geq \ell + 1$ for every $\ell < n$.*

(ii) *Let v be an $\mathbb{N}$-word such that $|\mathcal{P}^k(v)| \leq k$ for some k. Then there are integers $m \leq n$ such that v is eventually m-periodic and $|\mathcal{P}^\ell(v)| = n$ for every $\ell \geq n - 1$.* $\square$

This theorem is best possible in the sense that there are words w such that $|\mathcal{P}^n(w)| = n + 1$ for every n. Indeed, a simple example is $\ldots 0001000 \ldots$, the word containing only one 1. A less trivial example is the *Fibonacci word*

$$01\,0\,01\,010\,01001\,01001010\,\ldots$$

constructed from the sequence of finite words $a_1, a_2, \ldots$ defined as follows: $a_1 = 0$, $a_2 = 01$ and for $k \geq 3$ the word a_{k+1} is the concatenation of a_k and a_{k-1} (in this order): $a_{k+1} = a_k a_{k-1}$. Equivalently, the Fibonacci sequence is obtained from 0 by repeatedly substituting 01 for 0, and 0 for 1. Thus $a_3 = 01\,0$, $a_4 = 010\,01$, $a_5 = 01001\,010$, and so on. In fact, much work has been done on the infinite words w with $|\mathcal{P}^n(w)| = n + 1$, called *Sturmian words*; see, e.g., Berthé [31] and Berthé.

To see that in (ii) we cannot demand that $n = m$, take the $\mathbb{N}$-word $w = 0001100110011\ldots$. Here $n = 2$ and $m = 5$, and $|\mathcal{P}^\ell| = 5$ for every $\ell \geq 4$.

The complexity of an infinite word has been studied in great detail in many papers (see, e.g., [56], [59], [70, 71], [82, 83], [117]), but general hereditary properties have hardly been considered. Here we shall present some results of Balogh and Bollobás [19] concerning the slow-growing functions that may arise as complexities of general hereditary properties.

As in [19], we define a *word graph* over an alphabet A as a *directed graph* with loops whose edges (including loops) are *decorated* with elements of A such that no two edges with the same initial vertex have the same decoration and all the edges with the same terminal vertex have the same decoration. (In particular, there is at most one loop at every vertex.)

Given $k \geq 2$, the *k-word graph* or *de Bruijn graph* $G_k(w)$ of a word w is defined as follows. The vertex set is the set of k-words in w, and a vertex u sends an edge of decoration i to a vertex v if w contains a $(k + 1)$-word ending in i whose first k-word is u and second (and last) k-word is v. Equivalently, v is obtained from u by omitting its first letter and adding i as its last letter. For a set W, the word graph of W is $G_k(W) = \bigcup_{w \in W} G_k(w)$. Clearly, $G_k(W)$ is a word graph as defined above. (To obtain the usual de Bruijn graph, take for W the set of all $(k+1)$-words.) These word graphs play an important role in the study of complexity. To illustrate this, we reproduce the proof of the Morse–Hedlund theorem from [19]. First, note the following simple characterization of k-word graphs.

Lemma 12.2 (i) *A word graph G is a k-word graph iff any two walks of length at most k ending in the same vertex have the same sequence of decorations, and any two walks of length k with the same sequence of decorations end in the same vertex. If every vertex is the terminal vertex of a walk of length k then the alphabet of the k-word graph is the set formed by the decorations of the edges.*

(ii) *A k-word graph is of the form $G_k(w)$ for some n-word w iff it has a (directed) walk of length $n - k + 1$ passing through all edges.* $\square$

Now, in proving Theorem 12.1(i), we may assume that $|\mathcal{P}^1(w)| = 2$. If two words in $\mathcal{P}^{k+1}(w)$ have different initial k-words then they are themselves different. Consequently, the complexity $|\mathcal{P}^k(w)|$ is a monotone increasing function of k, and so $|\mathcal{P}^{n-1}(w)| = |\mathcal{P}^n(w)| = n$ for some n, i.e., the $(n-1)$-word graph $G = G_{n-1}(w)$ has n vertices and n edges. Since G has a walk containing all the edges, and every vertex in G has indegree at least 1 and outdegree at least 1, the graph G is an (oriented) n-cycle. Therefore, the word w is n-periodic and $|\mathcal{P}^\ell(w)| = n$ for every $\ell \geq n$, as claimed, proving (i).

Part (ii) needs only a little more work. Indeed, proceeding as in (i), we find that in the word graph $G_{n-1}(v)$ of the word $v = v_1 v_2 \ldots$ every vertex has outdegree at least 1 and, with the exception of at most one vertex (the word $v_1 \ldots v_{n-1}$), all vertices have indegree at least 1. This implies that $G_{n-1}(v)$ is a cycle together with a path ending on the cycle. (This path may have length 0.) Consequently, v becomes periodic (of some period $m \leq n$) if we omit its initial segment formed by the letters not in the blocks corresponding to the vertices of $G_{n-1}(v)$ forming this cycle, and $|\mathcal{P}^\ell(w)| = n$ for every $\ell \geq n$. This completes the proof of Theorem 12.2.

Another tool in the study of the complexity of a property is the following fundamental theorem of Fine and Wilf [72] concerning periods of words. As usual, given natural numbers p and q, we write (p, q) for their greatest common divisor.

Theorem 12.3 *Let w be a word of length n with periods p and q. If (p, q) is not a period of w then $n \leq p + q - (p, q) - 1$, and this inequality is best possible.* $\square$

By a careful analysis of word graphs, Bollobás and Balogh [19] proved the following extension of Theorem 12.1.

Theorem 12.4 *Let $\mathcal{P}$ be a hereditary property of finite words over an alphabet A. Then the complexity $|\mathcal{P}^n|$ is either bounded, or at least $n + 1$ for every n.* $\square$

It is tempting to conjecture that, as in the Morse–Hedlund theorem, Theorem 12.1, if the speed of a hereditary property is bounded then it is eventually constant. In fact, this is far from the case: the complexity may oscillate considerably, even if it is bounded. Indeed, for $s \geq 1$ there is a hereditary property $\mathcal{P}$ of finite words such that

$$\limsup_{n \to \infty} |\mathcal{P}^n| = s^2 \quad \text{and} \quad \liminf_{n \to \infty} |\mathcal{P}^n| = 2s - 1;$$

also, $|\mathcal{P}^{4rs}| = s^2$ and $|\mathcal{P}^{(4r-2)s}| = 2s - 1$ for every $r \geq 1$.

Similarly, for $s \geq 1$ there is a hereditary property $\mathcal{P}$ of finite words such that

$$\limsup_{n \to \infty} |\mathcal{P}^n| = s(s + 1) \quad \text{and} \quad \liminf_{n \to \infty} |\mathcal{P}^n| = 2s;$$

also, $|\mathcal{P}^{4rs}| = s(s + 1)$ and $|\mathcal{P}^{(4r-2)s}| = 2s$ for every $r \geq 1$.

However, if $|\mathcal{P}^n| \leq n$ for some n, then the complexity is not only bounded, but cannot even be larger than in the examples above, even if we do not demand oscillation.

Theorem 12.5 *Let $\mathcal{P}$ be a hereditary property of finite words over an alphabet A such that $|\mathcal{P}^n| = m \leq n$. Then for all $k \geq n+m$ we have $|\mathcal{P}^k| \leq \lfloor (m+1)/2 \rfloor \cdot \lceil (m+1)/2 \rceil$. Furthermore, this inequality is sharp.* □

Turning to words of higher complexity, Ferenczi [71] constructed an $\mathbb{N}$-word w with $\liminf_{n\to\infty} |\mathcal{P}^n(w)|/n = 2$, and $\limsup_{n\to\infty} |\mathcal{P}^n(w)|/n^\beta = \infty$ for every β. In [19] it was shown that for a property $\mathcal{P}(w)$ with $\liminf_{n\to\infty} |\mathcal{P}^n(w)|/n = 2$ much wilder oscillation is possible: it may happen that, as $n \to \infty$ through some subsequence, $|\mathcal{P}^n(w)|$ grows at an almost exponential rate. More precisely, given a function $\alpha(n) = o(\log n)$ with $\alpha(n) \to \infty$, there is an $\mathbb{N}$-word w such that

$$\liminf_{n\to\infty} |\mathcal{P}^n(w)|/n = 2, \quad \text{and} \quad \liminf_{n\to\infty} |\mathcal{P}^n(w)|/2^{n/\alpha(n)} = \infty.$$

The function $n \mapsto |\mathcal{P}^n(w)|$ is not the only way of measuring the richness of the structure of an infinite word. For example, an interesting measure was introduced by Kamae and Zamboni [82]. Let $\tau = \{\tau_1, \ldots, \tau_k\} \subset \mathbb{N}$, and for a word $w = (w_i)$ let $\mathcal{P}^\tau(w)$ be the set of words of the form $w_{\tau_1+t} w_{\tau_2+t} \ldots w_{\tau_k+t}$. (Note that for $\tau = \{1, \ldots, k\}$ we have $\mathcal{P}^\tau(w) = \mathcal{P}^k(w)$.) Then $p_w(k) = \max_{|\tau|=k} |\mathcal{P}^\tau(w)|$ is the *maximal pattern complexity* of w.

Kamae and Zamboni [82, 83] proved that, in the analogue of the Morse–Hedlund theorem for maximal pattern complexity, the cut-off is at $2k - 1$, rather than k.

Theorem 12.6 *An infinite word w over a finite alphabet is eventually periodic if and only if $p_w(k) \leq 2k - 1$ for some k.* □

Rather than arranging the letters in a linear order, using $\mathbb{N}$ or $\mathbb{Z}$ as the index set, we may consider multi-dimensional patterns. Thus, a *d-dimensional infinite word* over an alphabet A is an element of $A^{\mathbb{Z}^d}$ (or $A^{\mathbb{N}^d}$). A word $w = (w_\mathbf{n})$ is *periodic* if there is a vector $\mathbf{p} \in \mathbb{Z}^d$ such that $w_{\mathbf{n}+\mathbf{p}} = w_\mathbf{n}$ for every $\mathbf{n} \in \mathbb{Z}^d$. A fair amount of work has been done on multi-dimensional words (see, e.g., [32], [54, 55], [60], [108], [109, 110], [115], [117]), although, not surprisingly, most of this work concerns the case $d = 2$.

The natural analogue of the notion of complexity of a word $\mathbf{w} \in A^{\mathbb{Z}^d}$ is the *box complexity function* $N_\mathbf{w}(b_1, \ldots, b_d)$, the number of distinct $b_1 \times \cdots \times b_d$-blocks in $\mathbf{w}$. What corresponds to the Morse–Hedlund theorem for box complexity? As shown by the word $\mathbf{w}$ consisting of a single 1 entry in a sea of 0s, we may have $N_\mathbf{w}(b_1, \ldots, b_d) = b_1 \ldots b_d + 1$ for all $(b_i)_1^d$ without $\mathbf{w}$ being periodic. For $d = 2$, Nivat [100] conjectured that this example is best possible and so the exact analogue of the Morse–Hedlund theorem holds: if there are positive integers b_1 and b_2 such that $N_\mathbf{w}(b_1, b_2) \leq b_1 b_2$ then $\mathbf{w}$ is periodic.

Epifanio, Koskas and Mignosi [60] were the first to prove a weak form of Nivat's conjecture when they showed that $\mathbf{w}$ is indeed periodic if the condition is strengthened to $b_1 b_2/144$. Quas and Zamboni [101] went further: they showed that a 2-dimensional word $\mathbf{w}$ is periodic if $N_\mathbf{w}(b_1, b_2) = b_1 b_2/12$ for some $b_1, b_2 \geq 1$. However, it seems that these results are still rather far from a proof of Nivat's beautiful conjecture.

References

[1] D. Achlioptas and A. Naor, The two possible values of the chromatic number of a random graph, *Ann. of Math. (2)* **162** (2005), 1335–1351.

[2] M.H. Albert, M. Elder, A. Rechnitzer, P. Westcott and M. Zabrocki, On the Stanley-Wilf limit of 4231-avoiding permutations and a conjecture of Arratia, *Adv. in Appl. Math.* **36** (2006), 96–105.

[3] V.E. Alekseev, Hereditary classes and coding of graphs (in Russian), *Probl. Cybern.* **39** (1982), 151–164.

[4] V.E. Alekseev, On the entropy values of hereditary classes of graphs (in Russian), *Discrete Math. Appl.* **3** (1993), 191–199.

[5] V.E. Alekseev, S.V. Sorochan, On the entropy of hereditary classes of oriented graphs (in Russian), International Conference DAOR'2000, *Diskretn. Anal. Issled. Oper. Ser. 1*, **7** (2000), 20–28.

[6] G.R. Allan, An inequality involving product measures, in *Radical Banach Algebras and Automatic Continuity* (J.M. Bachar et al., eds.), Lecture Notes in Mathematics **975**, Springer-Verlag, 1981, pp. 277–279.

[7] N. Alon and K.A. Berman, Regular hypergraphs, Gordon's lemma and invariant theory, *J. Combinatorial Theory Ser. A* **43** (1986), 91-97.

[8] N.Alon, E. Fischer, M. Krivelevich and M. Szegedy, Efficient testing of large graphs, *40th Annual Symposium on Foundations of Computer Science* (New York, 1999), 656–666, IEEE Computer Soc., Los Alamitos, CA, 1999.

[9] N.Alon, E. Fischer, M. Krivelevich and M. Szegedy, Efficient testing of large graphs, *Combinatorica* **20** (2000), 451–476.

[10] N. Alon and E. Friedgut, On the number of permutations avoiding a given pattern, *J. Combin. Theory A* **89** (2000), 133–140.

[11] N. Alon and M. Krivelevich, The concentration of the chromatic number of random graphs, *Combinatorica* **17** (1997), 303–313.

[12] N. Alon and A. Shapira, A characterization of the (natural) graph properties testable with one-sided error, *Proc. of the 46 IEEE FOCS, IEEE* (2005), 429–438.

[13] Alon, Shapira and Sudakov, Additive approximation for edge-deletion problems, *Proc. of the 46 IEEE FOCS, IEEE* (2005), 419–428.

[14] N. Alon and U. Stav, What is the furthest graph from a hereditary property?, to appear

[15] N. Alon and U. Stav, The maximum edit distance from hereditary graph properties, to appear

[16] R. Arratia, On the Stanley-Wilf conjecture for the number of permutations avoiding a given pattern, *Electron. J. Combin.* **6** (1999), Note, N1 4 pp. (electronic)

[17] M. Axenovich, A. Kézdy and R. Martin, On editing distance in graphs, *J. Graph Theory*, submitted.

[18] M. Axenovich and R. Martin, Avoiding patterns in matrices via a small number of changes, *SIAM J. Discrete Math.* **20** (2006), 49–54.

[19] J. Balogh and B. Bollobás, Hereditary properties of words, *Theor. Inform. Appl.* **39** (2005), 49–65.

[20] J. Balogh, B. Bollobás and R. Morris, Hereditary properties of ordered graphs, in *Topics in Discrete Mathematics*, pp. 179–213, Algorithms Combin. **26**, Springer, Berlin, 2006.

[21] J. Balogh, B. Bollobás and R. Morris, Hereditary properties of partitions, ordered graphs and ordered hypergraphs, *European J. Combin.* **27** (2006), 1263–1281.

[22] J. Balogh, B. Bollobás and R. Morris, Hereditary properties of combinatorial structures: posets and oriented graphs, to appear

[23] J. Balogh, B. Bollobás and R. Morris, Hereditary properties of tournaments, to appear

[24] J. Balogh, B. Bollobás, M. Saks and V.T. Sós, The unlabelled speed of a hereditary graph property, to appear

[25] J. Balogh, B. Bollobás and M. Simonovits, The number of graph without forbidden subgraphs, *J. Combinatorial Theory B* **91** (2004), 1–24.

[26] J. Balogh, B. Bollobás and M. Simonovits, On the structure of almost all graphs without fixed forbidden subgraphs, to appear

[27] J. Balogh, B. Bollobás and D. Weinreich, The speed of hereditary properties of graphs, *J. Combinatorial Theory B* **79** (2000), 131–156.

[28] J. Balogh, B. Bollobás and D. Weinreich, The penultimate rate of growth for graph properties, *European J. Combin.* **22** (2001), 277–289.

[29] J. Balogh, B. Bollobás and D. Weinreich, Measures on monotone properties of graphs, *Discrete Appl. Math.* **116** (2002), 17–36.

[30] J. Balogh, B. Bollobás and D. Weinreich, A jump to the Bell number for hereditary graph properties, *J. Combin. Theory B* **95** (2005), 29–48.

[31] V. Berthé, Sequences of low complexity: automatic and Sturmian sequences, in *Topics in Symbolic Dynamics and Applications (Temuco, 1997)*, London Math. Soc. Lecture Note Series **279**, Cambridge University Press, Cambridge, 2000, pp. 1–34.

[32] V. Berthé and L. Vuillon, Tilings and rotations: a two-dimensional generalization of Sturmian sequences, *Discrete Math.* **223** (2000), 27–33.

[33] B. Bollobás, *Random Graphs*, Academic Press, London, 1985, xvi + 447 pp. Second edition: Cambridge Studies in Advanced Mathematics **73**, Cambridge University Press, Cambridge, 2001, xviii + 498 pp.

[34] B. Bollobás, The chromatic number of random graphs, *Combinatorica* **8** (1988), 49–55.

[35] B. Bollobás, Martingales, isoperimetric inequalities and random graphs, in *Combinatorics (Eger, 1987)* (A. Hajnal, L. Lovász and V.T. Sós, eds.), Colloq. Math. Soc. János Bolyai **52**, North-Holland, Amsterdam (1988), pp. 113–139.

[36] B. Bollobás, *Extremal Graph Theory*, Academic Press (1978), xx + 488 pp. Reprint of the 1978 original: Dover Publications, Inc., Mineola, NY, 2004.

[37] B. Bollobás, *Modern Graph Theory*, Graduate Texts in Mathematics **184**, Springer-Verlag, New York, 1998. xiv + 394 pp.

[38] B. Bollobás and P. Erdős, On the structure of edge graphs, *Bull. London Math. Soc.* **5** (1973), 317–321.

[39] B. Bollobás and P. Erdős, Cliques in random graphs, *Math. Proc. Cambridge Philos. Soc.* **80** (1976), 419–427.

[40] B. Bollobás, P. Erdős and M. Simonovits, On the structure of edge graphs II, *J. London Math. Soc.* **12** (1976), 219–224.

[41] B. Bollobás and Y. Kohayakawa, An extension of the Erdős-Stone theorem, *Combinatorica* **14** (1994), 279–286.

[42] B. Bollobás and A. Thomason, Projections of bodies and hereditary properties of hypergraphs, *Bull. London Math. Soc.* **27** (1995), 417–424.

[43] B. Bollobás and A. Thomason, Generalized chromatic numbers of random graphs, *Random Structures and Algorithms* **6** (1995), 353–356.

[44] B. Bollobás and A. Thomason, Hereditary and monotone properties of graphs, in *The Mathematics of Paul Erdős, II*, Algorithms Combin. **14**, Springer, Berlin, 1997, pp. 70–78.

[45] B. Bollobás and A. Thomason, The structure of hereditary properties and colourings of random graphs, *Combinatorica* **20** (2000), 173–202.

[46] M. Bóna, Exact enumeration of 1342-avoiding permutations: a close link with labeled trees and planar maps, J. Comb. Theory, Ser. A, 80 (1997), 257–272.

[47] M. Bóna, Permutations avoiding certain patterns: The case of length 4 and some generalizations, *Discrete Math.* **175** (1997), 55–67.

[48] M. Bóna, The solution of a conjecture of Stanley and Wilf for all layered patterns, *J. Combin. Theory A* **85** (1999), 96–104.

[49] M. Bóna, A simple proof for the exponential upper bound for some tenacious patterns, *Adv. in Appl. Math.* **33** (2004), 192–198.

[50] M. Bóna, The limit of a Stanley–Wilf sequence is not always rational, and layered patterns beat monotone patterns, *J. Combin. Theory A* **110** (2005), 223–235.

[51] G. Brightwell, D.A. Grable and H.J. Prömel, Forbidden induced partial orders, *Discrete Math.* **201** (1999), 53–80.

[52] G. Brightwell, H.J. Prömel and A. Steger, The average number of linear extensions of a partial order, *J. Combin. Theory A* **73** (1996), 193–206.

[53] Yu.D. Burago and V.A. Zalgaller, *Geometric Inequalities*, Springer-Verlag, 1988, xiv + 331 pp.

[54] J. Cassaigne, Double sequences with complexity $mn+1$, *J. Auto. Lang. Comb.* **4** (1999), 153–170.

[55] J. Cassaigne, Subword complexity and periodicity in two or more dimensions, in *Developments in Language Theory. Foundations, Applications, and Perspectives* (DLT'99), Aachen, Germany, World Scientific, Singapore, 2000, pp. 14–21.

[56] J. Cassaigne, S. Ferenczi and L. Q. Zamboni, Imbalances in Arnoux-Rauzy sequences, *Ann. Inst. Fourier (Grenoble)* **50** (2000), 1265–1276.

[57] V. Chvátal and E. Szemerédi, On the Erdős–Stone theorem, *J. London Math. Soc.* **23** (1981), 207–214.

[58] F.R.K. Chung, R.L. Graham, P. Frankl and J.B. Shearer, Some intersection theorems for ordered sets and graphs, *J. Combinatorial Theory Ser. A* **43** (1986), 23–37.

[59] E. Coven and G.A. Hedlund, Sequences with minimal block growth, *Math. Syst. Theor.* **7** (1973), 138–153.

[60] C. Epifanio, M. Koskas and F. Mignosi, On a conjecture on bi-dimensional words, *Theoret. Comput. Sci.* **299** (2003) 123–150.

[61] P. Erdős, Some recent results on extremal problems in graph theory, in *Theory of Graphs*, (Internat. Sympos., Rome, 1966), pp. 117–123 (English); pp. 124–130 (French), Gordon and Breach, New York; Dunod, Paris.

[62] P. Erdős, On some new inequalities concerning extremal properties of graphs, in *Theory of Graphs* (Proc. Colloq., Tihany, 1966), Academic Press, New York, pp. 77–81.

[63] P. Erdős, P. Frankl and V. Rödl, The asymptotic enumeration of graphs not containing a fixed subgraph and a problem for hypergraphs having no exponent, *Graphs and Combinatorics* **2** (1986), 113–121.

[64] P. Erdős, D.J. Kleitman and B.L. Rothschild, Asymptotic enumeration of K_n-free graphs, in *International Coll. Comb.*, Atti dei Convegni Lincei (Rome) **17** (1976), 3–17.

[65] P. Erdős and A. Rényi, On random graphs I, *Publ. Math. Debrecen* **6** (1959), 290–297.

[66] P. Erdős and A. Rényi, On the evolution of random graphs, *Publ. Math. Inst. Hungar. Acad. Sci.* **5** (1961), 17–61.

[67] P. Erdős and M. Simonovits, A limit theorem in graph theory, *Studia Sci. Math. Hungar.* **1** (1966), 51–57.

[68] P. Erdős and M. Simonovits, Some extremal problems in graph theory, in *Combinatorial Theory and its Applications*, (Proc. Colloq., Balatonfüred, 1969), North-Holland, Amsterdam, 1970, pp. 377–390.

[69] P. Erdős and A.H. Stone, On the structure of linear graphs, *Bull. Amer. Math. Soc.* **52** (1946), 1087–1091.

[70] S. Ferenczi, Rank and symbolic complexity, *Ergodic Theory Dyn. Systems* **16** (1996), 663–682.

[71] S. Ferenczi, Complexity of sequences and dynamical systems, *Discrete Mathematics* **206** (1999), 145–154.

[72] N.J. Fine and H.S. Wilf, Uniqueness theorems for periodic functions, *Proc. Amer. Math. Soc.* **16** (1965), 109–114.

[73] A.M. Frieze and T. Luczak, On the independence and chromatic numbers of random regular graphs, *J. Combinat. Theory B* **54** (1992), 123–132.

[74] Z. Füredi, Indecomposable regular graphs and hypergraphs, *Discrete Mathematics* **101** (1992), 59-64.

[75] Z. Füredi and P. Hajnal, Davenport–Schinzel theory of matrices, *Discrete Math.* **103** (1992), 231–251.

[76] W.T. Gowers, Lower bounds of tower type for Szemerédi's uniformity lemma, *Geom. Funct. Anal.* **7** (1997), 322-332.

[77] J.E. Graver, A survey of the maximum depth problem for indecomposable exact covers, in *Infinite and Finite Sets* (Keszthely, 1973), Proc. Colloq. Math. Soc. J. Bolyai **10**, North-Holland, Amsterdam, 1976, pp. 731-743.

[78] G.R. Grimmett and C.J.H. McDiarmid, On colouring random graphs, *Math. Proc. Cambridge Philos. Soc.* **77** (1975), 313–324.

[79] H. Hadwiger, *Vorlesungen über Inhalt, Oberfläche und Isoperimetrie*, Springer-Verlag, 1957, xiii + 312 pp.

[80] Y. Ishigami, Proof of a conjecture of Bollobás and Kohayakawa on the Erdős–Stone theorem, *J. Combin. Theory B* **85** (2002), 222–254.

[81] T. Kaiser and M. Klazar, On growth rates of closed permutation classes, *Electron. J. Combin.* **9** (2002/03), no. 2, Research paper 10, 20 pp. (electronic)

[82] T. Kamae and L. Zamboni, Sequence entropy and the maximal pattern complexity of infinite words, *Ergod. Th. and Dynam. Sys.* **22** (2002), 1191–1199.

[83] T. Kamae and L. Zamboni, Maximal pattern complexity for discrete systems, *Ergod. Th. and Dynam. Sys.* **22** (2002), 1201–1214.

[84] M. Klazar, The Füredi–Hajnal conjecture implies the Stanley–Wilf conjecture, in *Formal Power Series and Algebraic Combinatorics* (Proc. 12th International conference FPSAC'00, Moscow, 2000), Springer, Berlin 2000; pp. 250–255.

[85] M. Klazar, Counting pattern-free set partitions I. A generalization of Stirling numbers of the second kind, *Eur. J. Comb.*, **21** (2000), 367–378.

[86] M. Klazar, Counting pattern-free set partitions II. Noncrossing and other hypergraphs, *Electron. J. Combin.* **7** (2000), Research Paper 34, 25 pp. (electronic)

[87] M. Klazar and A. Marcus, Extensions of the linea bound in the Füredi–Hajnal conjecture, *Advances in Applied Math.* **38** (2006), 258–266.

[88] D.J. Kleitman and B.L. Rothschild, The number of finite topologies, *Proc. Amer. Math. Soc.* **25** (1970), 276–282.

[89] D.J. Kleitman and B.L. Rothschild, Asymptotic enumeration of partial orders on a finite set, *Trans. Amer. Math. Soc.* **205** (1975), 205–220.

[90] Ph.G. Kolaitis, H.J. Prömel and B.L. Rothschild, K_{l+1}-free graphs: asymptotic structure and a 0-1-law, *Trans. Amer. Math. Soc.* **303** (1987), 637–671.

[91] T. Kővári, V. T. Sós, and P. Turán, On a problem of Zarankiewicz, Colloq. Math. **3** (1954) 50–57.

[92] L.H. Loomis and H. Whitney, An inequality related to the isoperimetric inequality, *Bull. Amer. Math. Soc.* **55** (1949), 961–962.

[93] L. Lovász and B. Szegedy, Graph limits and testing hereditary graph properties, to appear

[94] T. Łuczak, The chromatic number of random graphs, *Combinatorica* **11** (1991), 45–54.

[95] T. Łuczak, A note on the sharp concentration of the chromatic number of random graphs, *Combinatorica* **11** (1991), 295–297.

[96] W. Mantel, Problem 28, Winkundige Opgaven **10** (1907), 60–61.

[97] A. Marcus and G. Tardos, Excluded permutation matrices and the Stanley-Wilf conjecture, *J. Combin. Theory A* **107** (2004), 153–160.

[98] D.W. Matula, Expose-and-merge exploration and the chromatic number of a random graph, *Combinatorica* **7** (1987), 275–284.

[99] M. Morse and A. G. Hedlund, Symbolic dynamics, *Amer. J. Math*, **60** (1938), 815–866.

[100] M. Nivat, Invited talk at ICALP, Bologna, 1997.

[101] A. Quas and L. Zamboni, Periodicity and local complexity, *Theoret. Comput. Sci.* **319** (2004), 229–240.

[102] H.J. Prömel and A. Steger, Excluding induced subgraphs: quadrilaterals, *Random Structures and Algorithms* **2** (1991), 55–71.

[103] H.J. Prömel and A. Steger, Excluding induced subgraphs II: Extremal graphs, *Discrete Applied Mathematics* **44** (1993), 283–294.

[104] H.J. Prömel and A. Steger, Excluding induced subgraphs III: a general asymptotic, *Random Structures and Algorithms* **3** (1992), 19–31.

[105] H.J. Prömel and A. Steger, The asymptotic structure of H-free graphs, in *Graph Structure Theory* (N. Robertson and P. Seymour, eds), Contemporary Mathematics **147**, Amer. Math. Soc., Providence, 1993, pp. 167-178.

[106] H.J. Prömel and A. Steger, Almost all Berge graphs are perfect, *Combinatorics, Probability and Computing* **1** (1992) 53-79.

[107] F.P. Ramsey, On a problem of formal logic, *Proc. London Math. Soc.* **30** (1992), 264–286.

[108] M. Regnier and L. Rostami, A unifying look at d-dimensional periodicities and space coverings, in *Proc. Fourth Symp. on Combinatorial Pattern Matching*, 1993, pp. 215–227.

[109] J.W. Sander and R. Tijdeman, The complexity function on lattices, *Theoret. Comput. Sci.* **246** (2000), 195–225.

[110] J.W. Sander and R. Tijdeman, The rectangular complexity of functions on two-dimensional lattices, *Theoret. Comput. Sci.* **270** (2002), 857–863.

[111] E.R. Scheinerman, Generalized chromatic numbers of random graphs, *SIAM J. Discrete Math.* **5** (1992), 74–80.

[112] E.R. Scheinerman and J. Zito, On the size of hereditary classes of graphs, *J. Combinatorial Theory Ser. B*, **61** (1994), 16–39.

[113] E. Shamir and J. Spencer, Sharp concentration of the chromatic number on random graphs $G_{n,p}$, *Combinatorica* **7** (1987), 121–129.

[114] M. Simonovits, A method for solving extremal problems in graph theory; stability problems, in *Theory of Graphs* (Proc. Colloq., Tihany, 1966), Academic Press, New York, pp. 279–319.

[115] R.J. Simpson and R. Tijdeman, Multi-dimensional versions of a theorem of Fine and Wilf and a Formula of Sylvester, *Proc. Amer. Math. Soc.* **131** (2003), 1661–1671.

[116] E. Szemerédi, Regular partitions of graphs, in *Problèmes Combinatoires et Théorie des Graphes*, (Proc. Colloq. Internat. CNRS, Univ. Orsay, Orsay, 1976; J.-C. Bermond, J.-C. Fournier, M. las Vergnas and D. Sotteau, eds), Colloq. Internat. CNRS **260**, CNRS, Paris, 1978, pp. 399–401.

[117] R. Tijdeman, Periodicity and almost-periodicity, in *More Sets, Graphs and Numbers*, Bolyai Soc. Math. Stud. **15**, Springer, Berlin, 2006, pp. 381–405.

[118] P. Turán, On an extremal problem in graph theory (in Hungarian), *Mat. Fiz. Lapok* **48** (1941) 436–452.

[119] H. S. Wilf, The patterns of permutations, *Discrete Math.* **257** (2002), 575–583.

Department of Pure Mathematics and Mathematical Statistics
University of Cambridge, Cambridge, UK
and
Department of Mathematical Sciences
University of Memphis, Memphis, TN, USA.
b.bollobas@dpmms.cam.ac.uk

Ordering Classes of Matrices of 0's and 1's

Richard A. Brualdi

Abstract

In this article we consider various ways in which certain subclasses of $(0,1)$-matrices may be ordered. In the case of general $(0,1)$-matrices, this is equivalent to ordering bipartite graphs; in the case of symmetric $(0,1)$-matrices with zero trace, this is equivalent to ordering graphs. Except for the Bruhat order and its generalizations, these orders are only quasiorders and we emphasize the extremal cases.

1 Introduction

Let $R = (r_1, r_2, \ldots, r_m)$ and $S = (s_1, s_2, \ldots, s_n)$ be positive integral vectors satisfying

$$r_1 + r_2 + \cdots + r_m = s_1 + s_2 + \cdots + s_n.$$

The class $\mathcal{A}(R,S)$ consists of all m by n $(0,1)$-matrices whose row sum vector is R and whose column sum vector is S. This class may be empty without further restrictions on R and S; in particular, $\mathcal{A}(R,S) \neq \emptyset$ implies that $r_i \leq n$ for each i and $s_j \leq m$ for each j. The nonemptiness is not affected by reordering the entries of R and S, and so we assume that R and S are nonincreasing:

$$n \geq r_1 \geq r_2 \geq \cdots \geq r_m \text{ and } m \geq s_1 \geq s_2 \geq \cdots \geq s_n.$$

The *conjugate* of R is the vector $R^* = (r_1^*, r_2^*, \ldots, r_n^*)$ where

$$r_j^* = |\{i : 1 \leq i \leq m, r_i \geq j\}| \quad (i = 1, 2, \ldots n).$$

The vector R^* satisfies

$$r_1^* \geq r_2^* \geq \cdots \geq r_n^* \text{ and } r_1^* + r_2^* + \cdots + r_n^* = r_1 + r_2 + \cdots + r_m.$$

The vector R^* is the column sum vector of the *perfectly nested matrix* $A(R,n)$ with row sum vector R such that the 1's in each row occur in its initial positions (and so the 1's in each column also occur in the initial positions). The matrix $A(R,n)$ is the unique matrix in the class $\mathcal{A}(R, R^*)$. Clearly, each matrix in $\mathcal{A}(R,S)$, if nonempty, can be obtained from $A(R,n)$ by shifting 1's to the right in each row. The class $\mathcal{A}(R,S)$ can be identified with the class of bipartite graphs with a bipartition into sets X and Y of sizes m and n with R and S being the degree sequences of vertices in X and Y, respectively.

The *dominance order* (or *majorization order*) is the partial order defined on nonincreasing vectors $U = (u_1, u_2, \ldots, u_n)$ and $V = (v_1, v_2, \ldots, v_n)$ of the same length n by $U \preceq V$ if and only if

$$u_1 + u_2 + \cdots + u_k \leq v_1 + v_2 + \cdots + v_k \quad (k = 1, 2, \ldots, n)$$

with equality for $k = n$. The *Gale-Ryser theorem* asserts that $\mathcal{A}(R,S)$ is nonempty if and only if $S \preceq R^*$. One important property of $\mathcal{A}(R,S)$ is a certain connectivity

41

property due to Ryser: Given matrices A_1 and A_2 in $\mathcal{A}(R,S)$, then A_1 can be transformed into A_2 by a sequence of *interchanges*

$$L_2 = \begin{bmatrix} 0 & 1 \\ 1 & 0 \end{bmatrix} \leftrightarrow \begin{bmatrix} 1 & 0 \\ 0 & 1 \end{bmatrix} = I_2$$

each of which replaces a submatrix of A_1 equal to L_2 with I_2, or the other way around. Such a sequence of interchanges applied to a matrix in $\mathcal{A}(R,S)$ always results in a matrix in $\mathcal{A}(R,S)$.

Let k and n be positive integers with $k \leq n$, and let $R = S$ be the constant vector $(k, k, \ldots, k)$ of length n. Then $\mathcal{A}(R,S)$ is clearly nonempty and we denote this class by $\mathcal{A}(n,k)$. The class $\mathcal{A}(n,1)$ is the class of permutation matrices of order n, and hence can be identified with the symmetric group $\mathcal{S}_n$, that is, with the group of permutations of $\{1, 2, \ldots, n\}$.

For later reference we introduce several other classes of matrices.

(1) The class $\mathcal{A}^s(R)$ of all symmetric $(0,1)$-matrices with row and column sum vector R. (The class of graphs with a loop possible at each vertex having degree sequence R, but note that loops will only contribute 1 to degrees.)

(2) The class $\mathcal{A}_0^s(R)$ of all symmetric $(0,1)$-matrices with row and column sum vector R and with zero trace. (The class of graphs with degree sequence R.)

(3) The class $\mathcal{A}_0^s(n,k)$ of all symmetric $(0,1)$-matrices of order n with zero trace having k 1's in each row and column. (The class of regular graphs with n vertices having degree k.) Note that if n is odd, then k must be even for this class to be nonempty.

(4) The class $\mathcal{A}(n|\tau)$ of all $(0,1)$-matrices of order n with exactly τ 1's ($\tau \leq n^2$).

(5) The class $\mathcal{A}_0^s(n|\tau)$ of all symmetric $(0,1)$-matrices of order n with zero trace and exactly τ 1's above the main diagonal ($\tau \leq n(n-1)/2$). (The class of graphs with n vertices and τ edges.)

(6) The class $\mathcal{A}_0^s(n|\tau, \mathrm{irr})$ of all symmetric, irreducible $(0,1)$-matrices of order n with zero trace and exactly τ 1's above the main diagonal ($n-1 \leq \tau \leq n(n-1)/2$). (The class of connected graphs with n vertices and τ edges.)

(7) The class $\mathcal{Z}^+(R,S)$ of all nonnegative integral matrices (matrices each of whose entries is a nonnegative integer) with row sum vector R and column sum vector S. This class is nonempty if and only if the sum of the entries of R equals the sum of the entries of S.

2   Bruhat order on $\mathcal{A}(R,S)$

Let π and τ be permutations in $\mathcal{S}_n$. Then π precedes τ in the *Bruhat order*, written $\pi \preceq_B \tau$, provided π can be obtained from τ by a sequence of *inversion-reducing transpositions* of the form

$$(i_1, \ldots, i_k, \ldots, i_l, \ldots, i_n) \to (i_1, \ldots, i_l, \ldots, i_k, \ldots, i_n) \text{ where } i_k > i_l.$$

In terms of the corresponding permutation matrices P and Q, we have $P \preceq_B Q$ if and only if P can be obtained from Q by a sequence of (one-sided) interchanges $L_2 \to I_2$. The Bruhat order is a partial order on the symmetric group $\mathcal{S}_n$ with the identity permutation $1, 2, \ldots, n$ (with no inversions) as the unique minimal permutation and the anti-identity permutation $n, n-1, \ldots, 2, 1$ (with $n(n-1)/2$ inversions) as the unique maximal permutation. The partially ordered set $(\mathcal{S}_n, \preceq_B)$ is graded by the number of inversions. In terms of matrices, the minimal (in the Bruhat order) permutation matrix of order n is the identity matrix I_n, and the maximal permutation matrix is the permutation matrix L_n with 1's in positions $(1, n), (2, n-2), \ldots, (n-1, 2), (n, 1)$.

There are many equivalent ways to define the Bruhat order on permutations [3, 15, 19, 26]. One of these (see [3]) goes like this. Let $\sigma = i_1, i_2, \ldots, i_n$ and $\tau = j_1, j_2, \ldots, j_n$ be permutations of $\{1, 2, \ldots, n\}$. For each k with $1 \leq k \leq n-1$, let $i_{k1}, i_{k2}, \ldots, i_{kk}$ be the increasing rearrangement of $i_1, i_2, \ldots, i_k$. Let $j_{k1}, j_{k2}, \ldots, j_{kk}$ be defined in a similar way. Then $\sigma \preceq_B \tau$ if and only if $i_{kp} \leq j_{kp}$ for all p and k with $1 \leq p \leq k \leq n-1$. For example, if $\sigma = 2, 1, 4, 5, 3$ and $\tau = 3, 1, 5, 4, 2$, then $\sigma \preceq_B \tau$ because of the entrywise inequalities satisfied by the arrays

$$
\begin{array}{llll}
1 & 2 & 4 & 5 \\
1 & 2 & 4 & \\
1 & 2 & & \\
2 & & &
\end{array}
\quad \text{and} \quad
\begin{array}{llll}
1 & 3 & 4 & 5 \\
1 & 3 & 5 & \\
1 & 3 & & \\
3 & & &
\end{array}.
$$

The number of comparisons in this criterion equals $\binom{n}{2}$ and this was reduced in [3].

The above characterization of the Bruhat order on $\mathcal{S}_n$ can be rephrased allowing for the possibility of extension to more general classes $\mathcal{A}(R, S)$. For an m by n matrix $A = [a_{ij}]$, let

$$
\sigma_{ij}(A) = \sum_{k=1}^{i} \sum_{l=1}^{j} a_{kl} \quad (i = 1, 2, \ldots, m; j = 1, 2, \ldots, n),
$$

the sum of the entries of A in its leading i by j submatrix. Define an m by n matrix by

$$
\Sigma_A = [\sigma_{ij}(A); i = 1, 2, \ldots, m; j = 1, 2, \ldots, n].
$$

For real matrices $X = [x_{ij}]$ and $Y = [y_{ij}]$ of the same size, we write $X \geq Y$ provided $x_{ij} \geq y_{ij}$ for all i, j, and $X > Y$ provided $X \geq Y$ but $X \neq Y$. Then (see e.g. [26, 5]) for permutation matrices P and Q of order n,

$$
P \preceq_B Q \text{ if and only if } \Sigma_P \geq \Sigma_Q.
$$

The Bruhat order on permutations (the class $\mathcal{A}(n, 1)$) was extended to general nonempty classes $\mathcal{A}(R, S)$ in [9]. Each of the two equivalent ways to define the Bruhat order on $\mathcal{A}(n, 1)$ makes sense for $\mathcal{A}(R, S)$: For $A_1, A_2 \in \mathcal{A}(R, S)$,

(B) *(Bruhat order on $\mathcal{A}(R, S)$)* $A_1 \preceq_B A_2$ provided that $\Sigma_{A_1} \geq \Sigma_{A_2}$.

($\widehat{B}$) *(Secondary Bruhat order on $\mathcal{A}(R, S)$)* $A_1 \preceq_{\widehat{B}} A_2$ provided that A_1 can be obtained from A_2 by a sequence of $L_2 \to I_2$ interchanges.

On $\mathcal{A}(n,1)$, $A_1 \preceq_B A_2$ if and only if $A_1 \preceq_{\widehat{B}} A_2$. There was an implicit conjecture in [9] that these two partial orders are identical on $\mathcal{A}(R,S)$ as they are on $\mathcal{A}(n,1)$. It is straightforward to verify that

$$A_1 \preceq_{\widehat{B}} A_2 \text{ implies that } A_1 \preceq_B A_2,$$

that is, the Bruhat order is a refinement of the secondary Bruhat order. The following example from [7] shows that this conjecture is false. Consider the class $\mathcal{A}(6,3)$ and three of its matrices

$$A = \begin{bmatrix} 1 & 0 & 0 & 0 & 1 & 1 \\ 1 & 0 & 1 & 1 & 0 & 0 \\ 1 & 1 & 0 & 1 & 0 & 0 \\ 0 & 0 & 0 & 1 & 1 & 1 \\ 0 & 1 & 1 & 0 & 1 & 0 \\ 0 & 1 & 1 & 0 & 0 & 1 \end{bmatrix}, \ C = \begin{bmatrix} 0 & 0 & 0 & 1 & 1 & 1 \\ 1 & 0 & 1 & 1 & 0 & 0 \\ 1 & 1 & 0 & 1 & 0 & 0 \\ 1 & 0 & 0 & 0 & 1 & 1 \\ 0 & 1 & 1 & 0 & 1 & 0 \\ 0 & 1 & 1 & 0 & 0 & 1 \end{bmatrix}, \qquad (2.1)$$

and

$$D = \begin{bmatrix} 0 & 0 & 0 & 1 & 1 & 1 \\ 1 & 1 & 0 & 1 & 0 & 0 \\ 1 & 0 & 1 & 1 & 0 & 0 \\ 1 & 0 & 0 & 0 & 1 & 1 \\ 0 & 1 & 1 & 0 & 1 & 0 \\ 0 & 1 & 1 & 0 & 0 & 1 \end{bmatrix}. \qquad (2.2)$$

Then

$$\Sigma_A = \begin{bmatrix} 1 & 1 & 1 & 1 & 2 & 3 \\ 2 & 2 & 3 & 4 & 5 & 6 \\ 3 & 4 & 5 & 7 & 8 & 9 \\ 3 & 4 & 5 & 8 & 10 & 12 \\ 3 & 5 & 7 & 10 & 13 & 15 \\ 3 & 6 & 9 & 12 & 15 & 18 \end{bmatrix}, \ \Sigma_C = \begin{bmatrix} 0 & 0 & 0 & 1 & 2 & 3 \\ 1 & 1 & 2 & 4 & 5 & 6 \\ 2 & 3 & 4 & 7 & 8 & 9 \\ 3 & 4 & 5 & 8 & 10 & 12 \\ 3 & 5 & 7 & 10 & 13 & 15 \\ 3 & 6 & 9 & 12 & 15 & 18 \end{bmatrix},$$

and

$$\Sigma_D = \begin{bmatrix} 0 & 0 & 0 & 1 & 2 & 3 \\ 1 & 2 & 2 & 4 & 5 & 6 \\ 2 & 3 & 4 & 7 & 8 & 9 \\ 3 & 4 & 5 & 8 & 10 & 12 \\ 3 & 5 & 7 & 10 & 13 & 15 \\ 3 & 6 & 9 & 12 & 15 & 18 \end{bmatrix},$$

from which it follows that

$$\Sigma_A > \Sigma_D > \Sigma_C \text{ and so } A \prec_B D \prec_B C.$$

The following theorem from [7] characterizing the cover relation for the secondary Bruhat order shows that C covers both D and A in the secondary Bruhat order. This implies that D and A are incomparable in the secondary Bruhat order, and hence the Bruhat order and secondary Bruhat order are already different on $\mathcal{A}(6,3)$.

For A a matrix of size m by n, and $I \subseteq \{1,2,\ldots,m\}$ and $J \subseteq \{1,2,\ldots,n\}$, $A[I,J]$ denotes the submatrix of A determined by the row indices in I and column indices in J.

Theorem 2.1 *Let $A = [a_{ij}]$ be a matrix in $\mathcal{A}(R, S)$ where $A[\{i, j\}, \{k, l\}] = L_2$. Let $A' = [a'_{ij}]$ be the matrix obtained from A by the $L_2 \to I_2$ interchange that replaces $A[\{i, j\}, \{k, l\}] = L_2$ with I_2. Then A covers A' in the secondary Bruhat order on $\mathcal{A}(R, S)$ if and only if*

(i) $a_{pk} = a_{pl}$ $(i < p < j)$,

(ii) $a_{iq} = a_{jq}$ $(k < q < l)$,

(iii) $a_{pk} = 0$ and $a_{iq} = 0$ imply $a_{pq} = 0$ $(i < p < j, k < q < l)$, and

(iv) $a_{pk} = 1$ and $a_{iq} = 1$ imply $a_{pq} = 1$ $(i < p < j, k < q < l)$.

Theorem 2.1 generalizes the characterization of the cover relation on $\mathcal{A}(n, 1)$. In terms of permutations of $\{1, 2, \ldots, n\}$, this characterization asserts that if $\pi = (i_1, \ldots, i_p, \ldots, i_q, \ldots, i_n)$ is a permutation with $p < q$ and $i_p > i_q$, and if the permutation $\tau = (i_1, \ldots, i_q, \ldots, i_p, \ldots, i_n)$ is obtained from π by the transposition that interchanges i_p and i_q, then τ is covered by π in the Bruhat order if and only if each i_t with $p < t < q$ satisfies $i_t < i_q$ or $i_t > i_p$. In terms of the corresponding permutation matrices P and Q, respectively, this means that the submatrix $P[\{p, p + 1, \ldots, q\}, \{i_q, i_q + 1, \ldots, i_p\}]$ has exactly two 1's and these 1's are in the upper right and lower left corners. The corresponding submatrix of Q has its 1's in the upper left and lower right corners.

Although, in general, the Bruhat order is a proper refinement of the secondary Bruhat order on $\mathcal{A}(n, 3)$, we have the following theorem [7].

Theorem 2.2 *On $\mathcal{A}(n, 2)$ the Bruhat order and secondary Bruhat order are identical.*

The class $\mathcal{A}(n, n)$ contains a unique matrix. In the Bruhat order on the class $\mathcal{A}(n, 1)$, I_n is the unique minimal matrix and L_n is the unique maximal matrix. From this it follows that in the class $\mathcal{A}(n, n - 1)$ the unique minimal matrix is $J_n - L_n$ and the unique maximal matrix is $J_n - I_n$, where J_n denotes the matrix of order n each of whose entries equals 1. In general we have the following result [7].

Theorem 2.3 *Let n and k be integers with $1 \le k \le n$. Then in the secondary Bruhat order, the class $\mathcal{A}(n, k)$ has a unique minimal matrix if and only if $k = 1, n - 1, n$ or $n = 2k$.*

A characterization of the cover relation for the Bruhat order on $\mathcal{A}(R, S)$ is not known.

An algorithm is given in [9] to construct a minimal matrix in a nonempty class $\mathcal{A}(R, S)$. For classes $\mathcal{A}(n, k)$ it specializes as described below. Let $J_{p,q}$ denote the p by q matrix of all 1's; if $p = q$, this is abbreviated to J_p.

Algorithm to Construct a Minimal Matrix in the Bruhat order on $\mathcal{A}(n, k)$

1. Let $n = qk + r$ where $0 \le r < k$.

2. If $r = 0$, then $A = J_k \oplus \cdots \oplus J_k$, ($q$ J_k's) is a minimal matrix.

3. Else, $r \ne 0$.

(a) If $q \geq 2$, let

$$A = X \oplus J_k \oplus \cdots \oplus J_k, \quad (q-1 \ J_k\text{'s}, \ X \text{ has order } k+r),$$

and let $n \leftarrow k + r$.

(b) Else, $q = 1$, and let

$$A = \left[\begin{array}{c|c} J_{r,k} & O_k \\ \hline X & J_{k,r} \end{array} \right], \quad (X \text{ has order } k),$$

and let $n \leftarrow k$ and $k \leftarrow k - r$.

(c) Proceed recursively with the current values of n and k to determine X.

For example, the algorithm produces the minimal matrix

$$\left[\begin{array}{c|c} J_{7,11} & O_7 \\ \hline \begin{array}{c|c} \begin{array}{c|c} J_{3,4} & O_3 \\ \hline I_4 & J_{4,3} \end{array} & O_{7,4} \\ \hline O_{4,7} & J_4 \end{array} & J_{11,7} \end{array} \right]$$

in $\mathcal{A}(18, 11)$.

The minimal matrices in $\mathcal{A}(n, 2)$ and $\mathcal{A}(n, 3)$ have been characterized, but there does not appear to be a useful characterization of the minimal matrices in $\mathcal{A}(n, k)$ for $k \geq 4$.

3 Bruhat order on $\mathcal{Z}^+(R, S)$

Recall that $\mathcal{Z}^+(R, S)$ is the class of all nonnegative integral matrices with row sum vector $R = (r_1, r_2, \ldots, r_m)$ and column sum vector $S = (s_1, s_2, \ldots, s_n)$, and that this class is nonempty provided

$$r_1 + r_2 + \cdots + r_m = s_1 + s_2 + \cdots + s_n.$$

We can carry over the definitions of Bruhat order and secondary Bruhat order to $\mathcal{Z}^+(R, S)$. For matrices A_1 and A_2 in $\mathcal{Z}^+(R, S)$:

(B) $A_1 \preceq_B A_2$ provided that $\Sigma(A_1) \geq \Sigma(A_2)$ (entrywise), and

($\widehat{B}$) $A_1 \preceq_{\widehat{B}} A_2$ if and only if A_1 can be obtained from A_2 by a sequence of $L_2 \to I_2$ interchanges of the form

$$\begin{bmatrix} a & b \\ c & d \end{bmatrix} \to \begin{bmatrix} a+1 & b-1 \\ c-1 & d+1 \end{bmatrix} \tag{3.1}$$

where $b, c \geq 1$.

As with $\mathcal{A}(R, S)$, it is obvious that $A_1 \preceq_{\widehat{B}} A_2$ implies that $A_1 \preceq_B A_2$. That $A_1 \preceq_B A_2$ implies that $A_1 \preceq_{\widehat{B}} A_2$ follows in much the same way (but more easily) as for permutation matrices A_1 and A_2. Hence we have the following.

Theorem 3.1 *The Bruhat order and secondary Bruhat order coincide on classes* $\mathcal{Z}^+(R, S)$.

Let $A = [a_{ij}]$ be a matrix in $\mathcal{Z}^+(R, S)$ which is minimal in the Bruhat order. For each i and j with $1 \le i < m$ and $1 \le j < n$, the two submatrices of A weakly above and weakly to the right of a_{ij}, and weakly below and weakly to the left of a_{ij}, respectively, must, except for a_{ij}, be zero matrices; otherwise an interchange of the type (3.1) applied to A results in a matrix below A in the Bruhat order. This implies that $\mathcal{Z}^+(R, S)$ contains a unique minimal element and, similarly, a unique maximal element. The unique minimal element has a "snake-like pattern" starting in the upper left corner and ending in the lower right corner; the unique maximal element has a "snake-like pattern" starting in the upper right corner and ending in the lower left corner. For example, if $R = (3, 7, 1, 1, 3, 7)$ and $S = (2, 4, 2, 5, 5, 4)$, then the unique matrices in $\mathcal{Z}^+(R, S)$ which are minimal and maximal in the Bruhat order are, respectively,

$$
\begin{bmatrix}
2 & 1 & 0 & 0 & 0 & 0 \\
0 & 3 & 2 & 2 & 0 & 0 \\
0 & 0 & 0 & 1 & 0 & 0 \\
0 & 0 & 0 & 1 & 0 & 0 \\
0 & 0 & 0 & 1 & 2 & 0 \\
0 & 0 & 0 & 0 & 3 & 4
\end{bmatrix}
\quad \text{and} \quad
\begin{bmatrix}
0 & 0 & 0 & 0 & 0 & 3 \\
0 & 0 & 0 & 1 & 5 & 1 \\
0 & 0 & 0 & 1 & 0 & 0 \\
0 & 0 & 0 & 1 & 0 & 0 \\
0 & 0 & 1 & 2 & 0 & 0 \\
2 & 4 & 1 & 0 & 0 & 0
\end{bmatrix}.
$$

(In general, the snake-like patterns may be disconnected depending on the relationships between the components of R and S.)

4   Ordering $\mathcal{A}(R, S)$ by shape of insertion tableau

Let $\lambda = (\lambda_1, \lambda_2, \ldots, \lambda_p)$ and $\mu = (\mu_1, \mu_2, \ldots, \mu_q)$ be two partitions of the same integer τ. A *Young diagram* of shape λ is a left-justified arrangement of τ boxes in p rows where there are λ_i boxes in row i, $(i = 1, 2, \ldots, p)$. A *Young tableau* of shape λ and content μ results from a Young diagram of shape λ by inserting in each of its boxes one of the integers $1, 2, \ldots, q$ where (i) the elements in each row are weakly increasing, (ii) the elements in each column are strictly increasing, and (iii) the integer j occurs μ_j times, $(j = 1, 2, \ldots, q)$. For example,

$$
\begin{array}{ccccc}
1 & 1 & 1 & 2 & 3 \\
2 & 2 & 3 & 3 & \\
4 & 4 & 5 & & \\
5 & & & &
\end{array}
$$

is a Young tableau of shape $\lambda = (5, 4, 3, 1)$ and content $\mu = (3, 3, 3, 2, 2)$. The *Kostka number* $K_{\lambda,\mu}$ is the number of Young tableaux of shape λ and content μ. It is a basic fact [19] that $K_{\lambda,\mu} \ne 0$ if and only if $\mu \preceq \lambda$ where as before $\preceq$ denotes the dominance order.

Let $R = (r_1, r_2, \ldots, r_m)$ and $S = (s_1, s_2, \ldots, s_n)$ be two nonincreasing, positive integral vectors with

$$
\tau = r_1 + r_2 + \cdots + r_m = s_1 + s_2 + \cdots + s_n.
$$

Thus R and S are partitions of the integer τ. Let $\kappa(R,S)$ denote the number of matrices in $\mathcal{A}(R,S)$. Then (see [19, 4, 5])

$$\kappa(R,S) = \sum_{\lambda} K_{\lambda,R} K_{\lambda^*,S} = \sum_{\lambda} K_{\lambda^*,R} K_{\lambda,S},$$

where the summations extend over all partitions λ of τ.

We have $K_{\lambda^*,R} K_{\lambda,S} \neq 0$ if and only if $R \preceq \lambda^*$ and $S \preceq \lambda$. Using the fact that conjugation reverses a dominance order relation between two partitions of the same integer, we see that

$$\kappa(R,S) = \sum_{S \preceq \lambda \preceq R^*} K_{\lambda^*,R} K_{\lambda,S}.$$

These facts are consequences of the *Knuth correspondence* [24] (see also [4, 5]), which is a bijection between m by n (0,1)-matrices and ordered pairs of Young tableaux of conjugate shape. Applied to $\mathcal{A}(R,S)$ it gives a bijection between matrices in $\mathcal{A}(R,S)$ and ordered pairs $(\mathcal{P}, \mathcal{Q})$ of Young tableaux, where $\mathcal{P}$ is a Young tableau of some shape λ and content S and $\mathcal{Q}$ is a Young tableau of shape λ^* and content R, and $S \preceq \lambda \preceq R^*$.

The Knuth correspondence is based on an operation called *column-bumping* which we illustrate by an example taken from [4]. Let

$$A = \begin{bmatrix} 1 & 0 & 1 \\ 1 & 1 & 0 \\ 1 & 0 & 0 \end{bmatrix} \in \mathcal{A}(R,S),$$

where $R = (2,2,1)$ and $S = (3,1,1)$. First describe A by a *generalized permutation array* in lexicographic order:

$$\Theta_A = \begin{pmatrix} 1 & 1 & 2 & 2 & 3 \\ 1 & 3 & 1 & 2 & 1 \end{pmatrix} = \begin{pmatrix} i_k \\ j_k \end{pmatrix} (k = 1, \cdots, 5), \tag{4.1}$$

where the ordered pairs (i_k, j_k) are the positions of A occupied by its 1's.

Start with $\mathcal{P}$ and $\mathcal{Q}$ as empty tableaux (corresponding to the unique (empty) partition of 0). We recursively construct $\mathcal{P}$ and $\mathcal{Q}$ simultaneously by inserting in $\mathcal{P}$ the element j_k of the second row of Θ_A by column-bumping, working from bottom to top and left to right, to maintain strict increasing in columns, and then inserting in $\mathcal{Q}$ the element i_k in the "conjugate square of the new square created." In the example given, this produces:

$$1 \rightarrow \begin{array}{c} 1 \\ 3 \end{array} \rightarrow \begin{array}{cc} 1 & 1 \\ 3 & \end{array} \rightarrow \begin{array}{cc} 1 & 1 \\ 2 & 3 \end{array} \rightarrow \begin{array}{ccc} 1 & 1 & 1 \\ 2 & 3 & \end{array}$$

(After starting with the first 1, the 3 is inserted at the end of column 1, since that does not violate the strict increasing requirement. In inserting the second 1, the second 1 bumps the first 1 which is then put in a new column. In inserting the 2, 2 bumps 3 and 3 is inserted at the end of the second column, since that does not violate the strict increasing requirement. Finally, the third 1 bumps the 1 in the first column, which then bumps the 1 in the second column, which then is inserted in a new column.)

and

$$1 \rightarrow \begin{array}{cc} 1 & 1 \end{array} \rightarrow \begin{array}{cc} 1 & 1 \\ 2 & \end{array} \rightarrow \begin{array}{cc} 1 & 1 \\ 2 & 2 \end{array} \rightarrow \begin{array}{cc} 1 & 1 \\ 2 & 2 \\ 3 & \end{array}.$$

Thus our matrix A corresponds to the pair $(\mathcal{P}, \mathcal{Q})$ of Young tableaux of conjugate shape and content $S = (3, 1, 1)$ and $R = (2, 2, 1)$, where

$$\mathcal{P} = \begin{array}{ccc} 1 & 1 & 1 \\ 2 & 3 & \end{array} \quad \text{and} \quad \mathcal{Q} = \begin{array}{cc} 1 & 1 \\ 2 & 2 \\ 3 & \end{array}.$$

$\mathcal{P}$ is the *insertion tableau* (of content S), and $\mathcal{Q}$ is the *recording tableau* (of content R) corresponding to the matrix A.

The insertion tableau has an important property [20] (a proof is also given in [5]). Consider the second row

$$j_1, j_2, \dots, j_p \tag{4.2}$$

of column indices in the generalized permutation array. The number of integers (occupied boxes) in the first column of the insertion array $\mathcal{P}$ is the maximal number of terms in a strictly increasing subsequence of (4.2). In general, the number of integers (occupied boxes) in the first k columns of the insertion array $\mathcal{P}$ equals the maximal number of terms in a subsequence of (4.2) which is the union of k strictly increasing subsequences ($k = 1, 2, \dots$).

In [4] the Knuth correspondence was used to define a (quasi) partial order on a class $\mathcal{A}(R, S)$. Let A and B be matrices in $\mathcal{A}(R, S)$ whose insertion tableaux in the Knuth correspondence have shapes λ_A and λ_B, respectively. Then A precedes B in the *Knuth order*, written $A \leq_K B$ provided that $\lambda_A \preceq \lambda_B$. (If in defining the Knuth partial order we use the shape of the recording tableau in place of the shape of the insertion tableau, then the dual order results.)

The Knuth correspondence is a bijection. Assume that $S \preceq \lambda \preceq R^*$. Then starting with a pair $(\mathcal{P}, \mathcal{Q})$ of Young tableaux of conjugate shapes λ and λ^*, respectively, where $\mathcal{P}$ has content S and $\mathcal{Q}$ has content R, we can invert the bumping operation (see [19, 5]) and obtain a matrix A in $\mathcal{A}(R, S)$ whose insertion tableau has shape λ. In the extreme cases of $\lambda = S$ and $\lambda = R^*$, direct algorithms are given in [4] to construct such an A that do not depend on choosing the pair $(\mathcal{P}, \mathcal{Q})$. These algorithms are variants of the well-known algorithm of Ryser to construct a canonical matrix in $\mathcal{A}(R, S)$.

Algorithm: $\lambda = R^*$

(0) Begin with the m by n matrix $A(R; n)$ with row sum vector R and column sum vector R^*.

(1) Shift s_n of the last 1's in s_n rows of $A(R; n)$ to column n, choosing those 1's in the rows with the largest sums but, in the case of ties, giving preference to the *topmost* rows.

(2) The matrix left in columns $1, 2, \ldots, n - 1$ of $A(R; n)$ is a matrix $A(R''; n - 1)$ with row sum vector R'' determined by R and the 1's chosen to be shifted. (In general, unlike the vector R' in Ryser's algorithm, the vector R'' will not be nondecreasing.) We now repeat with $A(R''; n-1)$ in place of $A(R; n)$ and s_{n-1} in place of s_n. We continue like this until we arrive at a matrix $\widetilde{A'}$ in $\mathcal{A}(R, S)$.

If $\mathcal{A}(R, S) \neq \emptyset$, this algorithm terminates with a matrix in $\mathcal{A}(R, S)$ whose insertion tableau has shape R^*.

Algorithm: $\lambda = S$

(0) Begin with the m by n matrix $A'(S; m)$ with row sum vector S^* and column sum vector S.

(1) Shift r_m of the last 1's in r_m columns of $A'(S; m)$ to row m, choosing those 1's in the columns with the largest sums but, in the case of ties, giving preference to the *rightmost* columns.

(2) The matrix left in rows $1, 2, \ldots, m - 1$ of $A'(S; m)$ is a matrix $A'(S''; m - 1)$ with column sum vector S'' determined by S and the 1's chosen to be shifted. (By choice of 1's to be shifted, the vector S'' is nondecreasing.) We now repeat with $A'(S''; m - 1)$ in place of $A'(S; m)$ and r_{m-1} in place of r_m. We continue like this until we arrive at a matrix $\widetilde{A''}$ in $\mathcal{A}(R, S)$.

This algorithm, the "transpose" of Ryser's algorithm, terminates with a matrix in $\mathcal{A}(R, S)$ whose insertion tableau has shape S.

We illustrate the algorithm $\lambda = R^*$ with the following example from [4]. Let $R = (4, 4, 3, 3, 2)$ and $S = (4, 3, 3, 3, 3)$, and let $\lambda = R^* = (5, 5, 4, 2)$. Applying the algorithm and using obvious notation, we get:

$$
\begin{bmatrix}
1 & 1 & 1 & 1 & 0 & \| & 4 \\
1 & 1 & 1 & 1 & 0 & \| & 4 \\
1 & 1 & 1 & 0 & 0 & \| & 3 \\
1 & 1 & 1 & 0 & 0 & \| & 3 \\
1 & 1 & 0 & 0 & 0 & \| & 2
\end{bmatrix}
\rightarrow
\begin{bmatrix}
1 & 1 & 1 & 0 & 1 & \| & 3 \\
1 & 1 & 1 & 0 & 1 & \| & 3 \\
1 & 1 & 0 & 0 & 1 & \| & 2 \\
1 & 1 & 1 & 0 & 0 & \| & 3 \\
1 & 1 & 0 & 0 & 0 & \| & 2
\end{bmatrix}
\rightarrow
$$

$$
\begin{bmatrix}
1 & 1 & 0 & 1 & 1 & \| & 2 \\
1 & 1 & 0 & 1 & 1 & \| & 2 \\
1 & 1 & 0 & 0 & 1 & \| & 2 \\
1 & 1 & 0 & 1 & 0 & \| & 2 \\
1 & 1 & 0 & 0 & 0 & \| & 2
\end{bmatrix}
\rightarrow
\begin{bmatrix}
1 & 0 & 1 & 1 & 1 & \| & 1 \\
1 & 0 & 1 & 1 & 1 & \| & 1 \\
1 & 0 & 1 & 0 & 1 & \| & 1 \\
1 & 1 & 0 & 1 & 0 & \| & 2 \\
1 & 1 & 0 & 0 & 0 & \| & 2
\end{bmatrix}
\rightarrow
$$

$$
\begin{bmatrix}
0 & 1 & 1 & 1 & 1 & \| & 0 \\
1 & 0 & 1 & 1 & 1 & \| & 1 \\
1 & 0 & 1 & 0 & 1 & \| & 1 \\
1 & 1 & 0 & 1 & 0 & \| & 1 \\
1 & 1 & 0 & 0 & 0 & \| & 1
\end{bmatrix}
\rightarrow
\begin{bmatrix}
0 & 1 & 1 & 1 & 1 & \| & 0 \\
1 & 0 & 1 & 1 & 1 & \| & 0 \\
1 & 0 & 1 & 0 & 1 & \| & 0 \\
1 & 1 & 0 & 1 & 0 & \| & 0 \\
1 & 1 & 0 & 0 & 0 & \| & 0
\end{bmatrix}.
$$

Thus the matrix obtained is

$$\widetilde{A'} = \begin{bmatrix} 0 & 1 & 1 & 1 & 1 \\ 1 & 0 & 1 & 1 & 1 \\ 1 & 0 & 1 & 0 & 1 \\ 1 & 1 & 0 & 1 & 0 \\ 1 & 1 & 0 & 0 & 0 \end{bmatrix},$$

and its corresponding generalized permutation array is

$$\begin{pmatrix} 1 & 1 & 1 & 1 & 2 & 2 & 2 & 2 & 3 & 3 & 3 & 4 & 4 & 4 & 5 & 5 \\ 2 & 3 & 4 & 5 & 1 & 3 & 4 & 5 & 1 & 3 & 5 & 1 & 2 & 4 & 1 & 2 \end{pmatrix}.$$

The corresponding insertion tableau $\mathcal{P}$, with intermediate results according to the five groups of column indices, is:

$$\begin{array}{c}
\begin{array}{c} 2 \\ 3 \\ 4 \\ 5 \end{array}
\rightarrow
\begin{array}{cc} 1 & 2 \\ 3 & 3 \\ 4 & 4 \\ 5 & 5 \end{array}
\rightarrow
\begin{array}{ccc} 1 & 1 & 2 \\ 3 & 3 & 3 \\ 4 & 4 & 5 \\ 5 & 5 & \end{array}
\rightarrow
\begin{array}{cccc} 1 & 1 & 1 & 2 \\ 2 & 3 & 3 & 3 \\ 4 & 4 & 4 & 5 \\ 5 & 5 & & \end{array}
\rightarrow
\begin{array}{ccccc} 1 & 1 & 1 & 1 & 2 \\ 2 & 2 & 3 & 3 & 3 \\ 4 & 4 & 4 & 5 & \\ 5 & 5 & & & \end{array}
\end{array}$$

which has shape $R^* = (5, 5, 4, 2)$. The recording tableau $\mathcal{Q}$, having shape $R^{**} = R$ and content R, must be of the form

$$\begin{array}{cccc} 1 & 1 & 1 & 1 \\ 2 & 2 & 2 & 2 \\ 3 & 3 & 3 & \\ 4 & 4 & 4 & \\ 5 & 5 & & \end{array}.$$

Canonical constructions for a matrix in $\mathcal{A}(R, S)$ whose corresponding insertion tableau has shape λ are not known in the nonextreme cases $S \prec \lambda \prec R^*$.

5 Bruhat order on symmetric matrices

Let $\mathcal{S}_n^*$ denote the set of all symmetric permutation matrices of order n. As a permutation of $\{1, 2, \ldots, n\}$, σ is a set of pairwise disjoint transpositions and fixed points. The number of symmetric permutation matrices of order n, enumerated according to the number of fixed points, equals

$$\sum_{k^*} \binom{n}{k} \frac{(n-k)!}{2^{(n-k)/2} \cdot ((n-k)/2)!} = \sum_{k^*} \frac{n!}{k! \cdot 2^{(n-k)/2} \cdot ((n-k)/2)!}$$

where the summation is over those k between 0 and n such that $n - k$ is even.

The Bruhat order on $\mathcal{S}_n$ induces a Bruhat order on $\mathcal{S}_n^*$. If $n = 3$, we have

$$(1, 2, 3) \preceq_B (1, 3, 2), (2, 1, 3) \preceq_B (3, 2, 1).$$

Since the identity permutation $(1, 2, \ldots, n)$ and anti-identity permutation $(n, n - 1, \ldots, 2, 1)$ are symmetric, they are the unique minimal and maximal permutations in the Bruhat order on $\mathcal{S}_n^*$.

A single interchange $L_2 \to I_2$ applied to a symmetric permutation matrix Q, where L_2 is a *principal* submatrix of order 2 of Q, results in a symmetric permutation matrix below Q in the Bruhat order. Otherwise, a sequence of two symmetrically situated $L_2 \to I_2$ interchanges (overlapping or disjoint) gives the following two types of interchanges:

$$L_3 = \begin{bmatrix} 0 & 0 & 1 \\ 0 & 1 & 0 \\ 1 & 0 & 0 \end{bmatrix} \to \begin{bmatrix} 1 & 0 & 0 \\ 0 & 0 & 1 \\ 0 & 1 & 0 \end{bmatrix} = I_1 \oplus L_2$$

and

$$L_4 = L_2 \otimes L_2 = \begin{bmatrix} 0 & 0 & 0 & 1 \\ 0 & 0 & 1 & 0 \\ 0 & 1 & 0 & 0 \\ 1 & 0 & 0 & 0 \end{bmatrix} \to \begin{bmatrix} 0 & 0 & 1 & 0 \\ 0 & 0 & 0 & 1 \\ 1 & 0 & 0 & 0 \\ 0 & 1 & 0 & 0 \end{bmatrix} = L_2 \otimes I_2$$

which, when applied to a principal submatrix of Q of orders 3 and 4, respectively, result in a symmetric permutation matrix below Q in the Bruhat order.

Theorem 5.1 *If P and Q are symmetric permutation matrices of order n, then $P \preceq_B Q$ if and only if P can be obtained from Q by a sequence of principal symmetric interchanges of the form $L_2 \to I_2$, $L_3 \to I_1 \oplus L_2$, and $L_2 \otimes L_2 \to L_2 \otimes I_2$. The matrix Q covers P in the Bruhat order on $\mathcal{S}_n^*$ if and only if the submatrix of Q of consecutive rows and columns determined by the L_2's in these interchanges contain no other ones other than those displayed.*

We remark that if we restrict ourselves to the set of symmetric permutation matrices with no fixed points, then only the $L_4 \to L_2 \otimes L_2$ interchanges are possible.

Now consider the classes $\mathcal{A}^s(R)$ and $\mathcal{A}_0^s(R)$ of symmetric $(0,1)$-matrices with row and column sum vector R with unrestricted trace and trace zero, respectively (equivalently, graphs with degree sequence equal to R in which loops may or may not be permitted). The Bruhat and secondary Bruhat orders on $\mathcal{A}(R,R)$ induce Bruhat and secondary Bruhat orders on $\mathcal{A}^s(R)$ and $\mathcal{A}_0^s(R)$, respectively. Recall the matrices A, C, and D in $\mathcal{A}(6,3)$ defined in (2.1) and (2.2) where A is smaller than D in the Bruhat order but incomparable in the secondary Bruhat order because C covers both A and D in the secondary Bruhat order. Let

$$A' = \begin{bmatrix} O_6 & A \\ A^T & O_6 \end{bmatrix}, C' = \begin{bmatrix} O_6 & C \\ C^T & O_6 \end{bmatrix}, \text{ and } D' = \begin{bmatrix} O_6 & D \\ D^T & O_6 \end{bmatrix} \tag{5.1}$$

be matrices in $\mathcal{A}_0^s(12,3)$ (and so in $\mathcal{A}^s(12,3)$). Since $\Sigma_A > \Sigma_D > \Sigma_C$, it follows that $\Sigma_{A'} > \Sigma_{D'} > \Sigma_{C'}$ and hence $A' \prec_B D' \prec_B C'$. Suppose there were a matrix X in $\mathcal{A}^s(12,3)$ such that $D' \prec_{\widehat{B}} X \prec_{\widehat{B}} C'$. Since the secondary Bruhat order is a refinement of the Bruhat order, X has the form

$$\begin{bmatrix} O_6 & U \\ U^T & O_6 \end{bmatrix}.$$

This implies that $D \prec_{\widehat{B}} U \prec_{\widehat{B}} C$ a contradiction. Thus C' covers D' in the secondary Bruhat order, and similarly C' covers A' in the secondary Bruhat order. Thus D' and A' are incomparable in the secondary Bruhat order on $\mathcal{A}_0^s(12,3)$ and $\mathcal{A}^s(12,3)$.

We now consider briefly the classes $\mathcal{A}_0^s(n, k)$. According to a result of Punnim [28], there exists a k-regular graph of order n having a complete subgraph K_{k+1} of order $k+1$ (so a connected component equal to K_{k+1}) if $n = k+1$ or $n \geq 2k+2$, and having a complete subgraph of order $m = \lfloor n/2 \rfloor$ if $k + 2 \leq n \leq 2k + 1$. In addition, no regular graph with n vertices of degree k can have a larger complete subgraph. Thus, if $n = k + 1$ or $n \geq 2k + 2$, $k + 1$ is the largest possible order of a principal submatrix of the form $J - I$ in a matrix belonging to $\mathcal{A}_0^s(n, k)$; if $k + 2 \leq n \leq 2k + 1$, then m is the largest such order. Note that if n is odd, then k must be even in order that $\mathcal{A}_0^s(n, k)$ be nonempty. It now follows that in the Bruhat order on $\mathcal{A}_0^s(n, k)$ with $n \geq 2k + 2$, there is a minimal matrix of the form $(J_{k+1} - I_{k+1}) \oplus X$ where X is a minimal matrix in $\mathcal{A}_0^s(n - k - 1, k)$. Thus we need only consider $k + 1 < n \leq 2k + 1$, and we now make this assumption. If n is even, say $n = 2m$, then there exists a matrix in $\mathcal{A}_0^s(2m, k)$ of the form

$$A = \begin{bmatrix} J_m - I_m & M \\ M^t & J_m - I_m \end{bmatrix},$$

where M is a matrix in $\mathcal{A}(m, k - m + 1)$. If we choose M to be a matrix that is minimal in the Bruhat order on the class $\mathcal{A}(m, k - m + 1)$, then A is clearly a minimal matrix in the Bruhat order on $\mathcal{A}_0^s(2m, k)$. Now consider the special case in which $n = 2k + 1$, where $k = m = 2h$. Then a minimal matrix in the Bruhat order can be constructed as follows:

$$\begin{bmatrix} J_k - I_k & \begin{matrix} j_h \\ o_h \end{matrix} & & \begin{matrix} & O_{h,h} & \\ I_h & & O_{h,h-1} \end{matrix} & o_k \\ \hline j_h^T \quad o_h^T & 0 & o_h^T & j_{h-1}^T & \\ \begin{matrix} & I_h & \\ O_{h,h} & & \\ & O_{h-1} & \end{matrix} & \begin{matrix} o_h^T \\ \\ j_{h-1}^T \end{matrix} & J_{k-1} - I_{k-1} & j_k \\ \hline o_k^T & & j_k^T & 0 \end{bmatrix},$$

where j_h is the column vector of h 1's and o_h is the column vector of h 0's.

6   Ordering by spectral radius (index)

In this section all matrices are square. Let A be a $(0, 1)$-matrix of order n (a bipartite graph with bipartition into two sets of size n). By the Perron-Frobenius theory of nonnegative matrices, A has a nonnegative eigenvalue $\rho(A)$ such that $|\lambda| \leq \rho(A)$ for every eigenvalue λ of A. The number $\rho(A)$ is the *spectral radius* or *index* of A. The spectral radius lies between the minimum $\tilde{r}$ and maximum $\bar{r}$ row sums. If the matrix is irreducible and $\tilde{r} \neq \bar{r}$, then there is strict inequality at both ends.

Let $R = (r_1, r_2, \ldots, r_n)$ and $S = (s_1, s_2, \ldots, s_n)$ be positive vectors such that $\mathcal{A}(R, S)$ is nonempty. One can attempt to order matrices in $\mathcal{A}(R, S)$ according to

spectral radius:
$$A_1 \leq_\rho A_2 \text{ provided that } \rho(A_1) \leq \rho(A_2).$$

Since the spectral radius depends on the entries of a matrix in a very complicated way it does not seem realistic to expect to be able to say anything general and substantial about this order when restricted to matrices in $\mathcal{A}(R, S)$. If $1 \leq k \leq n$ and $R = S = (k, k, \ldots, k)$, a constant vector of all k's, then the spectral radius is constant on $\mathcal{A}(R, S)$. This naturally leads to consideration of a *nearly-constant vector* with two different components k and $k+1$ $(1 \leq k \leq n-1)$. Since the spectral radius is invariant under simultaneous row and column permutations, without loss of generality we take

$$R = R_n(k; p) = (\underbrace{k+1, \ldots, k+1}_{p}, \underbrace{k, \ldots, k}_{n-p}) \quad (1 \leq p \leq n-1)$$

and S a rearrangement of R, that is, $S = R_n(k; p)Q$ for some permutation matrix Q of order n. The problem seems difficult enough already when $Q = I_n$, that is, $S = R_n(k; p)$. We call classes of the form $\mathcal{A}_n(k; p) = \mathcal{A}(R_n(k; p), R_n(k; p))$ *nearly-regular classes*.

A simple example reveals the difficulty even in this nearly-regular case. Consider $\mathcal{A}_n(1; p)$. If $n = 3$ and $p = 2$, then the spectral radius of a matrix with these parameters is one of $2, 1.8019, 1.7549, 1.6180$, respectively realized by

$$\begin{bmatrix} 1 & 1 & 0 \\ 1 & 1 & 0 \\ 0 & 0 & 1 \end{bmatrix}, \begin{bmatrix} 1 & 1 & 0 \\ 1 & 0 & 1 \\ 0 & 1 & 0 \end{bmatrix}, \begin{bmatrix} 1 & 1 & 0 \\ 0 & 1 & 1 \\ 1 & 0 & 0 \end{bmatrix}, \begin{bmatrix} 1 & 0 & 1 \\ 0 & 1 & 1 \\ 1 & 0 & 0 \end{bmatrix}.$$

If $p \geq 2$ and $n \geq 4$, the maximum spectral radius equals 2. This is because there is a matrix in $\mathcal{A}_n(1; p)$ with an irreducible component equal to J_2, that is, of the form $J_2 \oplus A'$ where $A' \in \mathcal{A}_{n-2}(1, p-2)$. If $p = 1$, the possibilities, up to simultaneous row and column permutations are

$$P'_k \oplus Q \quad (Q \text{ a permutation matrix}) \tag{6.1}$$

where P'_k is obtained from the full-cycle permutation matrix P_k of some order k by replacing the 0 in position $(1, 1)$ with a 1, and

$$\begin{bmatrix} 0 & 1 & 1 \\ 1 & 0 & 0 \\ 1 & 0 & 0 \end{bmatrix} \oplus Q \quad (Q \text{ a permutation matrix}). \tag{6.2}$$

For example,

$$P'_6 = \begin{bmatrix} 1 & 1 & 0 & 0 & 0 & 0 \\ 0 & 0 & 1 & 0 & 0 & 0 \\ 0 & 0 & 0 & 1 & 0 & 0 \\ 0 & 0 & 0 & 0 & 1 & 0 \\ 0 & 0 & 0 & 0 & 0 & 1 \\ 1 & 0 & 0 & 0 & 0 & 0 \end{bmatrix}.$$

The characteristic polynomial of P'_k is $(\lambda - 1)\lambda^{k-1} - 1$. The spectral radius of P'_k is between 1 and 2, and it is easy to see from the characteristic polynomial that the maximal spectral radius, namely $(1 + \sqrt{5})/2$, occurs when $k = 2$. The spectral radius of (6.2) is $\sqrt{2}$. Hence the maximum spectral radius is $(1+\sqrt{5})/2$ if $p = 2$ and $n \geq 4$.

Next we recall a theorem of Schwarz [30].

Theorem 6.1 *Let n^2 nonnegative numbers be given. Of all the matrices of order n whose n^2 entries are the given numbers, the largest spectral radius occurs among those matrices for which the entries in each row and in each column are weakly decreasing.*

The method used by Schwarz to prove this theorem is to first show that if a smaller entry precedes a larger entry in a row, the two entries can be interchanged without decreasing the spectral radius. In this way the entries in each row can be arranged to be weakly decreasing. Repeating on columns does not change the weakly decreasing property of the entries in each row.

Theorem 6.1 implies that the maximum spectral radius among all $(0,1)$-matrices with a specified row sum vector R but an unspecified column sum vector occurs at a matrix where the 1's are in the initial positions of each row; so, under the assumption that R is nonincreasing, the 1's in each column are also in the initial positions. If $R = R_n(2; p)$, the maximum spectral radius ρ for such matrices occurs for

$$J_3 = \begin{bmatrix} 1 & 1 & 1 \\ 1 & 1 & 1 \\ 1 & 1 & 1 \end{bmatrix} (p \geq 3; \rho = 3) \text{ and } J_3' = \begin{bmatrix} 1 & 1 & 1 \\ 1 & 1 & 1 \\ 1 & 1 & 0 \end{bmatrix} (p = 2; \rho = 1 + \sqrt{3}).$$

Now let $p = 1$. With the use of the method of Schwarz, it is not difficult to show that given a matrix A with row sum vector $R_n(2; 1)$, there is a $(0,1)$-matrix B of order 3 with two 0's whose spectral radius is at least as large as that of A. The largest spectral radius of such a B is $\rho = 1 + \sqrt{2}$ attained by both

$$J_3'' = \begin{bmatrix} 1 & 1 & 1 \\ 1 & 1 & 0 \\ 1 & 0 & 1 \end{bmatrix} \text{ and } \begin{bmatrix} 1 & 1 & 1 \\ 1 & 1 & 0 \\ 1 & 1 & 0 \end{bmatrix} (p = 1).$$

Since the column sum vector of the matrices J_3, J_3' and J_3'' is the same as its row sum vector, it remains to consider if and how they can be realized as a principal submatrix of a matrix in $\mathcal{A}_n(2; p)$. We have the following observations:

(i) $p \geq 4$ and $n \geq 6$: there is a matrix in $\mathcal{A}_n(2; p)$ having J_3 as a leading submatrix.

(ii) $p = 3$ and $n \geq 5$: there is a matrix having J_3 as a leading submatrix.

(iii) $p = 2$, and $n = 3$ or $n \geq 5$: there is a matrix having J_3' as a leading submatrix.

(iv) $p = 1$, and $n = 3$ or $n \geq 5$: there is a matrix having J_3'' as a leading submatrix.

Left to consider are $R_5(2; 4) = (3, 3, 3, 3, 2)$, $R_4(2; 3) = (3, 3, 3, 2)$, $R_4(2; 2) = (3, 3, 2, 2)$ and $R_4(2; 1) = (3, 2, 2, 2)$. Using MATLAB, one determines that the maximal spectral radius occurs, respectively, at the following matrices:

$$\begin{bmatrix} 1 & 1 & 1 & 0 & 0 \\ 1 & 1 & 1 & 0 & 0 \\ 1 & 1 & 0 & 1 & 0 \\ 0 & 0 & 1 & 1 & 1 \\ 0 & 0 & 0 & 1 & 1 \end{bmatrix}, \begin{bmatrix} 1 & 1 & 1 & 0 \\ 1 & 1 & 1 & 0 \\ 1 & 1 & 0 & 1 \\ 0 & 0 & 1 & 1 \end{bmatrix}, \begin{bmatrix} 1 & 1 & 1 & 0 \\ 1 & 1 & 0 & 1 \\ 1 & 0 & 1 & 0 \\ 0 & 1 & 0 & 1 \end{bmatrix}, \begin{bmatrix} 1 & 1 & 1 & 0 \\ 1 & 1 & 0 & 0 \\ 1 & 0 & 0 & 1 \\ 0 & 0 & 1 & 1 \end{bmatrix}.$$

We now consider $R_n(k;p)$ in general. Using the Schwarz method, the maximal spectral radius for this row sum vector (with the column sum vector unrestricted) equals the spectral radius of:

$$J_{k+1} \text{ if } p \geq k+1, \text{ and}$$

$$J_{k+1}^p = \left[\ J_{k+1,k} \mid \alpha_p\ \right] \text{ if } 1 \leq p \leq k. \tag{6.3}$$

where α is a column with p 1's followed by $k+1-p$ 0's.

Lemma 6.2 *Let p be a positive integer with $p \leq k$. The matrix J_{k+1}^p in (6.3) and the matrix*

$$L_{k+1}^p = \left[\begin{array}{c|c} J_p & J_{p,k+1-p} \\ \hline J_{k+1-p,p} & J_{k+1-p} - I_{k+1-p} \end{array}\right] \tag{6.4}$$

in $\mathcal{A}_{k+1}(k;p)$ have the same spectral radius.

Proof Each of the matrices (6.3) and (6.4) is irreducible. By the Perron-Frobenius theory each has a positive eigenvector (unique up to positive multiples) corresponding to its spectral radius. Let $x = (x_1, x_2, \ldots, x_{k+1})^T$ be a positive eigenvector for the spectral radius of J_{k+1}^p. Then $J_{k+1}^p x = \rho(J_{k+1}^p)x$ implies that $x_1 = \cdots = x_p$ and $x_{p+1} = \cdots = x_{k+1}$. From this it follows that $L_{k+1}^p x = \rho(J_{k+1}^p)x$ and $\rho(L_{k+1}^p) = \rho(J_{k+1}^p)$. $\qquad\square$

Thus provided we can realize J_{k+1} or the matrix (6.4) as a principal submatrix of a matrix in $\mathcal{A}_n(k;p)$, we will have determined which matrix in $\mathcal{A}_n(k;p)$ has the largest spectral radius. If $p = k+1$ and either $n = p$ or $n \geq 2k+1$, or if $p > k+1$ and $n \geq 2k+2$, then we have a matrix in $\mathcal{A}_n(k;p)$ of the form $J_{k+1} \oplus A$ and so $\rho = k+1$. If $p < k$ and either $n = k+1$ or $n \geq 2k-1$, then we have a matrix in $\mathcal{A}(p;k)$ of the form

$$L_{k+1}^p \oplus A = \left[\begin{array}{c|c} J_p & J_{p,k+1-p} \\ \hline J_{k+1-p,p} & J_{k+1-p} - I_{k+1-p} \end{array}\right] \oplus A$$

which therefore has the largest spectral radius. We observe that the matrix L_{k+1}^p is a symmetric matrix. Thus the largest spectral radius is attained by a symmetric matrix in these cases.

The determination of the maximum spectral radius for matrices in $\mathcal{A}_n(k;p)$ has now been reduced to a finite problem: the three possibilities are

$$p = k+1 \text{ and } p < n \leq 2k,$$
$$p > k+1 \text{ and } n \leq 2k+1, \text{ and}$$
$$p < k \text{ and } k+1 < n \leq 2k.$$

It appears to be a very difficult problem to determine which matrix has the largest spectral radius.

One can relax the condition that the rows and columns contain a specified number of 1's by prescribing only the total number of 1's, that is, by considering the class $\mathcal{A}(n|\tau)$ of all $(0,1)$-matrices of order n with exactly τ 1's ($\tau \leq n^2$). By Theorem 6.1 the maximum spectral radius occurs among those matrices whose 1's are concentrated in the upper left corner and have a staircase pattern. Such a matrix corresponds to a partition of τ:

$$\tau = r_1 + r_2 + \cdots + r_n \text{ where } r_1, r_2, \ldots r_n \text{ and } n \geq r_1 \geq r_2 \geq \cdots \geq r_n \geq 0, \quad (6.5)$$

(here we include terms equal to 0 corresponding to row sums equal to 0). The column sum vector for such a matrix is the conjugate partition τ^* (again including 0's to correspond to zero column sums). In [8] the question was raised to determine which partition (6.5) of τ gives the largest spectral radius. It was shown in [8] that for $\tau = k^2$ and $k^2 + 1$, respectively, the partitions

$$(\underbrace{k, k, \ldots, k}_{k}, 0, \ldots, 0), \text{ and } (\underbrace{k+1, k, \ldots, k}_{k}, 0, 0, \ldots, 0) \text{ and } (\underbrace{k, k, \ldots, k}_{k}, 1, 0, \ldots, 0)$$

give the maximum spectral radii, and this spectral radius equals k in both cases. If $k = 2$ there is only one other partition that gives spectral radius k, namely, $(2, 1, 1)$. Friedland [18] proves that for $\tau = k^2 + 2k$, the maximal spectral radius occurs for the partition

$$(\underbrace{k+1, k+1, \ldots, k+1}_{k}, k, 0 \ldots, 0).$$

Following up on a counterexample of Coppersmith to a conjecture reported in [8], he also showed that for $\tau = k^2 + 2k - 3$, the maximum spectral radius occurs for the partition

$$(\underbrace{k+1, \ldots, k+1}_{k-1}, k-1, k-1, 0, \ldots, 0).$$

It was conjectured in [18] that for $\tau = k^2 + l$, where $1 \leq l \leq 2k$, there is a partition giving the maximum spectral radius that has $k+1$ positive components with each component at most $k+1$. By our comments above, this has been proved for $l = 1, 2k, 2k - 3$ and for fixed l when k is sufficiently large depending on l.

We now turn to the class $\mathcal{A}_0^s(n|\tau)$ of symmetric $(0,1)$-matrices of order n with zero trace and exactly τ 1's above the main diagonal (graphs with n vertices and τ edges). Using a biquadratic form, Brualdi and Hoffman [8] proved the analogue of Schwarz's result for this class: the maximum spectral radius for a class $\mathcal{A}^s(n|\tau)$ occurs for a matrix $A = [a_{ij}]$ with a *stepwise pattern*, that is, for $i < j$, $a_{ij} = 1$ implies that $a_{pq} = 1$ whenever $p < q \leq j$ and $p \leq i$. They showed that if $\tau = \binom{d}{2}$, then the maximum spectral radius of matrices in $\mathcal{A}_0^s(n, \tau)$ is $d-1$ and that a matrix $A \in \mathcal{A}_0^s(n|\tau)$ with spectral radius $d-1$ satisfies

$$PAP^T = (J_d - I_d) \oplus O_{n-d}$$

for some permutation matrix P. Resolving a conjecture in [8], Rowlinson [29] (see also [14]) proved the following result.

Theorem 6.3 *Let* $\tau = \binom{d}{2} + t$ *where* $0 < t < d$. *Let* $K_{d,t}$ *be the matrix in* $\mathcal{A}^s(d+1|\tau)$ *obtained from* $J_d - I_d$ *by attaching a new last column and row with* t *1's in their initial positions. Then a matrix* A *in* $\mathcal{A}^s(n|\tau)$ *has maximum spectral radius if and only if there is a permutation matrix* P *such that*

$$PAP^T = K_{d,t} \oplus O_{n-d-1}.$$

The determination of the maximal spectral radius of the class $\mathcal{A}^s(n|\tau, \mathrm{irr})$, $n-1 \leq \tau \leq n(n-1)/2$ (index of connected graphs with n vertices and τ edges) is more difficult and has been solved in only some cases. A matrix in the class $\mathcal{A}^s(n|n-1, \mathrm{irr})$ is the adjacency matrix of a tree on n vertices and it is well-known that the minimum spectral radius occurs uniquely for a path and the maximal spectral radius occurs uniquely for the star $K_{1,n-1}$. Brualdi and Solheid [11] showed that even in the irreducible case, the maximum spectral radius occurs only at a matrix whose rows and columns can be simultaneously permuted to a matrix with a stepwise pattern (so the connected graph with the maximal index contains a star $K_{1,n-1}$ as a spanning tree; in particular, only the star has maximum spectral radius when $\tau = n-1$). The cases with $\tau = n+k$, $1 \leq k \leq 5$ were considered in [11] where graphs $G_{n,k}$ and $H_{n,k}$ (see below for a general definition of these graphs) were identified such that $G_{n,k}$ uniquely gave the maximal spectral radius for $k = 0, 1, 2$, while for $k = 3, 4$, or 5, $G_{n,k}$ gives the maximal spectral radius for some small values of n and $H_{n,k}$ uniquely gives the maximal spectral radius for all sufficiently large n.

We now define the graphs $G_{n,k}$ and $H_{n,k}$ as graphs with adjacency matrices in $\mathcal{A}^s(n|\tau, \mathrm{irr})$, $n-1 \leq \tau \leq n(n-1)/2$. Let $k = \binom{d-1}{2} + t - 1$ where $0 \leq t \leq d-2$. Then $G_{n,k}$ is the graph on n vertices with $\tau = n+k$ edges, having a complete subgraph K_d and an independent set of $n-d$ vertices each of which is joined by an edge to the same vertex of K_d, with one of the vertices in the independent set joined by an edge to t other vertices of K_d. If $0 \leq t < n-d$, the graph $H_{n,k}$ is the graph on n vertices with $\tau = n+k$ edges, having a complete subgraph K_d and an independent set of $n-d$ vertices each of which is joined by an edge to the same vertex of K_d, with one vertex of the stable set joined to t other vertices of the stable set. Cvetković and Rowlinson [12] extended the above results by showing that for any $k \geq 6$ and n large enough, $H_{n,k}$ uniquely gives the maximum spectral radius of the class $\mathcal{A}^s(n|n+k, \mathrm{irr})$. Bell [2] considered the case of $k = \binom{d-1}{2} - 1$ for some $d > 4$ (so $\tau = n - 1 + \binom{d-1}{2}$), the number of edges of a tree on n vertices together with those of a complete graph K_{d-1}) and showed that there is a function

$$g(d) = \frac{1}{2}d(d+5) + 7 + \frac{32}{d-4} + \frac{16}{(d-4)^2}$$

such that if $n \leq g(d)$, then $G_{n,k}$ gives the maximum spectral radius, while if $n \geq g(d)$ then $H_{n,k}$ gives the maximum spectral radius. Some bounds on this maximum spectral radius are also given in [2].

It has been conjectured [1] (see also [31]) that there is a function $g(k)$ such that a graph with adjacency matrix in $\mathcal{A}^s(n|n+k, \text{irr})$ of maximum index is isomorphic to $G_{n,k}$ if $n < f(k)$, is isomorphic to $G_{n,k}$ or $H_{n,k}$ if $n = f(k)$, and is isomorphic to $H_{n,k}$ if $n > f(k)$.

Finally we note that Sections 1 and 3 of [13] contain a discussion of the ordering of particular types of graphs by spectral radius, by spectra, and by spectral moments.

7 Ordering by rank

One can also order matrices in a nonempty class $\mathcal{A}(R, S)$ by rank (over the real field):
$$A_1 \leq_r A_2 \text{ provided that } \text{rk}(A_1) \leq \text{rk}(A_2).$$
As with the spectral radius, it is difficult to say anything substantial about this order for general $\mathcal{A}(R, S)$. Let
$$\widetilde{\text{rk}}(R, S) = \min\{\text{rk}(A) : A \in \mathcal{A}(R, S)\}$$
and
$$\overline{\text{rk}}(R, S)) = \max\{\text{rk}(A) : A \in \mathcal{A}(R, S)\}$$
be, respectively, the minimum and maximum rank possible for a matrix in $\mathcal{A}(R, S)$. An interchange $I_2 \leftrightarrow L_2$ can alter the rank by at most one. This follows since if A_2 is obtained from A_1 by a single interchange, taking place e.g. in the leading submatrix of order 2, then the matrices B_1 and B_2 obtained from A_1 and A_2, respectively, by adding column 1 to column 2 differ only in column 1 and thus differ in rank by at most 1. Since A_1 and A_2 have the same ranks as B_1 and B_2, respectively, the ranks of A_1 and A_2 differ by at most 1. Thus, by Ryser's theorem on interchanges, all values between the minimum and maximum ranks are attainable by matrices in $\mathcal{A}(R, S)$.

We now confine our attention to the *regular classes* $\mathcal{A}(n, k)$ consisting of all matrices of order n with k 1's in each row and column, and let $\widetilde{\text{rk}}_n(k)$ and $\overline{\text{rk}}_n(k)$ denote the minimum and maximum ranks for these classes. The maximum rank problem for these classes has a very neat answer due to Newman [27] and Houck and Paul [21].

Theorem 7.1 *Let n and k be integers with $0 \leq k \leq n$. Then*
$$\overline{rk}_n(k) = \begin{cases} 0 & \text{if } k = 0, \\ 1 & \text{if } k = n, \\ 3 & \text{if } k = 2 \text{ and } n = 4, \\ n & \text{otherwise.} \end{cases}$$

Thus, except for trivial cases, only for $n = 4$ and $k = 2$ does there not exist a nonsingular matrix in $\mathcal{A}(n, k)$.

Determining the minimum rank of matrices in $\mathcal{A}(n, k)$ is more difficult and is resolved in only certain special cases [10].

Theorem 7.2 *For $n \geq 2$ we have*

$$\widetilde{rk}_n(2) = \begin{cases} n/2 & \text{if } n \text{ is even,} \\ (n+3)/2 & \text{if } n \text{ is odd.} \end{cases}$$

For $n \geq 3$ we have

$$\widetilde{rk}_n(3) = \begin{cases} n/3 & \text{if } n \text{ is divisible by 3,} \\ \lfloor n/3 \rfloor + 3 & \text{otherwise.} \end{cases}$$

An inequality that is used in the proof of Theorem 7.2 is

$$\widetilde{rk}_n(k) \geq \lceil n/k \rceil \tag{7.1}$$

with equality if and only if k divides n.

The upper bound for the minimum rank,

$$\widetilde{rk}_n(k) \leq \lfloor n/k \rfloor + k$$

is a consequence of a general construction of a matrix in $\mathcal{A}(n,k)$ [10] which we illustrate for $n = 11$ and $k = 4$. The matrix

$$\left[\begin{array}{c|cc} J_4 & \multicolumn{2}{c}{O_{4,7}} \\ \hline \multirow{2}{*}{$O_{7,4}$} & O_3 & J_{3,4} \\ & J_{4,3} & I_4 \end{array} \right]$$

has rank 6. Hence using the characterization of equality in (7.1), we see that $5 \leq \widetilde{rk}_{11}(4) \leq 6$.

As already mentioned, all possible ranks between the minimum and maximum ranks are attainable by matrices in $\mathcal{A}(n,k)$. Thus the number of possible ranks is $\overline{rk}_n(k) - \widetilde{rk}_n(k) + 1$. For a positive integer r, Jorgenson [22] defined $\mathcal{R}_r$ to be the set of all numbers k/n for which there exists a matrix in $\mathcal{A}(n,k)$ with rank equal to r and showed that $|\mathcal{R}_r| < 2^{r^2}$, in particular, $\mathcal{R}_r$ is a finite set.

Little seems to be known about the order of matrices in $\mathcal{A}^{s}(R)$ or $\mathcal{A}_0^{s}(R)$ with respect to rank. Under the assumption that a matrix in $\mathcal{A}_0^{s}(R)$ has no identical rows (the corresponding graph does not have two vertices joined to the same set of vertices), in [25] there is an implicit lower bound[1] on the rank, namely $2 \log_2 n - c$ where c is a constant.

One can also order matrices in $\mathcal{A}(R,S)$ by determinant, permanent, and term rank but we do not pursue this line of investigation further here, and instead refer the interested reader to [5].

8 Higher-dimensional permutation arrays

In this section we describe some work of Erikkson and Linusson [16, 17] concerning generalizations of permutation matrices and its Bruhat order.

The natural way to generalize permutation matrices to higher dimensions is the following. Let n and d be positive integers and let $P = [p_{i_1 i_2 \ldots i_d}]$ be a d-dimensional

---

[1] As pointed out by a referee.

$(0,1)$-array of order n. Let k be a positive integer with $0 \leq k \leq d$. Define a k-*dimensional flat* of P to be the k-dimensional array obtained from P by fixing $d - k$ coordinate positions at values between 1 and n and allowing the other k coordinate positions to independently vary between 1 and n; notationally,

$$P[i_1 = a_1, i_2 = a_2, \ldots, i_{d-k} = a_{d-k}]$$

is a k-dimensional flat obtained by fixing coordinate position i_j at a_j for $j = 1, 2, \ldots, d - k$. There are $\binom{d}{k}$ k-dimensional flats. A (d, n, k)-*permutation array* is a d-dimensional $(0,1)$-array of order n such that each k-dimensional flat contains exactly one 1. A $(d, n, 0)$-permutation array is an array of all 1's; a (d, n, d)-permutation array is an array with exactly one 1. The permutation matrices of order n are the $(2, n, 1)$-permutation arrays. Latin squares of order n are equivalent to $(3, n, 1)$-permutation arrays. We can picture a $(d, 2, 1)$-permutation array as a d-dimensional unit cube with a 0 or 1 at each of its 2^d vertices. Combinatorial constructs known as orthogonal arrays of strength k and index 1 are equivalent to $(d, n, d - k)$-permutation arrays. Thus the existence of (d, n, k)-permutation arrays is equivalent to the existence of certain orthogonal arrays. For more on this theme see [6, 23] and the many references contained therein.

Eriksson and Linusson [16, 17] have proposed another definition of a permutation array in higher dimensions based on the following three properties of permutation matrices:

(i) Every row and column contains (at least) one 1.

(ii) In each leading submatrix the number of rows containing a 1 equals the number of columns containing a 1.

(iii) The collection of 1's is minimal with respect to (i) and (ii), that is, replacing a 1 by a 0 leads to a contradiction of (i) or (ii).

To get to their definition requires some preliminary work.

Let $P = [p_{i_1 i_2 \ldots i_d}]$ be a d-dimensional $(0,1)$-array of order n. For each position $(j_1, j_2, \ldots, j_d)$ with $1 \leq j_1, j_2, \ldots, j_d \leq n$, let $P[(j_1, j_2, \ldots, j_d)]$ be the *leading subarray* of P consisting of all entries at positions $(i_1, i_2, \ldots, i_d) \leq (j_1, j_2, \ldots, j_d)$ (componentwise ordering). For each coordinate index l with $1 \leq l \leq d$, define the *rank of P along the lth axis*, denoted $\mathrm{rk}_l(P)$ to be the number of values x $(1 \leq x \leq n)$, of the index i_l such that there is at least one 1 in some position with lth coordinate equal to x. If $\mathrm{rk}_l(P) = r$ for all $l = 1, 2, \ldots, d$, then P is *rankable* with rank r, denoted $\mathrm{rank} P = r$. For example, the 3-dimensional array of order 2 given by

$$\begin{vmatrix} 0 & 0 \\ 0 & 1 \end{vmatrix} \text{ (lower level)} \qquad \begin{vmatrix} 0 & 1 \\ 1 & 0 \end{vmatrix} \text{ (upper level)}$$

is rankable with rank 2. The array P is *totally rankable* provided that every leading subarray $P[(j_1, j_2, \ldots, j_d)]$ is rankable. If P is totally rankable, then the *rank array* of P is the d-dimensional nonnegative integral array rank P of order n whose entry in position $(j_1, j_2, \ldots, j_d)$ is $\mathrm{rank} P[(j_1, j_2, \ldots, j_d)]$. A $(d, n, d-1)$-permutation array is totally rankable. Another example of a totally rankable array taken from [16] is

$$
\begin{array}{ccc} 0 & 0 & 0 \\ 0 & 0 & 0 \\ 0 & 1 & 0 \end{array} \text{ (level 1)} \qquad
\begin{array}{ccc} 0 & 1 & 0 \\ 0 & 0 & 0 \\ 1 & 0 & 0 \end{array} \text{ (level 2)} \qquad
\begin{array}{ccc} 0 & 0 & 0 \\ 0 & 0 & 1 \\ 0 & 0 & 0 \end{array} \text{ (level 3)}
$$

with corresponding rank array

$$
\begin{array}{ccc} 0 & 0 & 0 \\ 0 & 0 & 0 \\ 0 & 1 & 1 \end{array} \text{ (level 1)} \qquad
\begin{array}{ccc} 0 & 1 & 1 \\ 0 & 1 & 1 \\ 1 & 2 & 2 \end{array} \text{ (level 2)} \qquad
\begin{array}{ccc} 0 & 1 & 1 \\ 0 & 1 & 2 \\ 1 & 2 & 3 \end{array} \text{ (level 3)}.
$$

Let P and Q be totally rankable d-dimensional $(0,1)$-arrays of order n. Then P and Q are *rank-equivalent* provided they have the same rank array, that is, for each $(j_1, j_2, \ldots, j_d)$ and each coordinate index l, we have

$$
\mathrm{rk}_l P[(j_1, j_2, \ldots, j_d)] = \mathrm{rk}_l Q[(j_1, j_2, \ldots, j_d)].
$$

Rank-equivalence partitions the set of totally rankable d-dimensional $(0,1)$-arrays of order n.

Theorem 8.1 [16] *If P is a totally rankable d-dimensional $(0,1)$-array of order n, then there exists a unique rankable d-dimensional $(0,1)$-array of order n that is rank-equivalent to P and has the smallest number of 1's. In fact, the rank-equivalence classes form intervals in the componentwise ordering.*

In view of Theorem 8.1, a *d-dimensional permutation array of order n* is defined to be a totally rankable d-dimensional $(0,1)$-array P of order n with the fewest number of 1's in its rank-equivalence class, equivalently, replacing any 1 of P with a 0 results in a permutation array that is not totally rankable. A 2-dimensional permutation array of order n is a permutation matrix. In [16] the rank arrays of permutation arrays (equivalently, totally rankable arrays) are characterized. In addition, an efficient algorithm is given to generate all d-dimensional permutation arrays of order n.

Let $\mathcal{P}_{d,n}$ denote the set of all d-dimensional permutation arrays of order n, and let $P, Q \in \mathcal{P}_{d,n}$. In [16, 17] the *Bruhat order*[2] is defined on $\mathcal{P}_{d,n}$ by

$$
P \preceq_B Q \text{ if and only if } \mathrm{rank}\, P[(i_i, i_2, \ldots, i_d)] \geq \mathrm{rank}\, Q[(i_i, i_2, \ldots, i_d)]
$$

for every position $(i_1, i_2, \ldots, i_n)$. If $d = 2$, this coincides with the Bruhat order on permutation matrices. The set $\mathcal{P}(d, 2)$, partially ordered using this Bruhat order, is isomorphic to the lattice of partitions of a set of d elements under refinement. The correspondence between permutation arrays P in $\mathcal{P}(d, 2)$ and partitions of $\{1, 2, \ldots, d\}$ is thus: Let there be q 1's in P, at positions $x^s = (i_1^s, i_2^s, \ldots, i_d^s)$, $1 \leq s \leq q$. Since $n = 2$, each i_j^s equals 1 or 2. Let X_s be the set of positions j in $\{1, 2, \ldots, d\}$ for which $i_j^s = 1$. Then $X_1, X_2, \ldots, X_q$ is a partition of $\{1, 2, \ldots, d\}$. The

---

[2]Actually we are reversing the order defined in [16, 17] in order to agree with the usual way, as adopted in this paper, of defining the Bruhat order on permutations.

largest permutation array in the Bruhat order on $\mathcal{P}(d, 2)$ is the "identity" permutation array with 1's in positions $(1, 1, \ldots, 1)$ and $(2, 2, \ldots, 2)$; this permutation array corresponds to the partition of $\{1, 2, \ldots, d\}$ with only one part, namely $\{1, 2, \ldots, d\}$ (the empty set corresponding to $(2, 2, \ldots, 2)$ is ignored). The smallest permutation array in this Bruhat order is the permutation array with d 1's where these 1's are in those positions $(1, 2, 2, \ldots, 2), (2, 1, 2, \ldots, 2), \ldots, (2, 2, \ldots, 1)$; this permutation array corresponds to the partition $\{1\}, \{2\}, \ldots, \{d\}$ of $\{1, 2, \ldots, d\}$.

References

[1] M. Aouchiche, F.K. Bell, D. Cvetković, P. Hansen, P. Rowlinson, S. Simić, and D. Stevanović, Variable neighborhood search for extremal graphs, 16. Some conjectures related to the largest eigenvalue of a graph, submitted.

[2] F.K. Bell, On the maximal index of connected graphs, *Linear Algebra Appl.* **144** (1991), 135–151.

[3] A. Björner and F. Brenti, An improved tableau criterion for Bruhat order, *Electron. J. Combin.* **3** (1996), #N22.

[4] R.A. Brualdi, Algorithms for constructing $(0, 1)$-matrices with prescribed row and column sum vectors, *Discrete Math.,* (to appear).

[5] R.A. Brualdi, *Combinatorial Matrix Classes*, Encyclopedia of Mathematicas and its Applications **108**, Cambridge University Press (2006).

[6] R.A. Brualdi and J. Csima, Small matrices of large dimension, *Linear Algebra Appl.* **150** (1991), 227–241.

[7] R.A. Brualdi and L. Deaett, More on the Bruhat order for $(0, 1)$-matrices, *Linear Algebra Appl.,* (to appear).

[8] R.A. Brualdi and A.J. Hoffman, On the spectral radius of $(0, 1)$-matrices, *Linear Algebra Appl.* **65** (1985), 133–146.

[9] R.A. Brualdi and S.-G. Hwang, A Bruhat order for the class of $(0, 1)$-matrices with row sum vector R and column sum vector S, *Electronic J. Linear Algebra* **12** (2004), 6–16.

[10] R.A. Brualdi, R. Manber, and J.A. Ross, On the minimum rank of regular classes of matrices of zeros and ones, *J. Combin. Theory, Ser. A* **41** (1986), 32–49.

[11] R.A. Brualdi and E. Solheid, On the spectral radius of connected graphs, *Publ. Inst. Math. (Beograd)* **39(53)** (1986), 45–54.

[12] D. Cvetković and P. Rowlinson, On connected graphs with maximal index, *Publ. Inst. Math. (Beograd)* **44(58)** (1988), 29–34.

[13] D. Cvetković and P. Rowlinson, The largest eigenvalue of a graph: a survey, *Linear Multilin. Algebra* **28** (1990), 3–33.

[14] D. Cvetković, P. Rowlinson, and S. Simić, *Eigenspaces of Graphs,* Encyclopedia of Mathematics and its Applications **66**, Cambridge University Press (1997), 69–72.

[15] B. Drake, S. Gerrish, and M. Skandera, Two new criteria for comparison in the Bruhat order, *Electron. J. Combin.*, **11** (2004), #N6.

[16] K. Eriksson and S. Linusson, A combinatorial theory of higher-dimensional permutation arrays, *Advances in Appl. Math.* **25** (2000), 194–211.

[17] K. Eriksson and S. Linusson, A decomposition of $FL(n)^d$ indexed by permutation arrays, *Advances in Appl. Math.* **25** (2000), 212–227.

[18] S. Friedland, The maximal eigenvalue of 0-1 matrices with prescribed number of ones, *Linear Algebra Appl.* **69** (1985), 33–69.

[19] W. Fulton, Young Tableaux With Applications to Representation Theory and Geometry, vol. 35 of London Mathematical Society of Student Texts, Cambridge University Press (1997).

[20] C. Greene, An extension of Schensted's theorem, *Advances in Math.* **14** (1974), 254–265.

[21] D.J. Houck and M.E. Paul, Nonsingular 0-1 matrices with constant row and column sums, *Linear Algebra Appl.* **50**, (1978), 143–152.

[22] L.K. Jorgenson, Rank of adjacency matrices of directed (strongly) regular graphs, *Linear Algebra Appl.* **407** (2005), 233–241.

[23] W.B. Jurkat and H.J. Ryser, Extremal configurations and decomposition theorems, *J. Algebra* **8** (1968), 194-222.

[24] D.E. Knuth, Permutation matrices and generalized Young tableaux, *Pacific J. Math.* **34** (1970), 709–727.

[25] A. Kotlov and L. Lovász, The rank and size of graphs, *J. Graph Theory* **23** (1996), 185–189.

[26] P. Magyar, Bruhat order for two flags and a line, *J. Algebraic Combin.* **21** (2005), 71–101.

[27] M. Newman, Combinatorial matrices with small determinants, *Canad. J. Math.* **30** (1978), 756-762.

[28] N. Punnim, The clique numbers of regular graphs, *Graphs Combin.* **18** (2002), 781–785.

[29] P. Rowlinson, On the maximal index of graphs with a prescribed number of edges, *Linear Algebra Appl.* **110** (1988), 43–53.

[30] B. Schwarz, Rearrangements of square matrices with non-negative elements, *Duke Math. J.*, **31** (1964), 45–62.

[31] D. Stevanović, Research problems from the Aveiro workshop on graph spectra, *Linear Algebra Appl.,* (to appear).

Richard A. Brualdi
Department of Mathematics
University of Wisconsin
Madison, WI 53706, USA
brualdi@math.wisc.edu

Cycle decompositions of complete graphs

Darryn Bryant

Abstract

The problem of decomposing complete graphs into cycles of specified lengths
dates back to the mid-nineteenth century when Kirkman solved the case where
all the cycles have length three. The case where all of the cycles are of an
arbitrary uniform length wasn't completely solved until just over one hundred
and fifty years later. The general problem where the specified cycle lengths vary
remains unsolved. This article gives an historical overview of the problem and
describes various cycle decomposition techniques that have been developed.

1 Introduction

1.1 Overview

A *decomposition* of a graph K is a set $\mathcal{D} = \{G_1, G_2, \ldots, G_t\}$ of subgraphs of K
such that $E(G_1) \cup E(G_2) \cup \cdots \cup E(G_t) = E(K)$ and $E(G_i) \cap E(G_j) = \emptyset$ for $i \neq j$. In
the case where the subgraphs are cycles we have a *cycle decomposition*. The purpose
of this article is to survey results on cycle decompositions and to illustrate some of
the techniques and ideas that have been used to obtain them. The most natural
graphs to decompose are complete graphs. The complete graph of order n is denoted
by K_n. There is no cycle decomposition of K_n when n is even, and so in this case it
is natural to consider cycle decompositions of $K_n - I$, the complete graph of order n
with the edges of a $1-$factor removed. Whenever the notation $K_n - I$ is used there
is an implication that n is even.

Example 1.1 A decomposition of $K_6 - I$ into a $3-$cycle, a $4-$cycle and a $5-$cycle
is shown on the left in the figure below, and a decomposition of K_7 into a $4-$cycle,
two $5-$cycles, and a $7-$cycle is shown on the right.

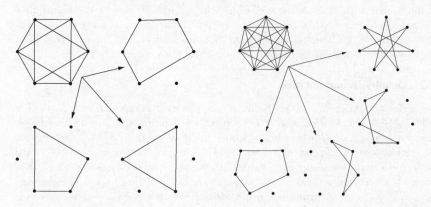

This article focuses on existence questions for cycle decompositions of K_n and
$K_n - I$. Other results on cycle decompositions can be found in the recent survey

[28], and results on decompositions into graphs other than cycles can be found in [17]. There are also several earlier surveys on decompositions into cycles [4, 14, 50] and a book [16] on graph decompositions.

An obvious necessary and sufficient condition for the existence of a cycle decomposition of a graph K is that each vertex of K has even degree. However, many very interesting and difficult problems arise if one asks questions about the existence of cycle decompositions where the lengths of the cycles are specified. That is, given a graph K and a sequence $m_1, m_2, \ldots, m_t$ of integers, determine whether there exists a cycle decomposition $\mathcal{D} = \{G_1, G_2, \ldots, G_t\}$ of K where G_i is an m_i-cycle for $i = 1, 2, \ldots, t$. The following lemma gives some obvious necessary conditions for the existence of such a decomposition.

Lemma 1.2 *Let K be a graph of order n, let $m_1, m_2, \ldots, m_t$ be a sequence of integers, and suppose there is a decomposition $\mathcal{D} = \{G_1, G_2, \ldots, G_t\}$ of K where G_i is an m_i-cycle for $i = 1, 2, \ldots, t$. Then*

1. $3 \le m_i \le n$ *for* $i = 1, 2, \ldots, t$;

2. *the number of edges in K is* $m_1 + m_2 + \cdots + m_t$; *and*

3. *each vertex of K has even degree.*

Of course, in many instances there are further obvious necessary conditions. For example, if K is bipartite then each m_i must be even. However, in the case of cycle decompositions of K_n and $K_n - I$ it seems likely that the necessary conditions given in Lemma 1.2 are also sufficient. The problem of proving this was posed by Alspach [3] in 1981 and remains unsolved. The special case of this problem where all the cycles have the same length is the subject of Section 2, and the general problem is discussed in Section 3.

It is worth mentioning that cycle decomposition problems are NP-complete in general. In fact, deciding whether an arbitrary graph has a decomposition into subgraphs each isomorphic to a given graph G is NP-complete if and only if G has a component with three or more edges [30, 38]. It should also be mentioned that Wilson [65] has shown that for any simple graph G, there exists an integer $N(G)$ such that for all $n \ge N(G)$ and satisfying certain obvious necessary numerical conditions, there is a decomposition of K_n into subgraphs each isomorphic to G.

1.2   Definitions and notation

A decomposition of a graph K into subgraphs each isomorphic to a graph G is called a G-*decomposition* of K. A graph (or set $\mathcal{S}$ of graphs) in a decomposition *uses* an edge xy if xy occurs in the graph (or in a graph in $\mathcal{S}$). The cycle with m vertices and m edges is called an m-*cycle* and is denoted by C_m. The path with $m + 1$ vertices and m edges is called an m-*path* and is denoted by P_{m+1}. The m-cycle with vertices $v_1, v_2, \ldots, v_m$ and edges $v_1 v_2, v_2 v_3, \ldots, v_{m-1} v_m, v_m v_1$ is denoted by $(v_1, v_2, \ldots, v_m)$. A connected graph with all vertices of even degree is a *closed trail*.

The complete multipartite graph with r parts of sizes $k_1, k_2, \ldots, k_r$ is denoted by $K_{k_1, k_2, \ldots, k_r}$. The graph obtained from K_n when the edges of a complete subgraph of order v are removed is called the *complete graph of order n with a hole of size v* and is denoted by $K_n - K_v$, the hole consisting of the vertices of K_v. For a graph G and set S of vertices not in G, $S \vee G$ denotes the graph with vertex set $S \cup V(G)$ and edge set $E(G) \cup \{xy : x \in S, y \in V(G)\}$.

We will often generate cycle decompositions via the action of permutations on *starter cycles*. For any graph G and any permutation π acting on the vertices of G, $\pi(G)$ is defined to be the graph with vertex set $V(\pi(G)) = \{\pi(v) : v \in V(G)\}$ and edge set $E(\pi(G)) = \{\pi(u)\pi(v) : uv \in E(G)\}$. For any set $\mathcal{S}$ of graphs and any permutation π acting on the vertices of the graphs in $\mathcal{S}$, $\pi(\mathcal{S}) = \{\pi(G) : G \in \mathcal{S}\}$.

Definition 1.3 The *circulant graph* of order n with *connection set* $S \subseteq \{1, 2, \ldots, \lfloor \frac{n}{2} \rfloor\}$, denoted $\mathrm{Circ}(n, S)$, has vertex set $\mathbf{Z}_n$ and edge set given by joining x to $x + s$ for each $x \in \mathbf{Z}_n$ and each $s \in S$.

Circulant graphs are a family of *Cayley graphs*, which are defined similarly but with an arbitrary underlying group rather than the cyclic group $\mathbf{Z}_n$. For n odd $K_n \cong \mathrm{Circ}(n, \{1, 2, \ldots, \frac{n-1}{2}\})$, and for n even $K_n - I \cong \mathrm{Circ}(n, \{1, 2, \ldots, \frac{n-2}{2}\})$.

In many of the constructions we discuss, the vertex set of K_n is either $\mathbf{Z}_n$ or $\mathbf{Z}_{n-1} \cup \{\infty\}$, and when n is even the vertex set of $K_n - I$ is either $\mathbf{Z}_n$ or $\mathbf{Z}_{n-2} \cup \{\infty_1, \infty_2\}$. In the latter case, the edges of the removed $1-$factor are $\infty_1 \infty_2$ and those edges joining x to $x + \frac{n-2}{2}$ for $x = 0, 1, 2 \ldots, \frac{n-4}{2}$. The permutation ρ of the following definition is then often used to generate cycles of the decomposition.

Definition 1.4 The permutations $(0, 1, 2, \ldots, n - 1)$, $(0, 1, 2, \ldots, n - 1)(\infty)$ and $(0, 1, 2, \ldots, n - 1)(\infty_1)(\infty_2)$ are each denoted by ρ_n. Where the value of n is clear from the context, we use just ρ rather than ρ_n.

When cycle decompositions involving the generation of cycles under the permutation ρ_n are illustrated in figures, the vertices of $\mathbf{Z}_n$ are arranged in cyclic order around the circumference of a circle with ∞, or ∞_1 and ∞_2 (if they are involved), in the interior of the circle. For any graph G with $\mathbf{Z}_n \subseteq V(G)$ and for any edge xy of G with $x, y \in \mathbf{Z}_n$, we define the *length* of xy to be the distance (length of the shortest path) between x and y in the cycle $(0, 1, 2, \ldots, n - 1)$.

As we shall see, the "doubling construction" given in the following definition has been used to prove several results, in particular results on decompositions of $K_n - I$. The construction is illustrated in the figure below.

Definition 1.5 Let K be a graph. The graph $K^{(2)}$ is defined by letting K' and K'' be vertex disjoint copies of K with corresponding vertices $v' \in V(K')$ and $v'' \in V(K'')$ for each $v \in V(K)$, and defining $K^{(2)}$ to have vertex set $V(K^{(2)}) = V(K') \cup V(K'')$ and edge set $E(K^{(2)}) = E(K') \cup E(K'') \cup \{u'v'', u''v' : uv \in E(K)\}$. Moreover, for any decomposition $\mathcal{D}$ of a graph K, define

$$\mathcal{D}^{(2)} = \{G^{(2)} : G \in \mathcal{D}\}.$$

Notice that $K_r^{(2)} \cong K_{2r} - I$ and that for any decomposition $\mathcal{D}$ of a graph K, $\mathcal{D}^{(2)}$ is a decomposition of $K^{(2)}$. The following figure shows the graph $P_8^{(2)}$.

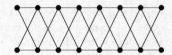

As a very simple example of the doubling construction, observe that since there is a trivial K_2−decomposition of K_r for any $r \geq 1$, and since $K_2^{(2)}$ is a 4−cycle, the doubling construction gives a C_4−decomposition of $K_n - I$ for any positive even integer n.

2   Uniform length cycles

This section deals with cycle decompositions of K_n and $K_n - I$ into cycles of uniform length and is divided into four subsections. Sections 2.1, 2.2 and 2.3 focus on decompositions of K_n into m−cycles, and decompositions of $K_n - I$ into m−cycles are discussed in Section 2.4. Section 2.1 introduces some notation and summarises results that were obtained up until about the end of the 1980s. Section 2.2 describes the results which reduced the problem to small values of n for each m, and then the solution to the problem is discussed in Section 2.3.

2.1   The spectrum problem for m-cycles

The problem of decomposing K_n into cycles of uniform length m is often phrased in terms of determining, for a given value of m, those values of n for which K_n can be decomposed into m−cycles. The set of such values of n is called the *spectrum* for m−cycles and is denoted by $\mathrm{Spec}(C_m)$. The problem of determining $\mathrm{Spec}(C_m)$ is called the *spectrum problem* for m−cycles. The obvious necessary conditions for $n \in \mathrm{Spec}(C_m)$ are $n \geq m$ for $n > 1$, n is odd, and m divides $\frac{n(n-1)}{2}$. For a given value of m, an integer satisfying these conditions is said to be m−*admissible*, or just *admissible* if the value of m is clear from the context. The problem of determining $\mathrm{Spec}(C_m)$ is now completely solved [5, 60]: for all $m \geq 3$, $\mathrm{Spec}(C_m)$ is precisely the set of all admissible n. We give here an historical overview of the problem including descriptions and examples of some of the constructions that have been devised.

It is worth making the following remarks concerning admissible values of n. For any given cycle length m, the admissible integers lie in certain residue classes modulo $2m$. For example, the admissible integers for cycles of length $m = 30$ are all $n \equiv 1, 21, 25, 45 \pmod{60}$ with $n = 1$ or $n \geq 45$. These admissible residue classes are sometimes called *fibers*, see [45]. For all $m \geq 3$, $n \equiv 1 \pmod{2m}$ is a fiber, and this is the only fiber when m is a power of 2. For odd $m \geq 3$, $n \equiv m \pmod{2m}$ is also a fiber, and when m is a power of an odd prime, $n \equiv 1, m \pmod{2m}$ are the only fibers. For $m \equiv 2 \pmod 8$ we have the fiber $n \equiv \frac{m}{2} \pmod{2m}$, for $m \equiv 6 \pmod 8$ we have the fiber $n \equiv \frac{3m}{2} \pmod{2m}$, and these together with the fiber $n \equiv 1 \pmod{2m}$ are the only fibers when m is twice a power of an odd prime.

Results on m−cycle decompositions of complete graphs date back to 1847 when Kirkman [47] proved that there is decomposition of K_n into 3-cycles if and only if $n \equiv 1, 3 \pmod 6$. That is, $\mathrm{Spec}(C_3) = \{n : n \equiv 1, 3 \pmod 6\}$. A decomposition of K_n into 3−cycles is of course a *Steiner triple system*. There is an entire text

[37] devoted to triple systems. Another result from the 1800s is that of Walecki (see Lucas [51]) concerning the decomposition of K_n and $K_n - I$ into $n-$cycles, or *Hamilton cycles*. Variations of Walecki's method were used over a hundred years later in the eventual solution to the spectrum problem for $m-$cycles. We include Walecki's construction here and illustrate it in the figure below.

For odd n, let $n = 2r + 1$, take $\mathbf{Z}_{2r} \cup \{\infty\}$ as the vertex set of K_n and let $\mathcal{D}$ be the orbit of the $n-$cycle

$$(\infty, 0, 1, 2r - 1, 2, 2r - 2, 3, 2r - 3, \ldots, r - 1, r + 1, r)$$

under the permutation ρ_{2r}. Then $\mathcal{D}$ is a decomposition of K_n into $n-$cycles. For even n, let $n = 2r + 2$, take $\mathbf{Z}_{2r} \cup \{\infty_1, \infty_2\}$ as the vertex set of $K_n - I$, and let $\mathcal{D}$ be the $n-$cycle decomposition given by the orbit under the permutation ρ_{2r} of the $n-$cycle

$$\left(\infty_1, 0, 1, 2r - 1, 2, 2r - 2, 3, 2r - 3, \ldots, \frac{r}{2}, \infty_2, \frac{3r}{2}, \ldots, r - 1, r + 1, r\right)$$

when r is even and

$$\left(\infty_1, 0, 1, 2r - 1, 2, 2r - 2, 3, 2r - 3, \ldots, \frac{3r + 1}{2}, \infty_2, \frac{r + 1}{2}, \ldots, r - 1, r + 1, r\right)$$

when r is odd. These decompositions are illustrated for the cases $n = 12, 13$ and 14 in the figure below.

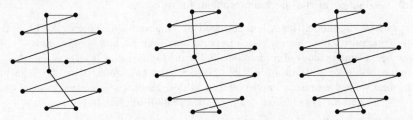

The next results on $m-$cycle decompositions of K_n appeared in papers of Kotzig and Rosa in the mid 1960s [48, 55, 56]. Kotzig [48] proved that $n \in \text{Spec}(C_m)$ whenever $n \equiv 1 \pmod{2m}$ and $m \equiv 0 \pmod 4$, thus settling the spectrum problem for $m-$cycles when m is a power of 2. Rosa [55] showed that $n \in \text{Spec}(C_m)$ for $n \equiv 1 \pmod{2m}$ and $m \equiv 2 \pmod 4$, and he also settled the spectrum problem for $5-$cycles and $7-$cycles [56]. The decompositions of Kotzig and Rosa were *cyclic*. That is, $\mathbf{Z}_n$ is taken as the vertex set of K_n and the decomposition $\mathcal{D}$ satisfies $\rho_n(\mathcal{D}) = \mathcal{D}$. An example of a cyclic decomposition of K_{25} into $6-$cycles is given by the orbit under ρ_{25} of the following two starter cycles.

$$(0, 1, 3, 7, 19, 8) \qquad (0, 3, 10, 20, 15, 9)$$

Other results on cyclic $m-$cycle systems can be found in [15, 18, 33, 34, 41, 53] and in the survey [28].

In the 1970s the spectrum problem for $m-$cycles was settled for several further small values of m as indicated in the following table, which also lists earlier results.

In each case, the spectrum consists of $n = 1$ and the set of all integers $n \geq m$ in the indicated residue classes modulo $2m$.

m	3	2^α	5, 7	6	10	12	14
Residues	1, 3	1	1, m	1, 9	1, 5	1, 9	1, 21
Year	1847	1965	1966	1975	1978	1978	1978
Reference	[47]	[48]	[56]	[59]	[13]	[13]	[13]

In 1980, Alspach and Varma [7] settled the spectrum problem for C_m for the case m is twice a power of an odd prime. They proved that for $m = 2p^\alpha$ where p is prime and $\alpha \geq 1$,

$$\mathrm{Spec}(C_m) = \{n : n \equiv 1, \tfrac{m}{2} \ (\mathrm{mod}\ 2m), n \neq \tfrac{m}{2}\} \qquad \text{for } m \equiv 2 \ (\mathrm{mod}\ 8),$$

and

$$\mathrm{Spec}(C_m) = \{n : n \equiv 1, \tfrac{3m}{2} \ (\mathrm{mod}\ 2m)\} \qquad \text{for } m \equiv 6 \ (\mathrm{mod}\ 8).$$

In 1988, Jackson [46] proved a similar result for cycles of odd length. He constructed $m-$cycle decompositions of K_n for $n \equiv 1, m \ (\mathrm{mod}\ 2m)$ and all odd m, thus establishing the entire spectrum for $m-$cycles when m is a power of an odd prime. For $m = p^\alpha$ where p is an odd prime and $\alpha \geq 1$,

$$\mathrm{Spec}(C_m) = \{n : n \equiv 1, m \ (\mathrm{mod}\ 2m)\}.$$

2.2 Reduction of the spectrum problem for m-cycles

In 1978, Bermond, Huang and Sotteau [13] reduced, for each even value of m, the spectrum problem for $m-$cycles to one of decomposing K_n into $m-$cycles for a finite number of small values of n. In 1989, Hoffman, Lindner and Rodger [45] proved a similar result for the case m is odd. These results are contained in the following theorem. They represent important breakthroughs on the spectrum problem for $m-$cycles and are critical ingredients in the eventual solution to the problem.

Theorem 2.1 [13, 45] *If there exists a decomposition of K_n into $m-$cycles for all admissible n in the range $m < n < 3m$, then there exists a decomposition of K_n into $m-$cycles for all admissible n.*

For m even, Theorem 2.1 is an easy corollary of the above-mentioned results of Kotzig [48] and Rosa [55], which proved the existence of $m-$cycle systems of K_n for all $n \equiv 1 \ (\mathrm{mod}\ 2m)$, and the following result of Sotteau [63] on $m-$cycle decompositions of complete bipartite graphs.

Theorem 2.2 [63] (Sotteau's Theorem) *There exists an $m-$cycle decomposition of $K_{x,y}$ if and only if x, y and m are even, m divides xy, $x \leq \tfrac{m}{2}$ and $y \leq \tfrac{m}{2}$.*

Suppose m is even and there exists an $m-$cycle decomposition of K_v. To construct an $m-$cycle decomposition of K_{2mx+v} for any $x \geq 1$, decompose K_{2mx+v} into a copy of K_{2mx+1} and a copy of K_v which intersect in one vertex, and a copy of $K_{v-1,2mx}$. The subgraphs in this decomposition each have a decomposition into $m-$cycles: K_v by assumption, K_{2mx+1} by the results of Kotzig [48] and Rosa [55], and $K_{v-1,2mx}$

by Sotteau's Theorem [63]. Thus it follows that K_{2mx+v} has a decomposition into $m-$cycles. Since the smallest admissible value of n in each fiber is in the range $m \leq n < 3m$ (and since Walecki's construction gives an $m-$cycle decomposition of $K_m - I$) Theorem 2.1 is established for even m.

We now outline the construction of Hoffman et al [45] which proves Theorem 2.1 for the case m is odd. The two main ingredients in the proof are Lemma 2.3 and Lemma 2.5.

Lemma 2.3 [45] *Let $m \geq 3$ and $v \geq 1$ be odd. There exists an $m-$cycle decomposition of $K_{2m+v} - K_v$ provided $q \leq m + 2r - 1$ where q and r are given by $v = q\frac{m-1}{2} + r$ with $1 \leq r \leq \frac{m-1}{2}$.*

Note that for $m \geq 5$ and $v < 3m$ the condition $q \leq m + 2r - 1$ is always satisfied. A considerable amount of work has been done on decompositions of $K_n - K_v$ into $m-$cycles, see [19, 23, 26, 27, 29, 39, 52] and the survey [28]. Results of this kind are often called "Doyen-Wilson" type results, after the authors of the 1973 article [39] which considered the problem for $3-$cycles.

Lemma 2.3 is proved by taking $\mathbf{Z}_m \times \{0,1\} \cup \{\infty_1, \infty_2, \ldots, \infty_v\}$ as the vertex set of $K_{2m+v} - K_v$, with $\{\infty_1, \infty_2, \ldots, \infty_v\}$ being the hole. Various clever combinations of starter $m-$cycles are then used to generate the decomposition under the permutation $(x,i) \mapsto (x+1,i)$ for $x \in \mathbf{Z}_m$ and $i \in \{0,1\}$, and $\infty_j \mapsto \infty_j$ for $j = 1,2,\ldots,v$. We illustrate some of the features of the construction with the following example.

Example 2.4 *A $15-$cycle decomposition of $K_{55} - K_{25}$.*

For any odd integer m, any subsets D_0, D_1 of $\{1,2,\ldots,\frac{m-1}{2}\}$ and any subset M of $\{0,1,2,\ldots,m-1\}$, define $\langle D_0, M, D_1 \rangle_m$ to be the graph with vertex set $\mathbf{Z}_m \times \{0,1\}$ and edges defined as follows. For $i \in \{0,1\}$ and each d in D_i vertices (x,i) and (y,i) are joined if x and y are at distance d in the $m-$cycle $(0,1,2,\ldots,m-1)$. For each $d \in M$ vertices $(x,0)$ and $(y,1)$ are joined if $x + d \equiv y \pmod{m}$. So $K_{2m} \cong \langle \{1,2,\ldots,\frac{m-1}{2}\}, \{0,1,2,\ldots,m-1\}, \{1,2,\ldots,\frac{m-1}{2}\} \rangle_m$.

We decompose $K_{55} - K_{25}$ into the following six subgraphs and then show that each of these can be decomposed into $15-$cycles.

1. $\{\infty_1, \infty_2, \ldots, \infty_7\} \vee \langle \emptyset, \{12\}, \emptyset \rangle_{15}$

2. $\{\infty_8, \infty_9, \ldots, \infty_{14}\} \vee \langle \emptyset, \{13\}, \emptyset \rangle_{15}$

3. $\{\infty_{15}, \infty_{16}, \ldots, \infty_{21}\} \vee \langle \emptyset, \{14\}, \emptyset \rangle_{15}$

4. $\{\infty_{22}\} \vee \langle \{7\}, \{1,2,\ldots,11\}, \{7\} \rangle_{15}$

5. $\{\infty_{23}, \infty_{24}, \infty_{25}\} \vee \langle \{3,4,5,6\}, \{0\}, \{3,4,5,6\} \rangle_{15}$

6. $\langle \{1,2\}, \emptyset, \{1,2\} \rangle_{15}$

For any cycle C in $\{\infty_1, \infty_2, \ldots, \infty_{25}\} \vee K_{30}$ denote by $\psi(C)$ the orbit of C under the permutation $(x,i) \mapsto (x+1,i)$ for $x \in \mathbf{Z}_{15}$ and $i \in \{0,1\}$, and $\infty_j \mapsto \infty_j$ for $j = 1,2,\ldots,25$. Now, for any $d \in \{0,1,2,\ldots,14\}$, $\psi(C)$ is a decomposition of $\{\infty_1, \infty_2, \ldots, \infty_7\} \vee \langle \emptyset, \{d\}, \emptyset \rangle_{15}$ into $15-$cycles, where

$$C = ((0,0),(d,1),\infty_1,(x_1,0),\infty_2,(x_2,1),\infty_3,(x_3,0),\ldots,(x_6,1),\infty_7)$$

and $x_1, x_2, \ldots, x_6$ are arbitrary distinct elements of $\mathbf{Z}_{15} \setminus \{0, d\}$. Hence graphs 1, 2 and 3 in the above list each have 15−cycle decompositions.

Also, for $d \in \{1, 2\}$ the graphs $\langle\{d\}, \emptyset, \emptyset\rangle_{15}$ and $\langle\emptyset, \emptyset, \{d\}\rangle_{15}$ are 15−cycles (ignoring the isolated vertices) and so graph number 6 in the list can be decomposed into four 15−cycles. A 15−cycle decomposition of graph 4 is given by $\psi(C)$ where C is the 15−cycle on the left in the figure below, and a 15−cycle decomposition of graph 5 is obtained similarly from the 15−cycle on the right.

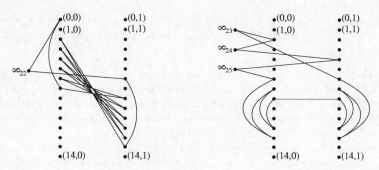

Lemma 2.5 [45] *Let $g \geq 3$ and $m \geq 3$ with m odd. There is an m−cycle decomposition of the complete multipartite graph $K_{2m,2m,\ldots,2m}$ with g parts of size $2m$.*

We give the construction used to prove Lemma 2.5. Let $Q = \{1, 2, \ldots, 2g\}$ and let $(Q, \circ)$ be a commutative quasigroup containing the subquasigroups $(\{1, 2\}, \circ)$, $(\{3, 4\}, \circ), \ldots, (\{2g - 1, 2g\}, \circ)$. Such quasigroups are well known to exist for all $g \geq 3$. Take $\mathbf{Z}_m \times Q$ as the vertex set of the complete multipartite graph with g parts of size $2m$; the g parts being $\mathbf{Z}_m \times \{1, 2\}, \mathbf{Z}_m \times \{3, 4\}, \ldots, \mathbf{Z}_m \times \{2g - 1, 2g\}$. For each pair $i, j \in Q$ with $i < j$ and $\{i, j\} \notin \{\{1, 2\}, \{3, 4\}, \ldots, \{2g - 1, 2g\}\}$ we include the orbit of the cycle

$$C = ((0, i), (\tfrac{m-3}{2}, j), (1, i), (\tfrac{m-5}{2}, j), \ldots, (\tfrac{m-3}{2}, i), (0, j), (\tfrac{m-1}{2}, i \circ j))$$

under the permutation $(x, i) \mapsto (x + 1, i)$ for each $x \in \mathbf{Z}_m$ and each $i \in Q$. The cycle C for the case $m = 9$ is shown in the figure below.

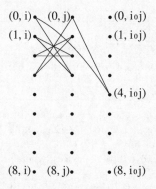

We now describe how one obtains Theorem 2.1 for the case m is odd from Lemmas 2.3 and 2.5. The above-mentioned result of Jackson [46] proves that the theorem holds for $m \leq 13$ so we may assume $m \geq 15$. Write $n = 2mx + v$ where $m \leq v < 3m$ and $x \geq 1$. Since n is admissible, it follows that v is also admissible, and hence by assumption that there is an $m-$cycle decomposition of K_v. For $x = 1$, decompose K_{2m+v} into one copy of $K_{2m+v} - K_v$ and one copy of K_v, and use the $m-$cycle decomposition of $K_{2m+v} - K_v$ given by Lemma 2.3. For $x = 2$, decompose K_{4m+v} into one copy of $K_{4m+v} - K_{2m+v}$ and one copy of K_{2m+v}, and use the $m-$cycle decomposition of K_{2m+v} just constructed and the $m-$cycle decomposition of $K_{4m+v} - K_{2m+v}$ given by Lemma 2.3. It is straightforward to check that the conditions of Lemma 2.3 are satisfied for $K_{4m+v} - K_{2m+v}$ when $m \geq 15$. For $x \geq 3$ decompose K_{2mx+v} into one copy of K_v, x copies of $K_{2m+v} - K_v$, and one copy of the complete multipartite graph $K_{2m,2m,\dots,2m}$ with x parts of size $2m$. By assumption K_v has an $m-$cycle decomposition, $K_{2m+v} - K_v$ has an $m-$cycle decomposition by Lemma 2.3, and $K_{2m,2m,\dots,2m}$ has an $m-$cycle decomposition by Lemma 2.5. Thus it follows that K_{2mx+v} has an $m-$cycle decomposition and Theorem 2.1 is established for odd m.

Hoffman et al [45] constructed $15-$cycle decompositions of K_{21} and K_{25}, and $21-$cycle decompositions of K_{49} and K_{57}, and thus by Theorem 2.1 settled the spectrum problem for $15-$cycles and $21-$cycles, the two smallest unresolved odd values of m:
$$\mathrm{Spec}(C_{15}) = \{n : n \equiv 1, 15, 21, 25 \ (\mathrm{mod} \ 30)\}$$
and
$$\mathrm{Spec}(C_{21}) = \{n : n \equiv 1, 7, 15, 21 \ (\mathrm{mod} \ 42), n \geq 21\}.$$

Their constructions also provide an alternative proof of the existence of $m-$cycle systems of K_n for all admissible n in the case m is a power of an odd prime, a case which was settled by Jackson [46]. Subsequently, Bell [11] used Theorem 2.1 to settle the spectrum problem for $m-$cycles for
$$m \in \{20, 24, 28, 30, 33, 35, 36, 39, 40, 42, 44, 45, 48\},$$
thus settling it for all $m \leq 50$, the spectrum for the other values of $m \leq 50$ having already been established by earlier results.

2.3   Solution to the spectrum problem for m-cycles

In [5] Alspach and Gavlas construct $m-$cycle decompositions of K_n for all odd m and all admissible n in the range $m < n < 3m$, which by Theorem 2.1 settles the spectrum problem for $m-$cycles when m is odd. Soon after, Šajna [60] extended their techniques and completed the solution by settling the case m is even. The same two papers also completely settle the existence problem for $m-$cycle decompositions of $K_n - I$, see Section 2.4.

Theorem 2.6 [5, 60] *Let $m \geq 3$. There exists an $m-$cycle decomposition of K_n if and only if n is odd, $n \geq m$ for $n > 1$, and m divides $\frac{n(n-1)}{2}$.*

We illustrate the techniques used by Alspach, Gavlas and Šajna to settle the spectrum problem for $m-$cycles with two examples: a decomposition of K_{21} into

15−cycles and a decomposition of K_{25} into 20−cycles. The general solution is quite lengthy and contains many techniques not illustrated by these two decompositions. For example, clever use was made of the result of Bermond et al [12] on decompositions of 4−regular Cayley graphs into Hamilton cycles (see Theorem 3.6) to link paths together into m−cycles. Nevertheless, the decompositions given in the following two examples serve to give some flavour of the constructions.

Example 2.7 *A 15−cycle decomposition of K_{21}.*

Take $\mathbf{Z}_{20} \cup \{\infty\}$ as the vertex set of K_{21}. Let C be the cycle on the left in the figure below and let C' be the cycle on the right. Then the orbits of C and C' under the permutation ρ form the required 15−cycle decomposition of K_{21}.

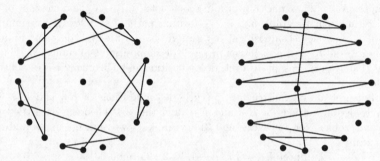

The orbit of C contains four 15−cycles and uses all the edges of lengths $1, 3$ and 6. The orbit of C' contains ten 15−cycles and uses all the edges of lengths $2, 4, 5, 7, 8, 9$ and 10 and all the edges incident with ∞.

Example 2.8 *A 20−cycle decomposition of K_{25}.*

Take $\mathbf{Z}_{24} \cup \{\infty\}$ as the vertex set of K_{25} and let C, C' and C'' be the cycles shown below on the left, in the centre, and on the right respectively.

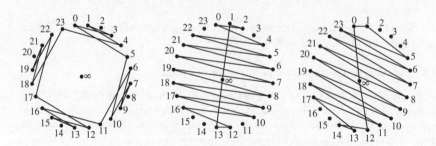

Define $\mathcal{D}$ to be the orbit of C under the permutation ρ^2, let $\mathcal{D}' = \{\rho^i(C') : i = 0, 2, 4, 6, 8, 10\}$, and similarly let $\mathcal{D}'' = \{\rho^i(C'') : i = 0, 2, 4, 6, 8, 10\}$. A 20−cycle decomposition of K_{25} is then given by $\mathcal{D} \cup \mathcal{D}' \cup \mathcal{D}''$. To see this, observe that

- $\mathcal{D}$ contains three 20−cycles, uses all the edges of lengths 3 and 5, and uses half the edges of length 2 (namely those joining vertex x to vertex $x + 2$ for $x = 1, 3, 5, \ldots, 23$);

- $\mathcal{D}'$ and $\mathcal{D}''$ each contain six $20-$cycles, together use all the edges of lengths $1, 4, 6, 7, 8, 9, 10, 11, 12$, and together use the other half of the edges of length 2 (namely those joining vertex x to vertex $x + 2$ for $x = 0, 2, 4, \ldots, 22$);

- the edges joining ∞ to $1, 3, \ldots, 23$ are used in $\mathcal{D}'$, and the edges joining ∞ to $0, 2, \ldots, 22$ are used in $\mathcal{D}''$.

More recently, Buratti [31] has provided an alternative solution to the spectrum problem for $m-$cycles in the case m is odd. A decomposition $\mathcal{D}$ of K_n into $m-$cycles is $1-rotational$ if there exists a labeling of the vertex set of K_n with the elements of $\mathbf{Z}_{n-1} \cup \{\infty\}$ such that $\rho_{n-1}(\mathcal{D}) = \mathcal{D}$. Walecki's $n-$cycle decompositions of K_n are examples of $1-$rotational decompositions, as are those in Example 2.7. Buratti constructed $1-$rotational $m-$cycle decompositions of K_n for all admissible n in the range $m \leq n \leq 3m$, with the exception that there is no $1-$rotational $m-$cycle decomposition of K_n when m is composite and $n = 2m + 1$. These exceptional values of m and n are covered by Jackson's result [46], so the spectrum problem for $m-$cycles with m odd is settled using Theorem 2.1. Other results on $1-$rotational cycle decompositions can be found in [32] and [54] and in the survey [28].

2.4   m-cycle decompositions of complete graphs minus a 1-factor

We extend the definition of the term *admissible* so that it applies to $m-$cycle decompositions of $K_n - I$, rather than just K_n. Thus, an odd integer $n > 1$ is *admissible* if $n \geq m$ and m divides $\frac{n(n-1)}{2}$, and an even integer $n > 2$ is admissible if $n \geq m$ and m divides $\frac{n(n-2)}{2}$.

Like the problem of decomposing K_n into $m-$cycles, the problem of decomposing $K_n - I$ into $m-$cycles was solved by first reducing the problem to small values of n for each value of m. For m even we have the following result, see [5] or [6].

Theorem 2.9 [5, 6] *Let m and n be even. If there exists a decomposition of $K_n - I$ into $m-$cycles for all admissible n in the range $m < n < 2m$ then there exists a decomposition of $K_n - I$ into $m-$cycles for all admissible n.*

Theorem 2.9 follows easily from Sotteau's Theorem (Theorem 2.2). Write $n = qm + r$ with $0 \leq r < m$ and decompose $K_n - I$ into $q - 1$ copies of $K_m - I$, one copy of $K_{m+r} - I$, $\binom{q-1}{2}$ copies of $K_{m,m}$ and $q - 1$ copies of $K_{m,m+r}$. Each of these graphs can be decomposed into $m-$cycles: $K_m - I$ by the result of Walecki, $K_{m+r} - I$ by assumption (since m divides $\frac{n(n-2)}{2}$ implies m divides $\frac{(m+r)(m+r-2)}{2}$), and $K_{m,m}$ and $K_{m,m+r}$ by Sotteau's Theorem. For the case m is odd we have the following result of Šajna [61]. A partial result of this kind had earlier been obtained by El-Zanati [40].

Theorem 2.10 [61] *Let m be odd and n be even. If there exists a decomposition of $K_n - I$ into $m-$cycles for all admissible n in the range $m < n < 3m$ then there exists a decomposition of $K_n - I$ into $m-$cycles for all admissible n.*

As mentioned above, the existence problem for $m-$cycle decompositions of $K_n - I$ was completely solved, using Theorem 2.9 and Theorem 2.10, by Alspach and Gavlas [5] and Šajna [60]. Again, the obvious necessary conditions given in Lemma 1.2 are also sufficient, and m−cycle decompositions of $K_n - I$ exist for all admissible even n.

Theorem 2.11 [5, 60] *Let $m \geq 3$ and let n be even. There exists an m−cycle decomposition of $K_n - I$ if and only if $n \geq m$ for $n > 2$, and m divides $\frac{n(n-2)}{2}$.*

The following examples illustrate some of the ideas involved in the proof of Theorem 2.11. Example 2.12 is for the case m is even which is the case settled by Alspach and Gavlas [5], and Example 2.13 is for the case m is odd which is the case settled by Šajna [60].

Example 2.12 *A 12−cycle decomposition of $K_{20} - I$.*

Take $\mathbf{Z}_{18} \cup \{\infty_1, \infty_2\}$ as the vertex set of $K_{20} - I$ and let the edges of the removed 1−factor be the edges of length 9 and $\infty_1 \infty_2$. Let C be the cycle on the left in the figure below and let C' be the cycle on the right. Then the orbits of C and C' under the permutation ρ form the required 12−cycle decomposition of $K_{20} - I$.

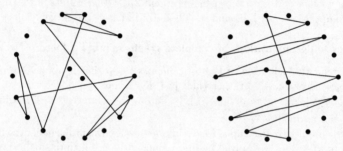

The orbit of C contains six 12−cycles and uses all the edges of lengths $1, 3, 4$ and 8. The orbit of C' contains nine 12−cycles and uses all the edges of lengths $2, 5, 6$, and 7 and all the edges incident with ∞_1 and ∞_2.

Example 2.13 *A 21−cycle decomposition of $K_{56} - I$.*

We begin with a decomposition of K_{28} into 21−cycles and 21−paths. The decomposition of $K_{56} - I$ is obtained from this by using a modification of the doubling construction given in Definition 1.5. Let K_{28} have vertex set $\mathbf{Z}_{28}$ and decompose it into two circulant graphs, namely $\mathrm{Circ}(28, S_C)$ and $\mathrm{Circ}(28, S_P)$ where $S_C = \{1, 3, 6\}$ and $S_P = \{2, 4, 5, 7, 8, 9, 10, 11, 12, 13, 14\}$. Thus, using the doubling construction we have a decomposition of $K_{56} - I$ into copies of $\mathrm{Circ}(28, S_C)^{(2)}$ and $\mathrm{Circ}(28, S_P)^{(2)}$ (see the remarks following Definition 1.5). We will obtain the required 21−cycle decomposition of $K_{56} - I$ by showing that each of these graphs has a decomposition into 21−cycles.

Let C be the cycle shown on the left in the figure below, let P be the path shown in the centre, and let P' be the path shown on the right. Then the orbit of C under the permutation ρ is a decomposition of $\mathrm{Circ}(28, S_C)$ into four 21−cycles. A decomposition of $\mathrm{Circ}(28, S_P)$ into fourteen 21−paths is given by the orbit of P under ρ, and a different decomposition of $\mathrm{Circ}(28, S_P)$ into fourteen 21−paths is given by the orbit of P' under ρ. The four bold edges in the figure indicate where P' differs from P.

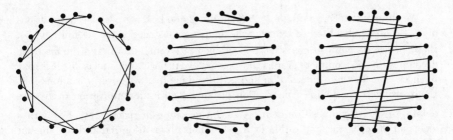

We now describe how the decomposition of $K_{56} - I$ is obtained from the above decompositions of K_{28}. It is easy to construct an $m-$cycle decomposition of $C_m^{(2)}$ for any integer $m \geq 3$. At the moment we are interested in odd m and such decompositions are obtained by generalising, in the obvious manner, the decomposition of $C_9^{(2)}$ shown in the figure below.

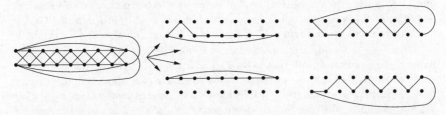

Thus, since we have a $21-$cycle decomposition of $\text{Circ}(28, S_C)$, we have a $C_{21}^{(2)}-$decomposition of $\text{Circ}(28, S_C)^{(2)}$ and hence a $21-$cycle decomposition of $\text{Circ}(28, S_C)^{(2)}$. We now only need a $21-$cycle decomposition of $\text{Circ}(28, S_P)^{(2)}$.

Unfortunately, there is no $m-$cycle decomposition of $P_m^{(2)}$ when m is odd. To see this observe that $P_m^{(2)}$ is bipartite. However, for $i = 0, 1, \ldots, 13$ there is an $m-$cycle decomposition of the graph, G_i say, which is obtained from $(\rho^i(P))^{(2)}$ by replacing each of the two original copies of $\rho^i(P)$ in $(\rho^i(P))^{(2)}$ with $\rho^i(P')$. This decomposition is illustrated in the figure below.

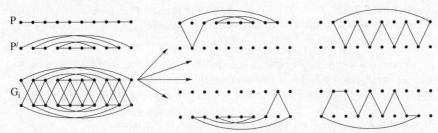

Since $\{\rho^i(P) : i = 0, 1, 2, \ldots, 13\}$ is a decomposition of $\text{Circ}(28, S_P)$, $\{(\rho^i(P))^{(2)} : i = 0, 1, 2, \ldots, 13\}$ is a decomposition of $\text{Circ}(28, S_P)^{(2)}$ and so it follows from the fact that $\{\rho^i(P') : i = 0, 1, 2, \ldots, 13\}$ is also a decomposition of $\text{Circ}(28, S_P)$ that $\{G_i : i = 0, 1, 2, \ldots, 13\}$ is also a decomposition of $\text{Circ}(28, S_P)^{(2)}$. Since we have a $21-$cycle decomposition of each G_i, we have the required $21-$cycle decomposition of $\text{Circ}(28, S_P)^{(2)}$.

We observe that for m odd, $n \geq 2m$ and $n \equiv 2 \pmod 4$, $m-$cycle decompositions of $K_n - I$ are easily obtained by using the doubling construction (see Definition 1.5), the $m-$cycle decompositions of $C_m^{(2)}$ mentioned in Example 2.13, and the $m-$cycle decompositions of complete graphs constructed by Alspach and Gavlas [5]. It is easy to check that for these values of m and n, $\frac{n}{2}$ is admissible whenever n is admissible. Hence using the doubling construction on an $m-$cycle decomposition of $K_{\frac{n}{2}}$ we obtain a $C_m^{(2)}-$decomposition of $K_n - I$, and decomposing each copy of $C_m^{(2)}$ thus yields an $m-$cycle decomposition of $K_n - I$. This observation reduced the amount of work that needed to be done in [60] to settle the problem of decomposing $K_n - I$ into $m-$cycles for m odd.

We mention two earlier results on decompositions of $K_n - I$ into $m-$cycles. The technique of Häggkvist [43], which also gives decompositions into cycles of varying lengths, in combination with a result of Tarsi [64] on decompositions into paths yields $m-$cycle decompositions of $K_n - I$ when m and $\frac{n(n-2)}{2m}$ (the required number of $m-$cycles for a decomposition) are both even. The construction for this result is described at the end of Section 3.2 and the result was used by Alspach and Gavlas [5] in their proof of Theorem 2.11 for even m. Indeed, when n and m are both even and $\frac{n(n-2)}{2m}$ is odd, it follows that $m \equiv 0 \pmod 4$. So the result significantly reduced the amount of work that needed to be done in [5] to settle the problem of decomposing $K_n - I$ into $m-$cycles for m even. Finally, it is also worth mentioning that Alspach and Marshall [6] used Häggkvist's technique in combination with other ideas to obtain results on $m-$cycle decompositions of $K_n - I$ in the case where m is even and $\frac{n(n-2)}{2m}$ is odd.

3   Cycles of varying lengths

We now turn our attention to the problem of decomposing complete graphs of odd order, and complete graphs of even order with the edges of a 1−factor removed, into cycles of varying specified lengths. The following definition introduces some useful notation.

Definition 3.1 Let $M = m_1, m_2, \ldots, m_t$ be a sequence of integers. An $(M)-$cycle decomposition of a graph K is a decomposition $\mathcal{D} = \{G_1, G_2, \ldots, G_t\}$ of K where G_i is an m_i-cycle for $i = 1, 2, \ldots, t$.

When K is a complete graph, or a complete graph of even order with the edges of a 1−factor removed, the obvious necessary conditions (see Lemma 1.2) for the existence of an $(M)-$cycle decomposition of K are given by the following Lemma.

Lemma 3.2 Let $n \geq 3$ be an integer and let $M = m_1, m_2, \ldots, m_t$ be a sequence of integers. If there exists an $(M)-$cycle decomposition of K_n, then n is odd, $3 \leq m_i \leq n$ for $i = 1, 2, \ldots, t$, and $m_1 + m_2 + \cdots + m_t = \frac{n(n-1)}{2}$. If there exists an $(M)-$cycle decomposition of $K_n - I$, then n is even, $3 \leq m_i \leq n$ for $i = 1, 2, \ldots, t$, and $m_1 + m_2 + \cdots + m_t = \frac{n(n-2)}{2}$.

The problem of showing that the necessary conditions given in Lemma 3.2 are also sufficient was posed by Alspach [3] in 1981.

Problem 3.3 (see [3])

(a) *Let n be odd and let $M = m_1, m_2, \ldots, m_t$ be a sequence of integers satisfying $3 \le m_i \le n$ for $i = 1, 2, \ldots, t$, and $m_1 + m_2 + \cdots + m_t = \frac{n(n-1)}{2}$. Prove that K_n has an (M)-cycle decomposition.*

(b) *Let n be even and let $M = m_1, m_2, \ldots, m_t$ be a sequence of integers satisfying $3 \le m_i \le n$ for $i = 1, 2, \ldots, t$, and $m_1 + m_2 + \cdots + m_t = \frac{n(n-2)}{2}$. Prove that $K_n - I$ has an (M)-cycle decomposition.*

Of course, the problem of decomposing K_n and $K_n - I$ into cycles of uniform length m, which was discussed in the previous section, is a special case of this general problem.

3.1 Summary of results

The following theorem summarises the results that have been obtained on Problem 3.3. Many further instances of the problem, not included in the theorem, are settled by results on other problems. For example, results on the 2−factorisations of K_n and $K_n - I$ including the *Oberwolfach Problem* and the *Hamilton-Waterloo Problem* yield (M)−cycle decompositions of K_n and $K_n - I$ for various M. See [28] and [58] for recent surveys of results on 2−factorisations. Another example is the result of Colbourn and Rosa [36] on *maximal partial triple systems with quadratic leaves*, which gives decompositions of K_n into 3−cycles and a relatively small number of cycles of other lengths. These are just a few of many such examples.

Theorem 3.4 *Let n be an integer and let $M = m_1, m_2, \ldots, m_t$ be a sequence of integers such that $3 \le m_i \le n$, for $i = 1, 2, \ldots, t$, $m_1 + m_2 + \cdots + m_t = \frac{n(n-1)}{2}$ when n is odd and $m_1 + m_2 + \cdots + m_t = \frac{n(n-2)}{2}$ when n is even. There exists an (M)−cycle decomposition of K_n if n is odd, or of $K_n - I$ if n is even, in each of the following cases.*

(1) $m_1 = m_2 = \cdots = m_t$ [5, 60].

(2) $n \equiv 2 \pmod 4$, t is even, $m_i \in \{4, 6, \ldots, n\} \setminus \{n - 2\}$ for $i = 1, 2, \ldots, t$, $m_i = m_{i+1}$ for $i = 1, 3, \ldots, t - 1$ [43].

(3) $n \ge N$ (N a large fixed constant) and $m_1, m_2, \ldots, m_t \le \lfloor (n - 112)/20 \rfloor$ [10].

(4) $\{m_1, m_2, \ldots, m_t\} \subseteq \{3, 4, 5\}$ [9].

(5) $n \le 14$ [10].

(6) $\{m_1, m_2, \ldots, m_t\} \subseteq \{n - 2, n - 1, n\}$ [44].

(7) $\{m_1, m_2, \ldots, m_t\} \subseteq \{3, 4, 6\}$ [44].

(8) $\{m_1, m_2, \ldots, m_t\} \subseteq \{2^k, 2^{k+1}\}$ for $k \ge 2$ [44].

(9) $\{m_1, m_2, \ldots, m_t\} \subseteq \{4, 10\}, \{6, 8\}, \{6, 10\}, \{8, 10\}$ [2].

(10) $\{m_1, m_2, \ldots, m_t\} \subseteq \{3, n\}$ [25].

In addition to the results in Theorem 3.4, there are also a few results on partial cycle decompositions and on problems which are closely related to Problem 3.3. Some such results are discussed in the sections that follow.

We now discuss some of the results included in Theorem 3.4. Result (1) is the case of uniform length cycles and was covered in Section 2. Result (2) is a corollary to Häggkvist's cycle decomposition lemma which will be described in Section 3.2. In particular, see Theorem 3.13. Results (3) and (4) are obtained using Balister's *trails of octahedra method* [9, 10] which will be discussed in Section 3.3, and result (5) is also due to Balister [10]. Result (4) supersedes the earlier result of Adams et al [1] which dealt with the case $\{m_1, m_2, \ldots, m_t\} \subseteq \{3, 5\}$. The case $\{m_1, m_2, \ldots, m_t\} \subseteq \{4, 5\}$ had also been dealt with previously [24]. Result (5) supersedes an earlier result of Rosa [57] which dealt with $n \leq 10$.

Results (6)-(8) of Theorem 3.4 were proven in a 1989 article of Heinrich, Horák and Rosa [44] and we now describe the constructions for each of these results in turn. For $\{m_1, m_2, \ldots, m_t\} \subseteq \{n - 2, n - 1, n\}$ the only possibilities are

- an n–cycle decomposition of K_n or $K_n - I$,

- an $(n - 2)$–cycle decomposition $K_n - I$, and

- a decomposition of K_n, n odd, into $\frac{n-1}{2}$ cycles of length $n - 2$ and one cycle of length $n - 1$.

So only in the last case do we have cycles of more than one length. We illustrate the solution for this case with a decomposition of K_{13} into six 11–cycles and a 12–cycle. The required decomposition $\mathcal{D}$ is obtained by taking $\mathbf{Z}_{12} \cup \{\infty\}$ as the vertex set of K_{13} and letting $\mathcal{D}$ consist of the orbits under the permutation ρ_{12} of the two cycles shown in the figure below. This construction generalises in an obvious manner.

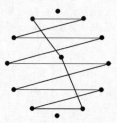 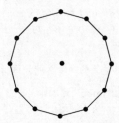

Heinrich et al [44] settled the case $\{m_1, m_2, \ldots, m_t\} \subseteq \{3, 4, 6\}$ as follows. Firstly, the result is proved for $n \leq 17$. For other even n, $K_n - I$ is decomposed into copies of $K_s - I$ with $s \in \{6, 8, 10, 12\}$. The method for doing this is illustrated shortly. The known decompositions of $K_s - I$ into 3–cycles, 4–cycles and 6–cycles for $s \in \{6, 8, 10, 12\}$ are then used in appropriate combinations to obtain all of the required decompositions of $K_n - I$. The proof for the case n is odd is simplified significantly by noting that if the number of 3–cycles is at least $\frac{n-1}{2}$, then the required decomposition can be obtained from the result for $K_{n-1} - I$. Simply form $\frac{n-1}{2}$ cycles of length 3 by using the edges of I and the vertex in $V(K_n) \backslash V(K_{n-1} - I)$, and then decompose $K_{n-1} - I$ into the required number of 4–cycles, 6–cycles and additional 3–cycles. When n is odd and the number of 3–cycles is less than $\frac{n-1}{2}$, the constructions are more complicated and use induction on n.

We now illustrate how $K_n - I$ is decomposed into copies of $K_s - I$ with $s \in \{6, 8, 10, 12\}$ by considering the case $n \equiv 10 \pmod{12}$. In this case $K_n - I$ is decomposed into copies of $K_6 - I$ and $K_{10} - I$ only (in fact there is only one copy of $K_{10} - I$). For $n \equiv 10 \pmod{12}$, $n = 2r$ where $r \equiv 5 \pmod{12}$ and there thus exists a decomposition of K_r into $\frac{1}{3}(r(r-1)/2 - 10)$ copies of C_3 and one copy of K_5. This result was proven by Mendelsohn and Rosa [52] in 1983. The required decomposition of $K_n - I$ into copies of $K_6 - I$ and a $K_{10} - I$ is obtained from this decomposition of K_r by using the doubling construction, see Definition 1.5: each copy of C_3 gives rise to a copy of $K_6 - I$ and the K_5 gives rise to a copy of $K_{10} - I$. The other residue classes of n modulo 12 are dealt with using similar ideas.

Heinrich et al [44] also settled the case $\{m_1, m_2, \ldots, m_t\} \subseteq \{2^k, 2^{k+1}\}$, $k \geq 2$. Briefly, they used several results on cycle decompositions of complete bipartite graphs, which they proved using Sotteau's Theorem, and then used the *cycle switching* technique illustrated in the figure below to obtain the required decompositions (of course, the two top left cycles must be vertex disjoint, but the bottom left cycle may have several vertices other than w, x, y, z in common with each of the two top left cycles).

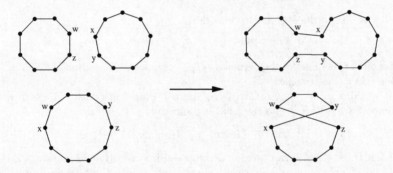

Result (9) of Theorem 3.4 is obtained by constructing the required small examples and applying the following general result from [2].

Theorem 3.5 [2] *Let m' and m'' be even integers with $4 \leq m' < m''$ and let $M = m_1, m_2, \ldots, m_t$ with $m_i \in \{m', m''\}$ for $i = 1, 2, \ldots, t$. If the necessary conditions given in Lemma 3.2 for the existence of $(M)-$cycle decompositions of K_n and $K_n - I$ are sufficient for $n < 7m''$ then they are sufficient for all n.*

Theorem 3.5 relies heavily on Doyen-Wilson type results for its proof. Essentially, the proof is by induction on n. For various values of v, K_n is decomposed into a copy of $K_n - K_v$ and a copy of K_v. Then $m'-$cycle decompositions of $K_n - K_v$, or $m''-$decompositions of $K_n - K_v$, are combined with decompositions of K_v into $m'-$cycles and $m''-$cycles which exist by the inductive hypothesis. This approach was developed in [24] and used to settle the problem of decomposing K_n and $K_n - I$ into 4$-$cycles and 5$-$cycles.

Result (10) of Theorem 3.4 concerns decompositions of K_n and $K_n - I$ into 3$-$cycles and $n-$cycles, or triangles and Hamilton cycles. The result was proved by considering K_n or $K_n - I$ as circulant graphs. In general, the decompositions into triangles and Hamilton cycles were obtained by partitioning the connection set into two parts S_T and S_H, decomposing $\mathrm{Circ}(n, S_T)$ into triangles using *Skolem Sequences* and their generalisations (see [62]), and decomposing $\mathrm{Circ}(n, S_H)$ into Hamilton cycles using the following theorem of Bermond et al [12].

Theorem 3.6 [12] *Every connected 4−regular Cayley graph on a finite abelian group can be decomposed into two Hamilton cycles.*

The following example illustrates one of these decompositions. In some cases the constructions were more complicated. They involved decomposing some 4− and 6−regular circulant graphs into combinations of triangles and Hamilton cycles.

Example 3.7 *A decomposition of K_{39} into 156 triangles and 7 Hamilton cycles.*
We first decompose $K_{39} \cong \mathrm{Circ}(39, \{1, 2, \dots, 19\})$ into the two circulant graphs $\mathrm{Circ}(39, S_T)$ and $\mathrm{Circ}(39, S_H)$ where $S_T = \{1, 2, \dots, 12\}$ and $S_H = \{13, 14, \dots, 19\}$. Using a Skolem sequence of order 4, namely $1, 1, 4, 2, 3, 2, 4, 3$, we obtain the four *difference triples*

$$(1, 5, 6) \qquad (2, 8, 10) \qquad (4, 7, 11) \qquad (3, 9, 12)$$

which give rise to the four *starter* triangles

$$(0, 1, 6) \qquad (0, 2, 10) \qquad (0, 4, 11) \qquad (0, 3, 12).$$

The orbits of these under the permutation ρ_{39} yield a decomposition of $\mathrm{Circ}(39, S_T)$ into $4 \times 39 = 156$ triangles.

It remains to decompose $\mathrm{Circ}(39, S_H)$ into 7 Hamilton cycles. To do this, we decompose $\mathrm{Circ}(39, S_H)$ into the following four circulant graphs.

$$\mathrm{Circ}(39, \{13, 14\}) \qquad \mathrm{Circ}(39, \{15, 16\}) \qquad \mathrm{Circ}(39, \{17, 18\}) \qquad \mathrm{Circ}(39, \{19\})$$

The first three of these can each be decomposed into two Hamilton cycles by Theorem 3.6. Since consecutive integers are relatively prime, these graphs are indeed connected. The last of the four circulant graphs is a Hamilton cycle. Note that $\frac{n-1}{2}$ is relatively prime to n so that $\mathrm{Circ}(n, \{\frac{n-1}{2}\})$ is an n−cycle for all odd $n \geq 3$.

3.2   Häggkvist's cycle decomposition lemma

In 1985 Häggkvist [43] described a technique for the construction of cycle decompositions and 2−factorisations containing cycles of even lengths. The following lemma is the critical ingredient.

Lemma 3.8 [43] *Let K be either an m−path or an m−cycle and let G be any 2−regular graph with $2m$ vertices where each component of G is a cycle of even length. Then there exists a G−decomposition of $K^{(2)}$.*

The proof of the lemma is obvious from the figure below. On the left, we have an example where K is a path with 12 edges, and a decomposition of $K^{(2)}$ into two 2−regular graphs each consisting of a 4−cycle, a 6−cycle, and a 14−cycle is shown. On the right, we have an example where K is a 15−cycle, and a decomposition of $K^{(2)}$ into two 2−regular graphs each consisting of a 6−cycle, a 10−cycle, and a 14−cycle is shown.

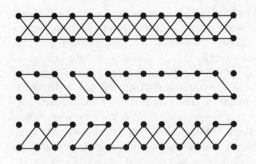 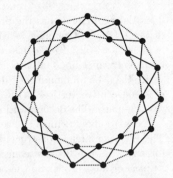

Lemma 3.8 can be used as follows. Suppose $\mathcal{D} = \{G_1, G_2, \ldots, G_t\}$ is a decomposition of K_r where G_i is either an m_i-cycle or an m_i-path for $i = 1, 2, \ldots, t$. Applying the doubling construction (see Definition 1.5) to $\mathcal{D}$ we obtain a decomposition $\{G_1^{(2)}, G_2^{(2)}, \ldots, G_t^{(2)}\}$ of $K_{2r} - I$. By Lemma 3.8, $G_i^{(2)}$ can be decomposed into two $2m_i$-cycles for $i = 1, 2, \ldots, t$, and we thus obtain a $(2m_1, 2m_1, 2m_2, 2m_2, \ldots, 2m_t, 2m_t)$-cycle decomposition of $K_{2r} - I$. This construction gives us the following lemma.

Lemma 3.9 [43] *Suppose there exists a decomposition $\{G_1, G_2, \ldots, G_t\}$ of K_r where G_i is either an m_i-cycle or an m_i-path for $i = 1, 2, \ldots, t$. Then there exists a $(2m_1, 2m_1, 2m_2, 2m_2, \ldots, 2m_t, 2m_t)$-cycle decomposition of $K_{2r} - I$.*

Of course, in order to use Lemma 3.9, we first require decompositions of complete graphs into paths and cycles. A 1983 paper of Tarsi [64] contains the following two results (and some others) on path decompositions.

Theorem 3.10 [64] *There exists an m-path decomposition of K_n if and only if m divides $\frac{n(n-1)}{2}$ and $m \leq n - 1$ for $n > 1$.*

Theorem 3.11 [64] *For any positive odd integer n and any sequence $m_1, m_2, \ldots, m_t$ satisfying $1 \leq m_i \leq n - 3$ for $i = 1, 2, \ldots, t$ and $m_1 + m_2 + \cdots, m_t = \frac{n(n-1)}{2}$, there exists a decomposition $\{G_1, G_2, \ldots, G_t\}$ of K_n where G_i is an m_i-path for $i = 1, 2, \ldots, t$.*

Tarsi [64] conjectures that Theorem 3.11 also holds if n is even and if the upper bound on the number of edges in the paths is increased to $n - 1$. Tarsi proved Theorem 3.11 using the following construction, which was also given in Häggkvist's paper [43]. Similar methods were used in the proof of Theorem 3.10.

Let n be odd and consider Walecki's decomposition of K_n into n-cycles. That is $\{C, \rho(C), \rho^2(C), \ldots, \rho^{k-1}(C)\}$ where

$$C = (\infty, 0, 1, 2k - 1, 2, 2k - 2, 3, 2k - 3, \ldots, k - 1, k + 1, k)$$

$\rho = \rho_{2k}$ and $k = \frac{n-1}{2}$ (see Section 2.1). We define an Eulerian circuit in K_n which starts at ∞ and traverses the n-cycles of Walecki's decomposition in the natural order, that is in the order

$$C, \rho(C), \rho^2(C), \ldots, \rho^{k-1}(C).$$

This Eulerian circuit has the property that travelling along the path, the shortest distance (smallest number of edges) between two occurrences of the same vertex is at least $n - 2$. To see this it is sufficient to consider the distance between the occurrences of x in C and $\rho(C)$ for each $x \in \mathbf{Z}_n \cup \{\infty\}$, and it is straightforward to verify that the minimum such distance is $n - 2$. Thus, for $s \leq n - 2$ any sequence of s consecutive vertices in the Eulerian circuit are distinct and hence define an $(s - 1)$−path. The decomposition required for Theorem 3.11 is now immediate.

Now, to apply Lemma 3.9 we require a decomposition into either paths or cycles. The Eulerian circuit constructed in the preceding paragraph has the property that for any $s \leq n - 2$, any sequence of $s + 1$ consecutive vertices defines either an s−path or an s−cycle. So the only "missing" values of s are $s = n - 1$ and $s = n$. But we can obtain n−cycles, say $\alpha \leq \frac{n-1}{2}$ of them, by removing the first α cycles of Walecki's n−cycle decomposition from the Eulerian circuit and including them in the decomposition. The Eulerian circuit that remains of course has the same property in terms of distance between occurrences of the same vertex. Thus we have the following Lemma [43]. We use r instead of n in the lemma so that it is more convenient to apply.

Lemma 3.12 [43] *Let r be odd and let $m_1, m_2, \ldots, m_t$ be any sequence of integers with $m_i \in \{1, 2, \ldots, r\} \setminus \{r - 1\}$ and $m_1 + m_2 + \cdots + m_t = \frac{k(k-1)}{2}$. Then there is a decomposition $\{G_1, G_2, \ldots, G_t\}$ of K_r where G_i is either an m_i−path or an m_i−cycle for $i = 1, 2, \ldots, t$.*

Combining Lemma 3.9 with Lemma 3.12 we obtain the following result.

Theorem 3.13 [43] *Let $n \equiv 2 \pmod 4$ and let $M = m_1, m_2, \ldots, m_t$ be any sequence of integers such that*

- $m_1 + m_2 + \cdots + m_t = \frac{n(n-2)}{2}$;

- $m_i \in \{4, 6, \ldots, n\} \setminus \{n - 2\}$; and

- $m_i = m_{i+1}$ for $i = 1, 3, \ldots, t - 1$.

Then $K_n - I$ can be decomposed into t cycles of lengths $m_1, m_2, \ldots, m_t$.

Finally, we consider the implications of Lemma 3.8 for m−cycle decompositions of $K_n - I$. Combining Lemma 3.9 and Theorem 3.10 we immediately obtain an m−cycle decomposition of $K_n - I$ whenever m and $\frac{n(n-2)}{2m}$ (the number of m−cycles in the decomposition) are both even with $m < n$. The case $m = n$ is of course covered by Walecki's construction.

3.3   Balister's trails of octahedra method

In 2001, Balister [10] obtained strong results on Problem 3.3, and on a related problem concerning decompositions of K_n and $K_n - I$ into closed trails [9]. Here we give a brief description of his method, focusing on cycle decompositions, to illustrate the central ideas. Thus we omit many nice constructions as well as the more technical details. The interested reader can find these in his articles [9, 10]. An essential component of his methods was the use of *trails of octahedra*. The graph of the octahedron, which is denoted by O and is isomorphic to $K_6 - I$ and $K_{2,2,2}$, is shown in the figure below.

Denote the graph shown below by O^v where v is the number of copies of O. In this case $v = 4$. Such graphs are referred to as *trails of octahedra*.

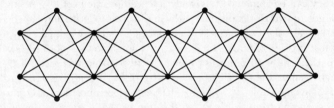

Balister proved the following theorem on cycle decompositions of O^v.

Theorem 3.14 [10] *Let v and L be integers and let $M = m_1, m_2, \ldots, m_t$ be a sequence of integers such that*

- $m_1 + m_2 + \cdots + m_t = 12v$;

- $72 \leq m_i \leq L$ *for* $i = 1, 2, \ldots, t$;

- $12v \geq 40L$.

Then there is an $M-$cycle decomposition of O^v.

The cycle decompositions of O^v are constructed by linking together path decompositions of each copy of O in appropriate combinations. The following seven path decompositions of O, together with others obtained by symmetry from these, are used.

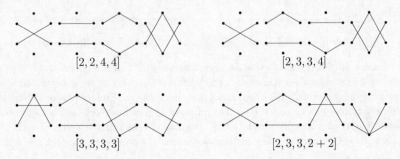

$[2, 2, 4, 4]$ $[2, 3, 3, 4]$

$[3, 3, 3, 3]$ $[2, 3, 3, 2 + 2]$

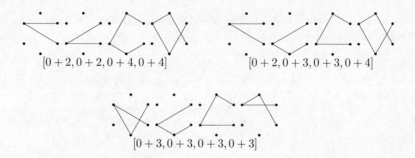

$$[0+2,0+2,0+4,0+4]$$ $$[0+2,0+3,0+3,0+4]$$

$$[0+3,0+3,0+3,0+3]$$

The notation $[s_1, s_2, s_3, s_4]$ (see the figure above) where s_i is either an integer or a sum $p_i + q_i$ of two integers is useful for describing the path decompositions of O, and how they can be combined to give cycle decompositions of O^v. A single integer s_i corresponds to two vertex disjoint paths in O which together have s_i edges, whose initial vertices are the two leftmost vertices of O, and whose final vertices are the two rightmost vertices of O. A sum $p_i + q_i$ corresponds to two (not necessarily vertex disjoint) paths: one with p_i edges joining the two leftmost vertices of O, and one with q_i edges joining the two rightmost vertices of O. When p_i or q_i is zero, no path is present. The quadruple $[s_1, s_2, s_3, s_4]$ denotes a decomposition of O into paths of types corresponding to s_1, s_2, s_3, s_4 (so the sum of the integers in the quadruple is always 12).

We illustrate how the above path decompositions of O can be combined to give cycle decompositions of O^v by constructing an (M)−cycle decomposition of O^3 with $M = 4, 4, 7, 10, 11$. For this, we can use $[0 + 2, 0 + 2, 0 + 4, 0 + 4], [2 + 2, 2, 3, 3]$ and $[2+0, 3+0, 3+0, 4+0]$. The paths corresponding to each coordinate in the quadruples generate the cycles. From the first coordinates of the three quadruples we obtain two 4−cycles, from the second a 7−cycle, from the third a 10−cycle, and from the fourth an 11−cycle. This decomposition of O^3 is shown in the figure below.

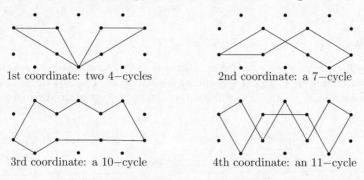

1st coordinate: two 4−cycles   2nd coordinate: a 7−cycle

3rd coordinate: a 10−cycle   4th coordinate: an 11−cycle

In order to make use of cycle decompositions of O^v, decompositions of $K_n - I$ into long trails of octahedra are required. Such decompositions are constructed from Steiner triple systems of order $r = \frac{n}{2}$. Of particular use are Steiner triple systems for which the triples can be ordered so that adjacent triples intersect, and non-adjacent intersecting triples are far apart in the ordering. Steiner triple systems with this property are constructed using Skolem sequences. Decompositions of $K_n - I$ into octahedra are obtained from these Steiner triple systems via the doubling construction. The resulting octahedra are placed in the same order as their corresponding triples, and hence are partitioned into long trails of octahedra. Balister [10] used Steiner triple systems of order 1 (mod 72) and thus obtained cycle decompositions of $K_n - I$ with $n \equiv 2$ (mod 144). In particular he proved the following theorem.

Theorem 3.15 [10] *Let $n \equiv 2$ (mod 144) and let $M = m_1, m_2, \ldots, m_t$ be a sequence of integers satisfying $72 \leq m_i \leq \lfloor \frac{n+37}{20} \rfloor$ and $m_1 + m_2 + \cdots + m_t = \frac{n(n-2)}{2}$. Then there exists an (M)−cycle decomposition of $K_n - I$.*

Balister points out that it is possible to reduce the lower bound of 72 on the cycle lengths in Theorem 3.14, and hence in Theorem 3.15, by using more complicated methods from his paper. However, as he also notes, Theorem 3.14 is not true without some restrictions on the values of m_i. He gives the example that for any v, O^v has no (M)−cycle decomposition for $M = 8, 4, 4, \ldots, 4$ (where the number of 4s in the sequence is $3v - 2$). Theorem 3.15 provides a platform from which the following general result is proved [10].

Theorem 3.16 [10] *There exists a (very large) constant N such that for any $n \geq N$ and any sequence $M = m_1, m_2, \ldots, m_t$ satisfying $3 \leq m_i \leq \lfloor \frac{n-112}{20} \rfloor$ and $m_1 + m_2 + \cdots + m_t = \frac{n(n-1)}{2}$ (n odd) or $m_1 + m_2 + \cdots + m_t = \frac{n(n-2)}{2}$ (n even), there exists an (M)−cycle decomposition of K_n (n odd) or of $K_n - I$ (n even).*

The proof of Theorem 3.16 makes critical use of the following result of Caro and Yuster [35], which in turn is proved using a result of Gustavsson [42].

Theorem 3.17 [35] *For a graph G, let $e(G)$ denote the number of edges in G and let $\gcd(G)$ denote the greatest common divisor of the degrees of the vertices in G. Let $H_1, H_2, \ldots, H_t$ be a sequence of graphs with the property that $\gcd(H_i) = \gcd(H_j)$ for $1 \leq i \leq j \leq t$. Then there exists an integer $N = N(H_1, H_2, \ldots, H_t)$ and a positive constant $\gamma = \gamma(H_1, H_2, \ldots, H_t)$ such that for any sequence $\alpha_1, \alpha_2, \ldots, \alpha_t$ of integers and any graph G with $n > N$ vertices and minimum degree $\delta(G) \geq n(1-\gamma)$ satisfying $\gcd(H_1)$ divides $\gcd(G)$ and $\alpha_1 e(H_1) + \alpha_2 e(H_2) + \cdots + \alpha_t e(H_t) = e(G)$ there is a decomposition of G consisting of exactly α_i copies of H_i for $i = 1, 2, \ldots, t$.*

Like Theorem 3.16, this theorem also provides a solution to Problem 3.3 when n is very large compared with the length of the longest required cycle. The difference is Theorem 3.16 says that provided n is large enough the cycle lengths can be up to $\lfloor \frac{n-112}{20} \rfloor$, whereas in Theorem 3.17, n is exponentially larger than the length of the longest cycle. Of course, Theorem 3.17 says a lot about decompositions into graphs other than cycles.

Balister also used his trails of octahedra method to completely settle a related problem on decompositions into closed trails [9]. Of course, a 2−regular closed trail is a cycle. In fact, Balister's result solves the more general question of *packing* closed trails in K_n.

Theorem 3.18 [9] *Let $m_1, m_2, \ldots, m_t \geq 3$ and let $e = m_1 + m_2 + \cdots + m_t$. There is a subgraph H of K_n such that there is a decomposition $\{G_1, G_2, \ldots, G_t\}$ of H where G_i is a closed trail with m_i edges for $i = 1, 2, \ldots, t$ if and only if*

- $e = \frac{n(n-1)}{2}$ or $e \leq \frac{n(n-1)}{2} - 3$ for n odd; and

- $e \leq \frac{n(n-2)}{2}$ for n even.

For $m \in \{3, 4, 5\}$ a closed trail with m edges is necessarily an m−cycle. Thus, Theorem 3.18 settles Problem 3.3 for the case $m_1, m_2, \ldots, m_t \subseteq \{3, 4, 5\}$ (see Result (5) in Theorem 3.4).

3.4 Cycle repacking

Some new results on cycle decompositions and related problems have recently been obtained by *repacking*. That is, the subgraphs in a decomposition $\mathcal{D} = \{G_1, G_2, \ldots, G_t\}$ are rearranged or *repacked* to give a new decomposition $\mathcal{D}' = \{G_1', G_2', \ldots, G_t'\}$ where $G_i' \cong G_i$ for $i = 1, 2, \ldots, t$. The initial decomposition $\mathcal{D}$ is a decomposition of some subgraph, K say, of K_n and the new decomposition $\mathcal{D}'$ is a decomposition of a different subgraph, K' say, of K_n. The goal is to repack so that a desired graph G occurs as a subgraph of the complement of K'. The graph G can then be added to the decomposition and the process repeated until a decomposition of K_n, or of some other desired subgraph of K_n, is obtained. In 1980, Andersen et al [8] used repacking of 3−cycles in their article on *embeddings of partial Steiner triple systems*. We now describe a generalisation of their technique which was given in [22] and which applies to decompositions into cycles of any lengths.

Let $\mathcal{D}$ be an (M)−cycle decomposition of a graph K. For any $\alpha, \beta \in V(K)$, an edge-coloured multigraph $G_{\alpha,\beta}$ is constructed as follows. The vertex set of $G_{\alpha,\beta}$ is $V(G_{\alpha,\beta}) = V(K) \cup J$ where J is a set of new vertices, disjoint from $V(K)$, that is given by the following construction.

- For each cycle $C \in \mathcal{D}$ that contains α and not β, a red edge joining the two neighbours of α in C is added.

- For each cycle $C \in \mathcal{D}$ that contains β and not α, a blue edge joining the two neighbours of β in C is added.

- If there is a cycle $C \in \mathcal{D}$ that contains the edge $\alpha\beta$, say $C = (\ldots, x, \alpha, \beta, y, \ldots)$, then a new vertex v is added to J, the edge xv is added and coloured red, and the edge yv is added and coloured blue (if C is a 3-cycle then $x = y$ and a double edge results).

- For each cycle $C \in \mathcal{D}$ that contains α and β at distance at least 2, say $C = (\ldots, x_1, \alpha, x_2, \ldots, y_2, \beta, y_1 \ldots)$, two new vertices u and v are added to J, the edges ux_1 and vx_2 are added and coloured red, and the edges uy_1 and vy_2 are added and coloured blue (if $x_1 = y_1$ or $x_2 = y_2$ then a double edge results).

The following figure shows the edges added to the graph $G_{\alpha,\beta}$ for various cycles. The edges in the cycles are shown as solid lines, the red edges in $G_{\alpha,\beta}$ are shown as dashed lines, and the blue edges in $G_{\alpha,\beta}$ are shown as dotted lines.

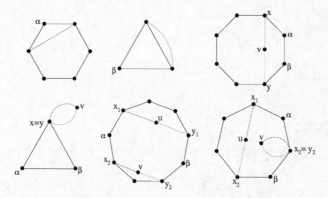

We note some properties of $G_{\alpha,\beta}$. Let A be the set of neighbours in K of α and let B be the set of neighbours in K of β. Then in $G_{\alpha,\beta}$ each $x \in A \setminus \{\beta\}$ is incident with exactly one red edge, each $x \in B \setminus \{\alpha\}$ is incident with exactly one blue edge, each $x \in J$ is incident with exactly one red and exactly one blue edge, and these are all the edges of $G_{\alpha,\beta}$. So each component of $G_{\alpha,\beta}$ is either an alternating red-blue path (this includes trivial paths consisting of an isolated vertex) or an alternating red-blue cycle.

Suppose that for some vertex $a_1 \in V(K) \setminus \{\alpha, \beta\}$, $\alpha a_1 \in E(K)$ and $\beta a_1 \notin E(K)$. Then there is a component of $G_{\alpha,\beta}$ which is a path, say $P = a_1, a_2, \ldots, a_r$, where $r \geq 2$ and $a_1 a_2$ is red. For $2 \leq i \leq r - 1$, if $a_i \in V(K)$ then both edges αa_i and βa_i are in $E(K)$, the edge $\alpha a_r \in E(K)$ if and only if $a_{r-1} a_r$ is red, and the edge $\beta a_r \in E(K)$ if and only if $a_{r-1} a_r$ is blue. So exactly one of αa_r and βa_r is in $E(K)$. Let e_1 be the one that is, let e_2 be the one that is not, and let K' be the graph obtained from K by replacing the edges αa_1 and e_1 with βa_1 and e_2. We now show how to modify the cycles of $\mathcal{D}$ to obtain an (M)−cycle decomposition of K'.

- In the cycle containing the edge αa_1, replace αa_1 with the edge βa_1.

- For $2 \leq i \leq r-1$, if $a_i \in V(K)$ then in the cycle containing the edge αa_i replace αa_i with the edge βa_i, and in the cycle containing the edge βa_i, replace βa_i with the edge αa_i.

- If $\alpha a_r \in E(K)$, then in the cycle containing the edge αa_r, replace αa_r with βa_r.

- If $\beta a_r \in E(K)$, then in the cycle containing the edge βa_r, replace βa_r with αa_r.

The figure below shows how various cycles are modified by this procedure under various scenarios. Shown on the top left is the case where a cycle contains α, does not contain β, and the red edge xy is in the path P. On the top right is the case where a cycle contains the edge $\alpha\beta$ and xvy $(v \in J)$ is in P. On the bottom left is the case where a cycle contains both α and β, the edge $\alpha\beta$ is not in the cycle, $x_1 v_1 y_1$ and $x_2 v_2 y_2$ $(v_1, v_2 \in J)$ are both in the path P. On the bottom right is the case where a cycle contains both α and β, the edge $\alpha\beta$ is not in the cycle, $x_1 v_1 y_1$ $(v_1 \in J)$ is in the path P, and $x_2 v_2 y_2$ $(v_2 \in J)$ is not in the path P.

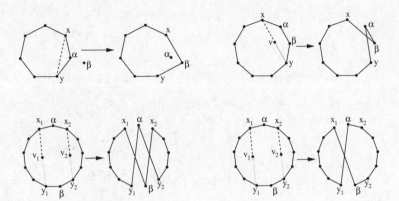

It is easy to see that this procedure results in an (M)−cycle decomposition of K'. The figure below illustrates a small example. At the top is a decomposition of a graph K into a 4-cycle, a 5-cycle and an 8−cycle. At the bottom is the decomposition of a graph K' into a 4-cycle, a 5-cycle and an 8−cycle. The decomposition of K' is obtained by repacking the decomposition of K. The dotted edges at the top are those in K' and not in K, and the dotted edges at the bottom are those in K and not K'.

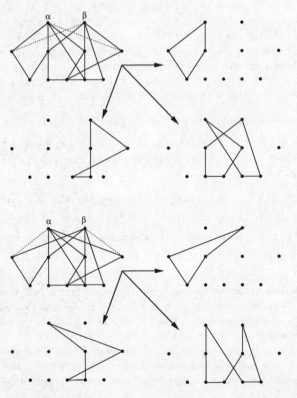

Note that if $G'_{\alpha,\beta}$ was constructed in the same manner using the new decomposition $\mathcal{D}'$, then it would differ from $G_{\alpha,\beta}$ only in that the colours on the path P have been interchanged. In this sense $\mathcal{D}'$ is constructed from $\mathcal{D}$ by interchanging the colours on P and changing the relevant cycles in the corresponding manner.

It is straightforward to prove the following theorem using this cycle repacking technique. A *partial* $(M)-$cycle decomposition of K_n is an $(M)-$cycle decomposition of some subgraph of K_n, and a partial $(M)-$cycle decomposition is said to be *equitable* if for any two vertices $\alpha, \beta \in V(K_n)$, the number of cycles containing α differs from the number containing β by at most one.

Theorem 3.19 [22] *If there is a partial $(M)-$cycle decomposition of K_n then there is an equitable partial $(M)-$cycle decomposition of K_n.*

A slight generalisation of the above-described cycle repacking technique was used in [22] to modify the closed trail decompositions of Balister (see Theorem 3.18) and obtain the following result. It gives a solution to a variant of Problem 3.3(a) in which the condition that the subgraphs are cycles is relaxed so that the subgraphs are arbitrary 2−regular graphs.

Theorem 3.20 [22] *Let $n \geq 3$ and $m_1, m_2, \ldots, m_t$ be integers. There exists a decomposition $\{G_1, G_2, \ldots, G_t\}$ of K_n where G_i is a 2−regular graph with m_i edges for $i = 1, 2, \ldots, t$ if and only if n is odd, $3 \leq m_1, m_2, \ldots, m_t \leq n$, and $m_1 + m_2 + \cdots + m_t = \binom{n}{2}$.*

New techniques for repacking 3−cycles were developed in [20] and used to prove Lindner's Conjecture [49] on embeddings of partial Steiner triple systems. In [21], some of these new techniques have been generalised to decompositions into cycles of arbitrary lengths and used to obtain the following result. It shows that one can get close to the cycle decompositions required to solve Problem 3.3. In fact the result in [21] is slightly stronger than that given below.

Theorem 3.21 [21] *Let n be an integer and let $M = m_1, m_2, \ldots, m_t$ be a sequence of integers satisfying $m_1 + m_2 + \cdots + m_t \leq \binom{n}{2} - 3\lfloor \frac{n}{2} \rfloor$ and $3 \leq m_i \leq n$ for $i = 1, 2, \ldots, t$. Then there exists a partial $(M)-$cycle decomposition of K_n.*

References

[1] P. Adams, D. Bryant and A. Khodkar, 3, 5−cycle decompositions, *J. Combin. Des.* **6** (1998), 91–110.

[2] P. Adams, D. Bryant and A. Khodkar, On Alspach's conjecture with two even cycle lengths, *Discrete Math.* **223** (2000), 1–12.

[3] B. Alspach, Research Problem 3, *Discrete Math.* **36** (1981), 333.

[4] B. Alspach, J. -C. Bermond and D. Sotteau, Decomposition into cycles I: Hamilton Decompositions, *Cycles and rays (Montreal, PQ, 1987), NATO Adv. Sci. Inst. Ser. C Math. Phys. Sci.* **301**, 9–18 Kluwer Acad. Publ., Dordrecht, (1990).

[5] B. Alspach and H. Gavlas, Cycle decompositions of K_n and $K_n - I$, *J. Combin. Theory Ser. B* **81** (2001), 77–99.

[6] B. Alspach and S. Marshall, Even cycle decompositions of complete graphs minus a 1−factor, *J. Combin. Des.* **2** (1994), 441–458.

[7] B. Alspach and B. N. Varma, Decomposing complete graphs into cycles of length $2p^e$, *Ann. Discrete Math.* **9** (1980), 155–162.

[8] L. D. Andersen, A. J. W. Hilton and E. Mendelsohn, Embedding partial Steiner triple systems, *Proc. London Math. Soc.* **41** (1980), 557–576.

[9] P. Balister, Packing Circuits into K_n, *Combin. Probab. Comput.* **10** (2001), 463–499.

[10] P. Balister, On the Alspach Conjecture, *Combin. Probab. Comput.* **10** (2001), 95–125.

[11] E. Bell, Decomposition of K_n into cycles of length at most fifty, *Ars Combin.* **40** (1995), 49–58.

[12] J. -C. Bermond, O. Favaron and M. Maheo, Hamiltonian decomposition of Cayley graphs of degree 4, *J. Combin. Theory Ser. B* **46** (1989), 142–153.

[13] J. -C. Bermond, C. Huang and D. Sotteau, Balanced cycle and circuit designs: even cases, *Ars Combin.* **5** (1978), 293–318.

[14] J. -C. Bermond and D. Sotteau, Graph decompositions and G−designs, Proceedings of the Fifth British Combinatorial Conference (Univ. Aberdeen, Aberdeen, 1975) *Congr. Numer.* **15** (1976), 53–72.

[15] A. Blinco, S. I. El-Zanati and C. Vanden Eynden, On the cyclic decomposition of complete graphs into almost-bipartite graphs, *Discrete Math.* **284** (2004), 71–81.

[16] J. Bosák, Decompositions of graphs, Kluwer, Dordrecht, (1990).

[17] D. Bryant and S. El-Zanati, Graph Decompositions, in The CRC Handbook of Combinatorial Designs, 2nd Edition, (J. H. Dinitz and C. J Colbourn Eds) CRC Press, Boca Raton, FL, (to appear).

[18] D. Bryant, H. Gavlas and A. C. H. Ling, Skolem-type difference sets for cycle systems, *Electron. J. Combin.* **10** (2003), Research Paper 38, 12 pp.

[19] D. Bryant, D. G. Hoffman and C. A. Rodger, 5−cycle systems with holes, *Des. Codes Cryptogr.* **8** (1996), 103–108.

[20] D. Bryant and D. Horsley, A proof of Lindner's conjecture on embeddings of partial Steiner triple systems, *Submitted*.

[21] D. Bryant and D. Horsley, Packing cycles in complete graphs, *Preprint*.

[22] D. Bryant, D. Horsley and B. Maenhaut, Decompositions into 2−regular subgraphs and equitable partial cycle decompositions, *J. Combin. Theory Ser. B* **93** (2005), 67–72.

[23] D. Bryant and A. Khodkar, 5−cycle systems of $K_v - F$ with a hole, *Util. Math.* **54** (1998), 59–73.

[24] D. Bryant, A. Khodkar and H. L. Fu, (m, n)−cycle systems, *J. Statist. Plann. Inference* **74** (1998), 365–370.

[25] D. Bryant and B. Maenhaut, Decompositions of complete graphs into triangles and Hamilton cycles, *J. Combin. Des.* **12** (2004), 221–232.

[26] D. Bryant and C. A. Rodger, The Doyen-Wilson theorem extended to 5−cycles, *J. Combin. Theory Ser. A* **68** (1994), 218–225.

[27] D. Bryant and C. A. Rodger, On the Doyen-Wilson theorem for m−cycle systems, *J. Combin. Des.* **2** (1994), 253–271.

[28] D. Bryant and C. A. Rodger, Cycle Decompositions, in The CRC Handbook of Combinatorial Designs, 2nd Edition, (J. H. Dinitz and C. J Colbourn Eds) CRC Press, Boca Raton, FL, (to appear).

[29] D. Bryant, C. A. Rodger and E. R. Spicer, Embeddings of m−cycle systems and incomplete m−cycle systems: $m \leq 14$, *Discrete Math.* **171** (1997), 55–75.

[30] K. Bryś and Z. Lonc, A complete solution of a Holyer problem, *Proc. 4th Twente Workshop on Graph and Combinatorial Optimization*, University of Twente, Enschede, The Netherlands, (1995).

[31] M. Buratti, Rotational k−cycle systems of order $v < 3k$; another proof of the existence of odd cycle systems, *J. Combin. Des.* **11** (2003), 433–441.

[32] M. Buratti, Existence of 1−rotational k−cycle systems of the complete graph, *Graphs Combin.* **20** (2004), 41–46.

[33] M. Buratti and A. Del Fra, Existence of cyclic k−cycle systems of the complete graph, *Discrete Math.* **261** (2003), 113–125.

[34] M. Buratti and A. Del Fra, Cyclic Hamiltonian cycle systems of the complete graph, *Discrete Math.* **279** (2004), 107–119.

[35] Y. Caro and R. Yuster, List decomposition of graphs, *Discrete Math.* **243** (2002), 67–77.

[36] C. J. Colbourn and A. Rosa, Quadratic leaves of maximal partial triple systems, *Graphs Combin.* **2** (1986), 317–337.

[37] C. J. Colbourn and A. Rosa, Triple Systems, Clarendon Press, Oxford (1999).

[38] D. Dor and M. Tarsi, Graph decomposition is NP-complete: A complete proof of Holyer's conjecture, *Siam J. Comput.* **26** (1997), 1166–1187.

[39] J. Doyen and R. M. Wilson, Embeddings of Steiner triple systems, *Discrete Math.* **5** (1973), 229–239.

[40] S. I. El-Zanati, Maximum packings with odd cycles. *Discrete Math.* **131** (1994), 91–97.

[41] H. L. Fu and S. L. Wu, Cyclically decomposing the complete graph into cycles, *Discrete Math.* **282** (2004), 267–273.

[42] T. Gustavsson, Decomposition of large graphs and digraphs with high minimum degree, Doctoral Dissertation, Dept. of Mathematics, Univ. of Stockholm (1991).

[43] R. Häggkvist, A lemma on cycle decompositions, *Ann. Discrete Math.* **27** (1985), 227–232.

[44] K. Heinrich, P. Horák and A. Rosa, On Alspach's conjecture, *Discrete Math.* **77** (1989), 97–121.

[45] D. G. Hoffman, C. C. Lindner and C. A. Rodger, On the construction of odd cycle systems, *J. Graph Theory* **13** (1989), 417–426.

[46] B. W. Jackson, Some cycle decompositions of complete graphs, *J. Combin. Inform. System Sci.* **13** (1988), 20–32.

[47] T. P. Kirkman, On a problem in combinations, *Cambridge and Dublin Math. J.* **2** (1847), 191–204.

[48] A. Kotzig, On decomposition of the complete graph into $4k$−gons, *Mat.–Fyz. Cas.* **15** (1965), 227–233.

[49] C. C. Lindner and T. Evans, Finite embedding theorems for partial designs and algebras, SMS 56, Les Presses de l'Université de Montréal, (1977).

[50] C. C. Lindner and C. A. Rodger, Decomposition in cycles II: cycle systems, in *Contemporary design theory: A collection of surveys* (J. H. Dinitz and D. R. Stinson Eds) Wiley, New York, 1992, 325–369.

[51] E. Lucas, "Récreations Mathématiqués," Vol II, Gauthier-Villars, Paris, 1892.

[52] E. Mendelsohn and A. Rosa, Embedding maximal packings of triples, *Cong. Numer.* **40** (1983), 235–247.

[53] R. Peltesohn, Eine Lösung der beiden Heffterschen Differenzenprobleme, *Composito Math.* **6** (1939), 251–257.

[54] K. T. Phelps and A. Rosa, Steiner triple systems with rotational automorphisms, *Discrete Math.* **33** (1981), 57–66.

[55] A. Rosa, On cyclic decompositions of the complete graph into $4m + 2$−gons, *Mat.–Fyz. Cas.* **16** (1966), 349–352.

[56] A. Rosa, On the cyclic decompositions of the complete graph into polygons with an odd number of edges, *Casopis Pest. Math.* **91** (1966), 53–63.

[57] A. Rosa, Alspach's conjecture is true for $n \leq 10$, Mathematical Reports, McMaster University.

[58] A. Rosa, Two-factorizations of the complete graph, *Rend. Sem. Mat. Messina Ser.* II **9**(25) (2003), 201–210.

[59] A. Rosa and C. Huang, Another class of balanced graph designs: balanced circuit designs, *Discrete Math.* **12** (1975), 269–293.

[60] M. Šajna, Cycle Decompositions III: Complete graphs and fixed length cycles. *J. Combin. Des.* **10** (2002), 27–78.

[61] M. Šajna, On decomposing $K_n - I$ into cycles of a fixed odd length. *Discrete Math.* **244** (2002), 435–444.

[62] N. Shalaby, Skolem and Langford sequences, in The CRC Handbook of Combinatorial Designs, 2nd Edition, (J. H. Dinitz and C. J Colbourn Eds) CRC Press, Boca Raton, FL, (to appear).

[63] D. Sotteau, Decomposition of $K_{m,n}$ ($K_{m,n}^*$) into cycles (circuits) of length $2k$. *J. Combin. Theory Ser. B* **30** (1981), 75–81.

[64] M. Tarsi, Decomposition of a complete multigraph into simple paths: Nonbalanced handcuffed designs, *J. Combin. Theory Ser. A* **34** (1983), 60–70.

[65] R. M. Wilson, Decomposition of complete graphs into subgraphs isomorphic to a given graph, Proceedings of the Fifth British Combinatorial Conference (Univ. Aberdeen, Aberdeen, 1975), *Congr. Numer.* **15** (1976), 647–659.

Darryn Bryant
Department of Mathematics
University of Queensland
QLD 4072, Australia
db@maths.uq.edu.au

Excluding induced subgraphs

Maria Chudnovsky and Paul Seymour

Abstract

In this paper we survey some results concerning the structure and properties of families of graphs defined by excluding certain induced subgraphs, including perfect graphs, claw-free graphs, even-hole-free graphs and others.

1 Introduction

All graphs in this paper are finite and simple. Given two graphs, G and H, we say that H is an *induced subgraph* of G if $V(H) \subseteq V(G)$, and two vertices of H are adjacent if and only if they are adjacent in G. Let $\mathcal{F}$ be a (possibly infinite) family of graphs. A graph G is called $\mathcal{F}$-free if no induced subgraph of G is isomorphic to a member of $\mathcal{F}$. A *clique* in a graph is a set of vertices all pairwise adjacent, and a *stable set* is a set of vertices all pairwise non-adjacent. The complement of a graph G is the graph $\overline{G}$, on the same vertex set as G, and such that two vertices are adjacent in G if and only if they are non-adjacent in $\overline{G}$.

It turns out that many interesting classes of graphs can be characterized as being $\mathcal{F}$-free for some family $\mathcal{F}$. The class of perfect graphs is, possibly, one of the most well-known examples. For a graph G, let us denote by $\chi(G)$ the chromatic number of G, and by $\omega(G)$ the size of the largest clique in G. A graph G is called *perfect* if for every induced subgraph H of G, $\chi(H) = \omega(H)$. In 1961 Claude Berge conjectured that being perfect is equivalent to the property of being $\mathcal{F}$-free for a certain infinite family $\mathcal{F}$ [2], and in 2002, in joint work with Neil Robertson and Robin Thomas, we were able to prove this conjecture [9]. More precisely, Berge conjectured that a graph is perfect if and only if no induced subgraph of it is a cycle of odd length at least five, or the complement of one. Today graphs with no induced subgraph isomorphic to a cycle of odd length at least five or its complement are called *Berge* graphs. The main part of our proof of the conjecture was a more general theorem, that describes the structure of all Berge graphs. More precisely, we proved that every Berge graph either belongs to one of a few well-understood families of basic graphs, or admits a certain decomposition (a version of this was conjectured earlier by Conforti, Cornuéjols and Vuškovié). Having obtained this explicit structural result for all Berge graphs, we were able to verify that all of them are perfect (the other direction of Berge's conjecture is easy, because odd cycles and their complements are not perfect, and every induced subgraph of a perfect graph is).

Theorems following the same general paradigm are known for $\mathcal{F}$-free graphs for other families $\mathcal{F}$. Some of them are easy—for example it is almost immediate to see that if $\mathcal{F}$ consists of a single graph which is a two-edge path, then every $\mathcal{F}$-free graph is either complete or disconnected. Others are difficult—take $\mathcal{F}$ to be the set of all even-length cycles, or the set of all cycles of odd length at least five (these are theorems of Conforti, Cornuéjols, Kapoor, and Vuškovié [21] and of Conforti, Cornuéjols, and Vuškovié [23], respectively).

One might then ask whether a structural theorem of that kind should exist for every family $\mathcal{F}$. This question is, of course, not well defined, because we do not know yet what graphs should be considered basic, and what kinds of decompositions should be allowed. However, it is of great interest, at least in our opinion, to understand to what extent forbidding an induced subgraph in a graph impacts the global structure of the graph. In the last few years, we have been studying $\mathcal{F}$-free graphs for different families $\mathcal{F}$, in an attempt to get some insight into this question. In this paper we will describe some of the theorems we came up with, and try to emphasize the similarities among them.

Let us now mention a conjecture of Erdős and Hajnal [27], that, in a sense, is concerned with the same question, namely whether forbidding a certain induced subgraph has a global effect on a graph:

Conjecture 1.1 *For every graph H, there exists $\delta(H) > 0$, such that if G is an $\{H\}$-free graph, then G contains either a clique or a stable set of size at least $|V(G)|^{\delta(H)}$.*

In Section 4 we will describe a structural result, that was used to solve a special case of 1.1, where H is a "bull" (we will give a precise definition later). The bull was one of the smallest subgraphs for which the conjecture was not known, and thus provided an interesting test case.

Finally, let us mention another problem concerning $\mathcal{F}$-free graphs, and that is the question of their recognition. We will focus on cases where $\mathcal{F}$ consists of all subdivisions of a given graph, possibly with parity conditions. It turns out that for some such families $\mathcal{F}$, there exist polynomial time algorithms to test whether a given graph is $\mathcal{F}$-free, while for others the recognition problem has been shown to be NP-complete. At the moment we do not understand what causes this difference, but in the last section of this paper we will survey some related results.

This paper is organized as follows. In Section 2 we describe the decomposition theorem for Berge graphs. Section 3 contains results about claw-free graphs; there we also try to explain the difference between a "composition" theorem and a "decomposition" theorem, and mention some results concerning colouring. Section 4 deals with bull-free graphs and the solution of the Erdős-Hajnal conjecture for them. In Section 5 we introduce the notion of a "trigraph", which is an object, slightly more general than a graph, which was quite useful to us on a number of occasions. Section 6 is about even-hole-free graphs; there we describe a solution to a conjecture of Reed, and a colouring property of even-hole-free graphs that it implies. Finally, in Section 7 we survey some results on testing for the presence of certain induced subgraphs in a given graph.

2   Perfect Graphs

We start with some definitions. A *hole* in a graph is an induced cycle with at least four vertices, and an *antihole* in a graph is an induced subgraph whose complement is a cycle with at least four vertices. The *length* of a hole is the number of edges in it (and the length of an antihole is the number of edges in its complement.) A *path* in G is an *induced* connected subgraph of G which is either a one-vertex graph, or such that exactly two of its vertices have degree one, and all the others have degree two (this definition is non-standard, but very convenient). An *antipath* is an induced subgraph whose complement is a path. The *length* of a path is the number of edges in it (and the length of an antipath is the number of edges in its complement). If P is a path, the set of internal vertices of P is called the *interior* of P; and similarly for antipaths. A path or a hole is called *even* if it has even length, and *odd* otherwise.

A graph is called *Berge* if every hole and antihole in it is even. The goal of this section is to describe a structural result about Berge graphs, that is used in [9] in order to prove Berge's Strong Perfect Graph Conjecture [2], which is now the following theorem:

Theorem 2.1 *A graph is perfect if and only if it is Berge.*

We first define the basic graphs. We say that G is a *double split graph* if $V(G)$ can be partitioned into four sets $\{a_1, \ldots, a_m\}$, $\{b_1, \ldots, b_m\}$, $\{c_1, \ldots, c_n\}$, $\{d_1, \ldots, d_n\}$ for some $m, n \geq 2$, such that:

- a_i is adjacent to b_i for $1 \leq i \leq m$, and c_j is non-adjacent to d_j for $1 \leq j \leq n$

- there are no edges between $\{a_i, b_i\}$ and $\{a_{i'}, b_{i'}\}$ for $1 \leq i < i' \leq m$, and all four edges between $\{c_j, d_j\}$ and $\{c_{j'}, d_{j'}\}$ for $1 \leq j < j' \leq n$

- there are exactly two edges between $\{a_i, b_i\}$ and $\{c_j, d_j\}$ for

  $1 \leq i \leq m$ and $1 \leq j \leq n$, and these two edges have no common end.

(The name is because such a graph can be obtained from what is called a "split graph" by doubling each vertex). The *line graph* $L(G)$ of a graph G has vertex set the set $E(G)$ of edges of G, and $e, f \in E(G)$ are adjacent in $L(G)$ if they share an end in G. In this section, we call a graph G *basic* if either G or $\overline{G}$ is bipartite or is the line graph of a bipartite graph, or is a double split graph. (Note that if G is a double split graph then so is $\overline{G}$.)

Now we turn to the various kinds of decomposition. If $X \subseteq V(G)$ we denote the subgraph of G induced on X by $G|X$. First, a special case of the "2-join" due to Cornuéjols and Cunningham [25] — a *proper 2-join* in G is a partition (X_1, X_2) of $V(G)$ such that there exist disjoint nonempty $A_i, B_i \subseteq X_i$ $(i = 1, 2)$ satisfying:

- every vertex of A_1 is adjacent to every vertex of A_2, and every vertex of B_1 is adjacent to every vertex of B_2,

- there are no other edges between X_1 and X_2,

- for $i = 1, 2$, every component of $G|X_i$ meets both A_i and B_i, and

- for $i = 1, 2$, if $|A_i| = |B_i| = 1$ and $G|X_i$ is a path joining the members of A_i and B_i, then it has odd length ≥ 3.

If $X \subseteq V(G)$ and $v \in V(G)$, we say v is *X-complete* if v is adjacent to every vertex in X (and consequently $v \notin X$), and v is *X-anticomplete* if v has no neighbours in X. If $X, Y \subseteq V(G)$ are disjoint, we say X is *complete* to Y (or the pair (X, Y) is *complete*) if every vertex in X is Y-complete; and being *anticomplete* to Y is defined similarly. Our second decomposition is a slight variation of the "homogeneous pair" of Chvátal and Sbihi [20]. Let A, B be two disjoint subsets of $V(G)$. The pair (A, B) is called a *homogeneous pair* in G if for every vertex $v \in V(G) \setminus (A \cup B)$, v is either A-complete or A-anticomplete and either B-complete or B-anticomplete. A *proper homogeneous pair* in G is a homogeneous pair (A, B) such that, if A_1, A_2 respectively denote the sets of all A-complete vertices and all A-anticomplete vertices in $V(G)$, and B_1, B_2 are defined similarly, then:

- every vertex in A has a neighbour in B and a non-neighbour in B, and vice versa

- the four sets $A_1 \cap B_1, A_1 \cap B_2, A_2 \cap B_1, A_2 \cap B_2$ are all nonempty.

Let A, B be disjoint subsets of $V(G)$. We say the pair (A, B) is *balanced* if there is no odd path between non-adjacent vertices in B with interior in A, and there is no odd antipath between adjacent vertices in A with interior in B. A set $X \subseteq V(G)$ is *connected* if $G|X$ is connected (so $\emptyset$ is connected); and *anticonnected* if $\overline{G}|X$ is connected. A *skew partition* in G (introduced by Chvátal [19]) is a partition (A, B) of $V(G)$ such that A is not connected and B is not anticonnected. The third kind of decomposition we use is a balanced skew-partition.

The main result of [9] is the following:

Theorem 2.2 *For every Berge graph G, either G is basic, or one of $G, \overline{G}$ admits a proper 2-join, or G admits a proper homogeneous pair, or G admits a balanced skew partition.*

Now, since all basic graphs are perfect (for bipartite graphs it is trivial; for line graphs of bipartite graphs it is a theorem of König [28]; for their complements it follows from a theorem of Lovász [29], although originally these were separate theorems of König; and for double split graphs we leave it to the reader); and none of the decompositions can occur in a minimum size counterexample to 2.1 (for 2-joins this is a result due to to Cornuéjols and Cunningham [25], for proper homogeneous pairs due to Chvátal and Sbihi [20], and for balanced skew partitions due to the authors together with Robertson and Thomas [9]), it follows that no graph is a minimum size counterexample to 2.1, and therefore 2.1 is true.

However, one can ask for more from a theorem of the kind of 2.2. While 2.2 provides enough insight into Berge graphs in order to prove 2.1, it does not give a recipe to build all Berge (or, equivalently, perfect) graphs, starting from easy pieces (unlike, say, the easy theorem we mentioned in the introduction, that says that every graph with no path of length two can be built by taking disjoint unions of complete graphs — we remind the reader that paths in this paper are induced subgraphs). The problem lies, unfortunately, in the most elegant of all the decompositions we used, the balanced skew-partition. We have tried, but failed, to "reverse" it, that is turn it into a way to combine two smaller perfect graphs together, to obtain a bigger perfect graph. This is also the reason why 2.2 does not immediately imply the existence of a polynomial time recognition algorithm for Berge graphs (we will come back to this in Section 7).

Another natural question to ask is whether all the basic classes and decompositions used in 2.2 are necessary. The answer to this question turns out to be "no", because the use of the proper homogeneous pair decomposition can be avoided and 2.2 can be strengthened as follows (this is the main result of [3]):

Theorem 2.3 *For every Berge graph G, either G is basic, or one of G, $\overline{G}$ admits a proper 2-join, or G admits a balanced skew partition.*

In Section 5 we will explain the main idea of the proof of 2.3, which was to consider more general objects, called "Berge trigraphs".

3 Claw-free Graphs

A *claw* is the complete bipartite graph $K_{1.3}$ (a vertex with three pairwise non-adjacent neighbours). A graph is called *claw-free* if it is $\{K_{1,3}\}$-free. One well-known class of claw-free graphs is the class of line graphs; some properties of line graphs have been generalized to all claw-free graphs. (For example, Edmonds' matching algorithm, that finds a maximum weight stable set in a line graph [26], was generalized by Minty to solve the maximum weight stable set problem in claw-free graphs [30].)

However, the question "what does a general claw-free graph look like?" remained open, and we are now in the process of writing a series of papers answering it [11, 12, 13, 14, 15]. Unlike in the case of perfect graphs, here we were able to prove a theorem that says: every claw-free graph can be built starting from graphs that belong to certain explicitly-constructed basic classes, and gluing them together by prescribed operations; and all graphs built in this way are claw-free. We do not have a formal way to tell what graphs should be allowed to count as basic (can the class of all claw-free graphs be basic?), or what operations are acceptable (is the operation "add a vertex to a graph that has already been constructed provided it does not introduce a claw" allowed?), but we do think that we managed to put our finger on an interesting structural property of claw-free graphs. Informally, all of our basic graphs are "explicit constructions", meaning graphs defined by a list of adjacencies, rather than properties (e.g. being claw-free). For the operations, our criterion was to "eliminate guessing". That means, roughly, that instead of constructing just all claw-free graphs, we constructed pairs (G, X), where G is a claw-free graph, and X is a "handle" (usually a subset of the vertex set of G, or, in some cases, a partition of the vertex set), that will be used when we combine G with another claw-free graph in the construction process. The question of formalizing these ideas is of great interest to us.

The first step in proving the theorem described in the previous paragraphs is obtaining a result similar to 2.2 for the class of claw-free graphs. First we need a number of definitions.

Let G be a graph. If $X \subseteq V(G)$, the graph obtained from G by deleting X is denoted by $G \setminus X$. A clique of size three is a *triangle*, and a stable set of size three is a *triad*. Distinct vertices u, v of G are *twins* (in G) if they are adjacent and have exactly the same neighbours in $V(G) \setminus \{u, v\}$.

Next, let us explain the decompositions. The first is just that there are two vertices in G that are twins, or briefly, "G admits twins". For the second, let (A, B) be a homogeneous pair, such that A, B are both cliques, and A is neither complete nor anticomplete to B. In these circumstances we call (A, B) a *W-join*. (Note that there is no requirement that $A \cup B \neq V(G)$. If the complement of G is bipartite, then G admits a W-join except in degenerate cases.) The pair (A, B) is *non-dominating* if some vertex of $G \setminus (A \cup B)$ has no neighbour in $A \cup B$; and it is *coherent* if the set of all $(A \cup B)$-complete vertices in $V(G) \setminus (A \cup B)$ is a clique.

Next, suppose that V_1, V_2 is a partition of $V(G)$ such that V_1, V_2 are nonempty and there are no edges between V_1 and V_2. We call the pair (V_1, V_2) a *0-join* in G. Thus G admits a 0-join if and only if it is not connected.

Next, suppose that V_1, V_2 partition $V(G)$, and for $i = 1, 2$ there is a subset $A_i \subseteq V_i$ such that:

- for $i = 1, 2$, A_i is a clique, and $A_i, V_i \setminus A_i$ are both nonempty

- A_1 is complete to A_2

- every edge between V_1 and V_2 is between A_1 and A_2.

In these circumstances, we say that (V_1, V_2) is a *1-join*.

Next, suppose that V_0, V_1, V_2 are disjoint subsets with union $V(G)$, and for $i = 1, 2$ there are subsets A_i, B_i of V_i satisfying the following:

- for $i = 1, 2$, A_i, B_i are cliques, $A_i \cap B_i = \emptyset$ and A_i, B_i and $V_i \setminus (A_i \cup B_i)$ are all nonempty

- A_1 is complete to A_2, and B_1 is complete to B_2, and there are no other edges between V_1 and V_2, and

- V_0 is a clique; and for $i = 1, 2$, V_0 is complete to $A_i \cup B_i$ and anticomplete to $V_i \setminus (A_i \cup B_i)$.

We call the triple (V_1, V_0, V_2) a *generalized 2-join*, and if $V_0 = \emptyset$ we call the pair (V_1, V_2) a *2-join*. (This is closely related to, but not the same as, the proper 2-join from the previous section.)

We use one more decomposition, the following. Let (V_1, V_2) be a partition of $V(G)$, such that for $i = 1, 2$ there are cliques $A_i, B_i, C_i \subseteq V_i$ with the following properties:

- For $i = 1, 2$ the sets A_i, B_i, C_i are pairwise disjoint and have union V_i

- V_1 is complete to V_2 except that there are no edges between A_1 and A_2, between B_1 and B_2, and between C_1 and C_2.

- V_1, V_2 are both nonempty.

In these circumstances we say that G is a *hex-join* of $G|V_1$ and $G|V_2$. Note that if G is expressible as a hex-join as above, then the sets $A_1 \cup B_2, B_1 \cup C_2$ and $C_1 \cup A_2$ are three cliques with union $V(G)$, and consequently no graph G with a stable set of size four is expressible as a hex-join.

Next, we list some basic classes of graphs.

- **Line graphs.** We say $G \in \mathcal{S}_0$ if G is isomorphic to a line graph.

- **The icosahedron.** This is the unique planar graph with twelve vertices all of degree five. For $0 \leq k \leq 2$, $icosa(-k)$ denotes the graph obtained from the icosahedron by deleting k pairwise adjacent vertices. We say $G \in \mathcal{S}_1$ if G is isomorphic to $icosa(0)$, $icosa(-1)$ or $icosa(-2)$.

- **The graphs $\mathcal{S}_2$.** Let G be the graph with vertex set $\{v_1, \ldots, v_{13}\}$, with adjacency as follows. v_1-$\cdots$-v_6 is a hole in G of length 6. Next, v_7 is adjacent to v_1, v_2; v_8 is adjacent to v_4, v_5, and possibly to v_7; v_9 is adjacent to v_6, v_1, v_2, v_3; v_{10} is adjacent to v_3, v_4, v_5, v_6, v_9; v_{11} is adjacent to $v_3, v_4, v_6, v_1, v_9, v_{10}$; v_{12} is adjacent to $v_2, v_3, v_5, v_6, v_9, v_{10}$; and v_{13} is adjacent to $v_1, v_2, v_4, v_5, v_7, v_8$. We say $H \in \mathcal{S}_2$ if H is isomorphic to $G \setminus X$, where $X \subseteq \{v_{11}, v_{12}, v_{13}\}$.

- **Circular interval graphs.** Let Σ be a circle and let $F_1, \ldots, F_k$ be subsets of Σ, each homeomorphic to the closed interval $[0, 1]$, and no three with union Σ. Let V be a finite subset of Σ, and let G be the graph with vertex set V in which $v_1, v_2 \in V$ are adjacent if and only $v_1, v_2 \in F_i$ for some i. Such a graph is called a *circular interval graph*. If $\bigcup_{i=1}^{k} F_i \neq \Sigma$, we say that G is a *linear interval graph*. We write $G \in \mathcal{S}_3$ if G is a circular interval graph. .

- **An extension of $L(K_6)$.** Let H be the graph with seven vertices $h_0, \ldots, h_6$, in which $h_1, \ldots, h_6$ are pairwise adjacent and h_0 is adjacent to h_1. Let G be the graph obtained from the line graph $L(H)$ of H by adding one new vertex, adjacent precisely to the members of $V(L(H)) = E(H)$ that are not incident with h_1 in H. Then G is claw-free. Let $\mathcal{S}_4$ be the class of all graphs isomorphic to induced subgraphs of G.

- **The graphs $\mathcal{S}_5$.** Let $n \geq 2$. Let $A = \{a_1, \ldots, a_n\}, B = \{b_1, \ldots, b_n\}$ and $C = \{c_1, \ldots, c_n\}$ be three cliques, pairwise disjoint. For $1 \leq i, j \leq n$, let a_i, b_j be adjacent if and only if $i = j$, and let c_i be adjacent to a_j, b_j if and only if $i \neq j$. Let d_1, d_2, d_3, d_4, d_5 be five more vertices, where d_1 is $A \cup B \cup C$-complete; d_2 is complete to $A \cup B \cup \{d_1\}$; d_3 is complete to $A \cup \{d_2\}$; d_4 is complete to $B \cup \{d_2, d_3\}$; d_5 is adjacent to d_3, d_4; and there are no more edges. Let the graph just constructed be G. We say $H \in \mathcal{S}_5$ if (for some n) H is isomorphic to $G \setminus X$ for some $X \subseteq A \cup B \cup C$.

- **2-simplicial graphs of antihat type.** Let $n \geq 0$. Let $A = \{a_0, a_1, \ldots, a_n\}$, $B = \{b_0, b_1, \ldots, b_n\}$ and $C = \{c_1, \ldots, c_n\}$ be three cliques, pairwise disjoint. For $0 \leq i, j \leq n$, let a_i, b_j be adjacent if and only if $i = j > 0$, and for $1 \leq i \leq n$ and $0 \leq j \leq n$ let c_i be adjacent to a_j, b_j if and only if $i \neq j \neq 0$. Let the graph just constructed be G. We say $H \in \mathcal{S}_6$ if (for some n) H is isomorphic to $G \setminus X$ for some $X \subseteq A \cup B \cup C$, and then H is said to be 2-*simplicial of antihat type*.

- **Antiprismatic graphs.** Let us say a graph is *antiprismatic* if for every triad u, v, w, every vertex different from u, v, w is adjacent to exactly two of them. Antiprismatic graphs are claw-free, and we gave a structural description of them in the first two papers of the series [11],[12]. We will not include it here for reasons of space.

We can now state the theorem:

Theorem 3.1 *Let G be claw-free. Then either*

- $G \in \mathcal{S}_0 \cup \cdots \cup \mathcal{S}_6$, *or*

- G *admits either twins, a non-dominating W-join, a coherent W-join, a 0-join, a 1-join, a generalized 2-join, or a hex-join, or*

- G *is antiprismatic.*

Similarly to 2.2, we call 3.1 a "decomposition" theorem. But, unlike 2.2, 3.1 can be converted into what we call a "composition theorem", meaning a theorem that allows us to build all claw-free graphs. This is done by "reversing" the decompositions, to obtain "compositions". For example, every claw-free graph that admits twins can be obtained from a smaller claw-free graph by adding a new adjacent copy of an existing vertex. Moreover, given a claw-free graph, one can do this operation, and the resulting graph will be claw-free, no matter what vertex has been replicated (so there is no need to guess the "right" vertex to replicate). Reversing other operations is more difficult, and the general result we obtain for claw-free graphs is quite complicated, and we will not include it here.

Instead, let us consider a subclass of claw-free graphs, the class of "quasi-line" graphs. These are graphs in which the vertex set of the neighbourhood of every vertex is the union of two cliques. Let W_i be the graph consisting of an antihole H of length i, and a $V(H)$-complete vertex v (therefore $v \notin V(H)$); and let $\mathcal{F}$ be the family of graphs consisting of the claw, together with all W_i with odd $i \geq 5$. Then G is a quasi-line graph if and only if G is $\mathcal{F}$-free. In particular, every line graph is a quasi-line graph.

Circular interval graphs are quasi-line graphs, but there is another way to construct quasi-line graphs, that we explain next. A vertex $v \in V(G)$ is *simplicial* if the set of neighbours of v is a clique. A *strip* (G, a, b) consists of a claw-free graph G together with two designated non-adjacent simplicial vertices a, b called the *ends* of the strip. For instance, if G is a linear interval graph, with vertices $v_1, \ldots, v_n$ in order and with $n > 1$, then v_1, v_n are simplicial, and so (G, v_1, v_n) is a strip, called a *linear interval strip*.

Suppose that (G, a, b) and (G', a', b') are two strips. We compose them as follows. Let A, B be the set of vertices of $G \setminus \{a, b\}$ adjacent in G to a, b respectively, and define A', B' similarly. Take the disjoint union of $G \setminus \{a, b\}$ and $G' \setminus \{a', b'\}$; and let H be the graph obtained from this by adding all possible edges between A and A' and between B and B'. Then H is claw-free.

This method of composing two strips is symmetrical between (G, a, b) and (G', a', b'), but we do not use it in a symmetrical way. We use it as follows. Start with a graph G_0 with an even number of vertices and which is the disjoint union of complete graphs, and pair the vertices of G_0. Let the pairs be $(a_1, b_1), \ldots, (a_n, b_n)$, say. For $i = 1, \ldots, n$, let (G_i', a_i', b_i') be a strip. For $i = 1, \ldots, n$, let G_i be the graph obtained by composing (G_{i-1}, a_i, b_i) and (G_i', a_i', b_i'); then the resulting graph G_n is called a *composition* of the strips (G_i', a_i', b_i') $(1 \leq i \leq n)$. For instance, if for each of the strips (G_i', a_i', b_i'), G_i' is a 3-vertex path with ends a_i', b_i', then the effect of composing with (G_i', a_i', b_i') is the identification of a_i, b_i; and so the graphs that are compositions of such 3-vertex path strips are precisely line graphs.

It is easy to check that every graph that is the composition of linear interval strips is a quasi-line graph, so this gives us a second construction for quasi-line graphs (and this includes line graphs, since the 3-vertex strip mentioned above is a linear interval strip).

We can prove the following decomposition theorem for quasi-line graphs [16]:

Theorem 3.2 *For every quasi-line graph G, either G is a circular interval graph, or G is a composition of linear interval strips, or G admits a 0-join, or a W-join.*

It is clear how to "reverse" the 0-join decomposition: all one needs to do is take a disjoint union. The W-join decomposition is trickier, but, it turns out, that one can avoid it at the expense of expanding the list of basic graphs. (In order to do that, we use the same idea as in eliminating proper homogeneous pairs from 2.2, and we will explain it later).

Let us now describe the expanded list of basic graphs. We say that a graph G is a *fuzzy circular interval graph* if:

• there is a map ϕ from $V(G)$ to a circle C (not necessarily injective), and

• there is a set of intervals from C, none including another, and such that no point of C is an end of more than one of the intervals, so that

• for u, v in G, if u, v are adjacent then $\{\phi(u), \phi(v)\}$ is a subset of one of the intervals, and if u, v are non-adjacent then $\phi(u) \neq \phi(v)$, and $\phi(u), \phi(v)$ are both ends of any interval including both of them.

(If also we required ϕ to be injective, this would be equivalent to the definition of a circular interval graph.) If x, y are ends of an interval and the sets $\phi^{-1}(x)$ and $\phi^{-1}(y)$ are not complete and not anticomplete to each other, then the pair $(\phi^{-1}(x), \phi^{-1}(y))$ is a W-join, and and these turn out to be the only kinds of W-joins that we need. (Fuzzy linear interval strips are defined analogously, with the additional condition that if a, b are the ends of the strip then $\phi(a), \phi(b)$ are different from $\phi(v)$ for all other vertices v of G.)

We prove [16]:

Theorem 3.3 *Every quasi-line graph G can be obtained by taking disjoint unions of fuzzy circular interval graphs and graphs that are compositions of fuzzy linear interval strips. Moreover, every graph obtained this way is a quasi-line graph.*

Finally, let us mention, that, similarly to the case of Berge graphs, the property of being claw-free implies that the chromatic number of a graph (and therefore all its induced subgraphs) is bounded by a function of the size of its largest clique. It is easy to see that for a claw-free graph G, $\chi(G) \leq \omega(G)^2$, and this is not far from being best possible because every graph with no triad is claw-free. Even if we insist that G has a triad, it may still have a large triad-free component controlling the chromatic number, and so $\chi(G)$ may still be super-linear in $\omega(G)$. However, if we restrict our attention to connected claw-free graphs with triads, a much better bound is true [17]:

Theorem 3.4 *Let G be a connected claw-free graph that contains a triad. Then $\chi(G) \leq 2\omega(G)$.*

The proof of 3.4 uses our structure theorem for claw-free graphs, but if we replace $\chi(G) \leq 2\omega(G)$ by $\chi(G) \leq 4\omega(G)$, there is an easy elementary proof. However, the factor of 2 is tight. 3.4 can be strengthened further if we assume that G is a quasi-line graph [8]:

Theorem 3.5 *Let G be a quasi-line graph. Then $\chi(G) \leq \frac{3}{2}\omega(G)$.*

The proof of 3.5 relies on 3.3, and the factor of $\frac{3}{2}$ is tight.

Curiously, we also a get a theorem similar to 3.4 for graphs whose complements are claw-free [17], and there the proof does not use any of the heavy machinery described earlier in this section.

Theorem 3.6 *Let G be the complement of a connected claw-free graph that contains a triad. Then $\chi(G) \leq 2\omega(G)$.*

4   Bull-free Graphs

The *bull* is the graph B with vertex set

$$\{x_1, x_2, x_3, y, z\}$$

and edge set

$$\{x_1x_2, x_2x_3, x_1x_3, x_1y, x_2z\}.$$

A graph is called *bull-free* if it is $\{B\}$-free. Obvious examples of bull-free graphs are graphs with no triangle and graphs with no triad; but there are others. Let us call a graph G an *ordered split graph* if there exists an integer n such that the vertex set of G is the disjoint union of a clique $\{k_1, \ldots, k_n\}$ and a stable set $\{s_1, \ldots, s_n\}$, and s_i is adjacent to k_j if and only if $i + j \leq n + 1$. It is easy to see that every ordered split graph is bull-free. A large ordered split graph contains a large clique and a large stable set, and therefore the three classes (triangle-free, triad-free and ordered split graphs) are significantly different.

It turns out, however, that, similarly to claw-free graphs, there is a composition theorem for bull-free graphs; all bull-free graphs can be built starting from graphs that belong to a few basic classes, gluing them together by certain operations [4]. The basic classes we need are triangle-free graphs, triad-free graphs, a certain generalization of the ordered split graphs, and a couple of others, that we will not describe here. Let $\mathcal{B}$ denote the set of all bull-free graphs that belong to one of the basic classes. Next we describe some operations, that are used to combine two smaller bull-free graphs together, to obtain a new, larger, bull-free graph. For a graph G and a vertex v of G, we denote by $\Gamma_G(v)$ the set of all vertices of $V(G) \setminus \{v\}$ that are adjacent to v.

Operation $\mathcal{O}_1$ is the operation of complementation. The input of $\mathcal{O}_1$ is a graph G_1, and the output is the complement of G_1.

Operation $\mathcal{O}_2$ is the operation of taking the disjoint union of two graphs. The input of $\mathcal{O}_2$ is a pair of graphs G_1, G_2, and the output is a new graph G_3, with $V(G_3) = V(G_1) \cup V(G_2)$ and $E(G_3) = E(G_1) \cup E(G_2)$.

Operation $\mathcal{O}_3$ is defined as follows. The input of $\mathcal{O}_3$ is a pair of graphs G_1, G_2, and disjoint ordered subsets A_1, B_1 of $V(G_1)$ and A_2, B_2 of $V(G_2)$, with the following properties:

- A_1, B_1, A_2, B_2 are stable sets, with $|A_1| = |A_2|$ and $|B_1| = |B_2|$.

- A_1 is complete to B_1, and A_2 to B_2.

- For $i = 1, 2$ let G_i' be the graph obtained from G_i by adding two new vertices a_i, b_i such that $\{a_i\}$ is complete to A_i and $\{b_i\}$ to B_i, and there are no other edges incident with a_i, b_i. Then both G_1' and G_2' are bull free.

- Let $\{i, j\} = \{1, 2\}$. Let $a_i, a_i' \in A_i$, and let a_j, a_j' be the corresponding vertices in A_j. If there exists an edge uv of G_i such that a_i is complete to $\{u, v\}$, and a_i' is adjacent to u and not to v, then $\Gamma_{G_j}(a_j) \subseteq \Gamma_{G_j}(a_j')$; and the same for B_i and B_j.

Under these circumstances, the result of applying $\mathcal{O}_3$ to $G_1, G_2, A_1, B_1, A_2, B_2$ is the graph G_3, obtained from the disjoint union of G_1 and G_2 by identifying the corresponding vertices of A_1 and A_2, and the corresponding vertices of B_1 and B_2.

We would like to remark that operation $\mathcal{O}_3$, the way it is defined here, is really just a decomposition in disguise, but it can be strengthened to get a real composition operation, see [4].

Operation $\mathcal{O}_4$ is the operation of substitution. The input of $\mathcal{O}_4$ is a pair of graphs G_1, G_2 and a vertex $v \in V(G_1)$. The output is a new graph G_3, with

$$V(G_3) = V(G_1) \cup V(G_2) \setminus \{v\}$$

and

$$E(G_3) = E(G_1 \setminus \{v\}) \cup E(G_2) \cup \{xy \ : \ x \in V(G_1) \setminus \{v\}, \ y \in V(G_2), \ \text{and} \ xv \in E(G_1)\}.$$

Please note that unlike all the previous operations, $\mathcal{O}_4$ is not symmetric between G_1 and G_2.

The main result of [4] is the following:

Theorem 4.1 *Let G be a bull-free graph. Then either $G \in \mathcal{B}$, or G can be obtained starting from graphs in $\mathcal{B}$, by repeated applications of operations $\mathcal{O}_1, \ldots, \mathcal{O}_4$. Conversely, every graph obtained in this way is bull-free.*

As in the case of claw-free graphs, we start by proving a "decomposition" theorem for bull-free graphs, that is, a theorem that says that every bull-free graph is either basic, or admits a decomposition. Reversing the decompositions yields the operations $\mathcal{O}_1, \ldots, \mathcal{O}_4$. Another similarity with claw-free graphs (and quasi-line graphs) is that one can state a decomposition theorem for bull-free graphs that uses very few basic classes, but needs a decomposition similar to a W-join. The conditions under which introducing a homogeneous pair in a bull-free graph produces another bull-free graph are quite complicated, and do not seem to be far from saying "add a vertex if it does not create a bull". But again, by considering the more general structure of "bull-free trigraphs", we were able to eliminate the use of homogeneous pairs, at the expense of expanding the list of basic classes.

In [10] Safra and the first author use 4.1 to settle the Erdős-Hajnal conjecture for the case when H is the bull, by proving the following:

Theorem 4.2 *Let G be a bull-free graph. Then G contains a stable set or a clique of size $|V(G)|^{\frac{1}{4}}$.*

In order to prove 4.2, it is shown inductively, using 4.1, that every bull-free graph G can be "fractionally" covered by at most $|V(G)|^{\frac{1}{2}}$ induced subgraphs of G, each of which is perfect. It follows that there exists an induced subgraph H of G, containing at least $|V(G)|^{\frac{1}{2}}$ vertices, and such that H is perfect. Consequently, H contains a stable set or a clique of size at least $|V(H)|^{\frac{1}{2}} \geq |V(G)|^{\frac{1}{4}}$, and 4.2 follows.

5   Trigraphs

The goal of this section is to introduce the notion of a "trigraph". A *trigraph* T is a 4-tuple $(V(T), E(T), S(T), N(T))$ where V is the *vertex set* of T and every unordered pair of vertices belongs to one of the three disjoint sets: the *strong edges* $E(T)$, the *strong non-edges* $N(T)$ and the *switchable pairs* $S(T)$, and such that every vertex of T belongs to at most one switchable pair. Let us say that two vertices u, v of T are *strongly adjacent* if $\{u, v\}$ is a strong edge, *strongly anti-adjacent* if $\{u, v\}$ is a strong non-edge, and *semi-adjacent* if $\{u, v\}$ is a switchable pair. In this notation a graph can be viewed as a trigraph with an empty set of switchable pairs. A *realization* of a trigraph $T = (V(T), E(T), S(T), N(T))$ is a graph $G = (V(G), E(G))$ such that $V(G) = V(T)$, and $E(T) \subseteq E(G) \subseteq E(T) \cup S(T)$.

Thus trigraphs are objects generalizing graphs, and on a number of occasions, when dealing with classes of graphs defined by forbidding certain induced subgraphs, considering trigraphs instead of graphs allowed us to prove stronger theorems for the class of *graphs* we were interested in.

We use trigraphs while dealing with Berge graphs, claw-free graphs, and bull-free graphs. In all three cases the situation is as follows: we were able to prove a theorem that said "every Berge (claw-free, bull-free) graph either belongs to one of a few basic classes, or admits one of a few decompositions", where one of the decompositions was some variety of a homogeneous pair decomposition, where the two sets of the homogeneous pair are not complete and not anticomplete to each other. In all cases, it is possible to define an operation that is the "reverse" of the homogeneous pair decomposition, let us call it a *thickening*. Given a list $\mathcal{L}$ of basic graphs, we call a *thickened basic* graph every graph that can be obtained from a graph in $\mathcal{L}$ by performing thickenings. Now we would like to strengthen the theorem, and prove that every Berge (claw-free, bull-free) graph is either a thickened basic graph, or admits one of a few decompositions (none of which is a homogeneous pair decomposition). The last step is to describe explicitly all thickened basic graphs, thus eliminating the use of homogeneous pairs.

The approach we use is as follows. Let $\mathcal{F}$ be a family of graphs. Let us say that the family $\mathcal{T}$ of trigraphs is $\mathcal{F}$-free, if every graph that is a realization of a trigraph in $\mathcal{T}$ is $\mathcal{F}$-free. Now, instead of considering Berge (claw-free, bull-free) graphs, we turn to Berge (claw-free, bull-free) trigraphs. For every decomposition we expect to use for the class of $\mathcal{F}$-free graphs, we define its trigraph analogue, in such a way that if two vertices of a graph were specified as being adjacent in the graph decomposition, they are specified as being strongly adjacent in the trigraph decomposition, and similarly for pairs that were specified to be non-adjacent. For example, the graph decomposition "G is disconnected", becomes the trigraph decomposition "$V(T)$ can be partitioned into two non-empty subsets V_1 and V_2, such that every vertex of V_1 is strongly anti-adjacent to every vertex of V_2". This has the useful consequence that if T is a trigraph admitting such a decomposition, and ab is a switchable pair in T, and we replace a and b by two sets of new vertices, A and B, respectively, making every vertex of A strongly adjacent to all the vertices of $V(T) \setminus \{a, b\}$ that were strongly adjacent to a, and strongly anti-adjacent to all the others, and similarly for B, forming a trigraph T', then T' admits the same decomposition (this is true with a few exceptions, but they are dealt with separately). We will refer to this as "property X".

For every basic class $\mathcal{C}$ of graphs, the corresponding basic class of trigraphs consists of all $\mathcal{F}$-free trigraphs T, such that some graph of $\mathcal{C}$ is a realization of T.

In each of the three cases (Berge, claw-free and bull-free) we were able to prove that every $\mathcal{F}$-free graph is either basic or admits one of the decompositions in some list, say, $D_1, \ldots, D_k$, or admits a homogeneous pair. In each case we can then prove that every $\mathcal{F}$-free trigraph is either basic (in the trigraph sense explained above) or admits (the trigraph analogue of) one of the decompositions $D_1, \ldots, D_k$, or admits (the trigraph analogue of) a homogeneous pair. So far this is the same theorem, only in slightly greater generality. It turns out, however, that this more general version allows us to prove the strengthened theorem for graphs that we are interested in.

It is enough to prove that every $\mathcal{F}$-free trigraph is either basic (in the trigraph sense) or admits (the trigraph analogue of) one of the decompositions $D_1, \ldots, D_k$. Here is the outline of the proof. Suppose this is false and let T be a trigraph that is not basic, and does not admit any of the decomposition $D_1, \ldots, D_k$, and subject to that has $|V(T)|$ as small as possible. By the theorem we know for trigraphs, T admits (the trigraph analogue of) a homogeneous pair (A, B). So every vertex of $V(T) \setminus (A \cup B)$ is either strongly adjacent to every vertex of A, or strongly anti-adjacent to every vertex of A, and the same for B. Let T' be the trigraph obtained from T by replacing the set A by a new vertex a, and the set B by a new vertex b, such that

- a is semi-adjacent to b in T'

- for every vertex $v \in V(T) \setminus (A \cup B)$, v is strongly adjacent to a in T' if v is strongly adjacent to every vertex of A in T, and v is strongly anti-adjacent to a in T' if v is strongly anti-adjacent to every vertex of A in T, and

- for every vertex $v \in V(T) \setminus (A \cup B)$, v is strongly adjacent to b in T' if v is strongly adjacent to every vertex of B in T, and v is strongly anti-adjacent to b in T' if v is strongly anti-adjacent to every vertex of B in T.

Then T' is $\mathcal{F}$-free (this requires some checking, but we ensure it by imposing some non-triviality conditions on the decompositions). By the minimality of $|V(T)|$, it follows that T' is either basic, or admits one of the decompositions $D_1, \ldots, D_k$. If T' is basic then so is T, with a few exceptions. So we may assume that T' admits some decomposition D_i. But then, since the pair $\{a, b\}$ is a switchable pair of T', by property X, T admits the same decomposition D_i. This, however, is a contradiction to the way T was chosen. This completes the proof.

At first it seems that instead of using trigraphs, one could redefine the decompositions and do the whole proof in terms of graphs only. We would like to remark that despite a certain amount of effort invested in this approach, we were unable to come up with a consistent set of definitions, and so the idea of using trigraphs seems crucial.

6 Even-hole-free Graphs

In this section we discuss the family of even-hole-free graphs; these are $\mathcal{F}$-free graphs where $\mathcal{F}$ is the family of all cycles of even length. (Similarly, *odd-hole-free graphs* are graphs with no induced odd cycles of length at least five). Unfortunately, for even-hole-free graphs we do not have a composition theorem similar to 3.3 or 4.1. The best known result of this kind is a theorem similar to 2.2, due to Conforti, Cornuéjols, Kapoor and Vušković [21], that states that every even-hole-free graph is either basic or admits a decomposition. This theorem was then used in [22] to design a polynomial time recognition algorithm for the class of even-hole-free graphs.

However, the following, conjectured by Reed [32], remained open for a while longer, and was proved only recently by Addario-Berry, Havet, Reed and the authors in [1] (a *bisimplicial* vertex in a graph is a vertex whose set of neighbours is the union of two cliques):

Theorem 6.1 *Every non-null even-hole-free graph has a bisimplicial vertex.*

At first we directed our effort to trying to find a composition theorem for even-hole-free graphs, but were unsuccessful. It still seemed, however, that proving a statement stronger than 6.1, that would contain some information about the location of the bisimplicial vertices in the graph, would allow us to apply induction and prove 6.1. This direction was a lot more fruitful, and eventually lead to a proof of 6.1, that we now outline.

Let us start with some definitions. Let G be a graph and let S be a subset of $V(G)$. The *neighbourhood* of S, denoted by $N_G(S)$, is S together with the set of all vertices of $V(G) \setminus S$ with a neighbour in S. The *non-neighbourhood* of S is the set $V(G) \setminus N_G(S)$. If S consists of a single vertex s, we write $N_G(s)$ instead of $N_G(\{s\})$. A set S of vertices in a graph G is called *dominating (in G)* if $N_G(S) = V(G)$, and *non-dominating* otherwise. An induced subgraph H of G is *dominating* if $V(H)$ is dominating, and *non-dominating* otherwise; we denote by $N_G(H)$ the set $N_G(V(H))$. The stronger statement we ended up proving is the following:

Theorem 6.2 *Let G be an even-hole-free graph. Then both the following statements hold:*

1. *If H is a non-dominating hole in G, then some vertex of $V(G) \setminus N_G(H)$ is bisimplicial in G.*

2. *If K is a non-dominating clique in G of size at most two, then some vertex of $V(G) \setminus N_G(K)$ is bisimplicial in G.*

Clearly the second statement of 6.2 with $K = \emptyset$ implies 6.1. We remark that the second statement of 6.2 would be false if we replace "at most two" by "at most three". The graph obtained from K_4 by choosing a vertex and subdividing once the edges incident with it is a counterexample.

Let us now describe the proof of 6.2. The proof uses induction. Let G be a graph such that 6.2 holds for all smaller graphs. First we suppose that G fails to satisfy the first statement, that is there is a non-dominating hole H in G, but there is no bisimplicial vertex in the non-neighbourhood of $V(H)$. Now the idea is to examine the neighbourhood of $V(H)$ and try to find what we call a "useful cutset" in G, that is, a subset C of $V(G)$ and an edge e with both ends in C such that

- $V(G) \setminus C$ is the disjoint union of two non-empty sets, L and R, anticomplete to each other

- $C \subseteq N(e)$ and the non-neighbourhood of e in the graph $G|(C \cup R)$ is a non-empty subset of the non-neighbourhood of $V(H)$ in G.

If we find such a cutset C, then it follows, from the minimality of G, that R contains a vertex v which is bisimplicial in $G|(C \cup R)$; and since L is anticomplete to R, it follows that v is a bisimplicial vertex of G, which is a contradiction.

Unfortunately, we do not always succeed in finding a useful cutset; sometimes we have to make do with a set C and a list $u_1, .., u_k, v_1, .., v_k$ of vertices of C (possibly with repetitions) where u_i is non-adjacent to v_i in G for every $1 \leq i \leq k$, such that:

- $V(G) \setminus C$ is the disjoint union of two non-empty sets, L and R, anticomplete to each other

- the graph G' obtained from $G|(R \cup C)$ by adding the edge $u_i v_i$ for every $1 \leq i \leq k$ is even-hole-free

- for some edge e of G', $C \subseteq N_{G'}(e)$, and the non-neighbourhood of e in G' is a non-empty subset of the non-neighbourhood of $V(H)$ in G

- if v is a bisimplicial vertex of G' contained in the non-neighbourhood of e, then v is bisimplicial in G.

Having found such a set C etc, the same argument as in the case of a "genuine" useful cutset leads to a contradiction.

So G satisfies the first statement of 6.2. Suppose it fails to satisfy the second. This means that there is a non-dominating clique K of size at most two in G with no bisimplicial vertex in its non-neighbourhood. An easy argument shows that there is a hole H of G such that K is included in $V(H)$. Since the first assertion of the theorem holds for G, we deduce that H is dominating in G. Now we can examine the structure of G relative to H, and again find variations on the idea of a useful cutset, such as the one described above, that lead to a contradiction. So G satisfies the second statement of 6.2 too. This completes the inductive proof.

A graph G is called *odd-signable* if there exists a function $f : E(G) \rightarrow \{0, 1\}$ such that $\sum_{e \in E(H)} f(e)$ is odd for every hole H of G. It is natural to ask whether 6.1 is true if we replace "even-hole-free" by "odd-signable". The answer to this question is "no", and the six vertex graph which is the 1-skeleton of the cube is a counterexample.

Finally, we would like to point out an easy corollary of 6.1, that, similarly to the case of perfect graphs, claw-free graphs and quasi-line graphs, establishes a connection between the property of being $\mathcal{F}$-free, and the fact the the chromatic number of the graph (and therefore of all induced subgraphs) is bounded by a function of the size of the largest clique.

Theorem 6.3 *Let G be a non-null even-hole free graph. Then $\chi(G) \leq 2\omega(G) - 1$.*

Proof. The proof is by induction on $|V(G)|$. By 6.1 there exists a bisimplicial vertex v in G. The graph G' obtained from G by deleting v is another even-hole free graph, we can assume G' is non-null, $\omega(G') \leq \omega(G)$, and, inductively, G' can be properly coloured with at most $2\omega(G) - 1$ colours. Let c be such a colouring of G'. Since v is bisimplicial in G, $|N_G(v)| \leq 2\omega(G) - 1$ and at least one of the $2\omega - 1$ colours does not appear in $N_G(v) \setminus \{v\}$ in c. Now v can be coloured with this colour, thus extending c to a proper colouring of G with at most $2\omega(G) - 1$ colours. This proves 6.3.

Unfortunately, we do not know if this theorem is sharp, the best example we know has chromatic number $\frac{5}{4}\omega$.

7 Detecting Induced Subgraphs

Given an infinite family $\mathcal{F}$ of graphs, it is natural to ask whether one can test in polynomial time if a given graph G is $\mathcal{F}$-free. In this section, will survey some known results in this direction. For brevity, let us say "testing for $\mathcal{F}$" when we mean "testing for being $\mathcal{F}$-free". In all cases the family $\mathcal{F}$ we consider consists of subdivisions of a given graph, possibly with some parity conditions. It turns out that even in this restricted setting, testing for $\mathcal{F}$ can be done in polynomial time for some families $\mathcal{F}$, and can be shown to be NP-complete for others. At the moment we do not know the reason for this difference in behaviour.

A *pyramid* is a graph consisting of a triangle $\{b_1, b_2, b_3\}$, called the *base*, a vertex $a \notin \{b_1, b_2, b_3\}$, called the *apex*, and three paths P_1, P_2, P_3, such that for $i, j \in 1, 2, 3$

- the ends of P_i are a and b_i,

- if $i \neq j$ then $V(P_i) \setminus \{a\}$ is disjoint from $V(P_j) \setminus \{a\}$ and the only edge between them is $b_i b_j$, and

- at most one of P_1, P_2, P_3 has length one.

In this case we say that the pyramid is *formed* by the paths P_1, P_2, P_3.

Let $\mathcal{P}$ be the family of all pyramids. It turns out that testing for $\mathcal{P}$ is relatively easy, and can be done in time $O(|V(G)|^9)$ [5]. The idea is as follows. If G contains a pyramid, then it contains a pyramid P with the number of vertices smallest. We are going to "guess" some of the vertices of P in G, then find shortest paths in G between pairs of vertices that we guessed that were joined by a path in P, and then test whether the subgraph of G formed by the union of these shortest paths is a pyramid. If the answer is "yes", then G contains a pyramid, and we stop.

Surprisingly, it turns out, that choosing the shortest paths with a little bit of care, we can guarantee that if the answer is "no", then there is no pyramid in P. We call this general strategy of testing for a family $\mathcal{F}$ a *shortest-paths detector* for $\mathcal{F}$.

Let us now be more precise. For $u, v \in V(G)$ we denote by $d_G(u, v)$ the length of the shortest path of G between u and v. If P is a pyramid, formed by three paths P_1, P_2, P_3, with apex a and base $\{b_1, b_2, b_3\}$, we say its *frame* is the 10-tuple

$$a, b_1, b_2, b_3, s_1, s_2, s_3, m_1, m_2, m_3,$$

where

- for $i = 1, 2, 3$, s_i is the neighbour of a in P_i

- for $i = 1, 2, 3$, $m_i \in V(P_i)$ satisfies $d_{P_i}(a, m_i) - d_{P_i}(m_i, b_i) \in \{0, 1\}$.

A pyramid P in G is *optimal* if there is no pyramid P' with $|V(P')| < |V(P)|$.

Theorem 7.1 *[5] Let P be an optimal pyramid, with frame $a, b_1, b_2, b_3, s_1, s_2, s_3, m_1, m_2, m_3$. Let S_1, T_1 be the subpaths of P_1 from m_1 to s_1, b_1 respectively. Let F be the set of all vertices non-adjacent to each of s_2, s_3, b_2, b_3.*

1. *Let Q be a path between s_1 and m_1 with interior in F, and with minimum length over all such paths. Then a-s_1-Q-m_1-T_1-b_1 is a path (say P_1'), and P_1', P_2, P_3 form an optimal pyramid.*

2. *Let Q be a path between m_1 and b_1 with interior in F, and with minimum length over all such paths. Then a-s_1-S_1-m_1-Q-b_1 is a path (say P_1'), and P_1', P_2, P_3 form an optimal pyramid.*

Analogous statements hold for P_2, P_3.

7.1 can be used to design an algorithm to test for $\mathcal{P}$:

- guess the frame $a, b_1, b_2, b_3, s_1, s_2, s_3, m_1, m_2, m_3$ of an optimal pyramid P of G, by trying all 10-tuples of vertices,

- find shortest paths between m_1 and b_1, and between m_1 and s_1, not containing any neighbours of s_2, s_3, b_2, and b_3; do the same for m_2, b_2, s_2 and m_3, b_3, s_3,

- test if the union of the six shortest paths, together with the vertex a forms a pyramid.

Now, by 7.1, the answer if "yes", if and only if G contains a pyramid. The algorithm in [5] is similar; it was modified a little to bring the running time down to $O(|V(G)|^9)$.

The main result of [5] is a polynomial time algorithm for testing if a graph is Berge (and therefore perfect). Since every pyramid contains an odd hole, it follows that every odd-hole-free, and therefore every Berge, graph is $\mathcal{P}$-free.

Even though the algorithm in [5] was found after 2.2 had been proved, it does not use 2.2. The idea in [5] is to use the shortest-path detector for odd holes. Unfortunately, there does not seem to be a theorem similar to 7.1 for odd holes, and so, first, the graph needs to be "prepared" for using a shortest-paths detector. The first step is to test for $\mathcal{P}$, and a few other families $\mathcal{F}$ that are easy to test for, and such that every Berge graph is $\mathcal{F}$-free. Now we can assume that the graph in question is $\mathcal{F}$-free for all these $\mathcal{F}$. The next step is applying "cleaning", a technique first proposed in [24]. The idea of cleaning is to find, algorithmically, polynomially many subsets $X_1, \ldots, X_k$ of $V(G)$, such that if G contains an odd hole, then for at least one value of $i \in \{1, \ldots, k\}$ the graph $G_i = G \setminus X_i$ contains an odd hole that can be found using a shortest-paths detector. Finally, applying a shortest-paths detector for odd holes to each of $G_1, \ldots, G_k$, we detect an odd hole if and only if G contains one.

In addition to the algorithm just described, [5] contains another algorithm to test for Bergeness, that instead of a shortest-paths detector for odd holes, uses a decomposition theorem for odd-hole-free graphs from [23], but we will not describe this algorithm here. We remark that both algorithms in [5] test for Bergeness, and not for the family of odd holes. The complexity of testing if a graph contains an odd hole is still unknown. On the other hand, the problem of testing if a graph contains an even hole can be solved in polynomial time. There are two known algorithms. One is due to Conforti, Cornuéjols, Kapoor, and Vušković [22], and the other to Kawarabayshi and the authors [7]. Both algorithms use cleaning, and then the former uses a decomposition theorem of [21] for even-hole-free graphs, and the latter a shortest-paths detector.

There are two other kinds of graphs that are somewhat similar to the pyramid, called a "theta" and a "prism". A *theta* is a graph consisting of two non-adjacent vertices s, t and three paths P_1, P_2, P_3, each between s and t, such that the sets $V(P_1) \setminus \{s, t\}$, $V(P_2) \setminus \{s, t\}$, and $V(P_3) \setminus \{s, t\}$ are pairwise disjoint, the union of every pair of P_1, P_2, P_3 is a hole. A *prism* is a graph consisting of two disjoint triangles $\{a_1, a_2, a_3\}$ and $\{b_1, b_2, b_3\}$ and three paths P_1, P_2, P_3, with the following properties:

- for $i = 1, 2, 3$, the ends of P_i are a_i and b_i,

- P_1, P_2, P_3 are pairwise disjoint , and

- for $1 \leq i < j \leq 3$, there are precisely two edges between $V(P_i)$ and $V(P_j)$, namely $a_i a_j$ and $b_i b_j$.

Let $\mathcal{T}$ be the family of all thetas, and $\mathcal{P}r$ the family of all prisms. Then every even-hole-free graph is $\mathcal{T} \cup \mathcal{P}r$-free, and so prisms and thetas play a similar role for even-hole-free graphs to the one that pyramids play for odd-hole-free graphs. It turns out, however, that, unlike $\mathcal{P}$, the problem of testing for $\mathcal{P}r$ is NP-complete (this is a theorem due to Maffray and Trotignon [31]). On the other hand, testing for $\mathcal{T}$ can be done in polynomial time [18]. The problems of testing for $\mathcal{P} \cup \mathcal{P}r$ and testing for $\mathcal{T} \cup \mathcal{P}r$ can also be solved in polynomial time (see [31] and [6], respectively).

All the algorithms mentioned above use variations on the ideas of cleaning and shortest paths detectors (or decomposition theorems), except one, and that is the algorithm for testing for $\mathcal{T}$. There our approach is different. In order to be able to test for $\mathcal{T}$, we study a slightly more general problem: given a graph G, and three vertices v_1, v_2, v_3 of G, does there exist an induced subgraph T of G, such that T is a tree and $v_1, v_2, v_3 \in V(T)$? We call this the *three-in-a-tree* problem. It turns out that the answer to this question is "no" if and only if the graph admits a certain structure. This fact allows us to design a polynomial time algorithm for the three-in-a-tree problem. Now, if $\{v_1, v_2, v_3\}$ is a triad with a common neighbour w in G, the degree of each of v_1, v_2, v_3 in $G \setminus \{w\}$ is one, and the degree of w in G is three, then the answer to the three-in-a-tree problem with input $(G \setminus \{w\}, v_1, v_2, v_3)$ is "yes" if and only if G contains a theta using v_1, v_2, v_3, w. On the other hand, if $\{v_1, v_2, v_3\}$ is a triangle and no vertex of G has two neighbours in it, then the answer to the three-in-a-tree problem with input (G, v_1, v_2, v_3) is "yes" if and only if G contains a pyramid with base $\{v_1, v_2, v_3\}$. Thus, our algorithm to solve the "three-in-a-tree" problem can be used, after some pre-processing, to test both for $\mathcal{P}$ and for $\mathcal{T}$ (and this is the only algorithm known to test for $\mathcal{T}$). This result is particularly pleasing from our point of view, because this is one of the few times that a composition theorem and an algorithm appear together in the study of graphs with forbidden induced subgraphs.

8 Acknowledgments

This research was conducted during the period the first author served as a Clay Mathematics Institute Research Fellow. The second author is supported by ONR grant N00014-01-1-0608 and NSF grant DMS-0070912.

References

[1] L. Addario-Berry, M. Chudnovsky, F. Havet, B. Reed, P. Seymour, Bisimplicial vertices in even-hole-free graphs, *manuscript*, 2006.

[2] C. Berge, Färbung von Graphen, deren sämtliche bzw. deren ungerade Kreisestarr sind, *Wiss. Z. Martin-Luther-Univ. Halle-Wittenberg Math.-Natur. Reihe* **10** (1961), 114.

[3] M. Chudnovsky, Berge trigraphs, *J. Graph Theory* **53** (2006), 1-55.

[4] M. Chudnovsky, The structure of bull-free graphs, *in preparation*.

[5] M. Chudnovsky, G. Cornuéjols, X. Liu, P. Seymour, K. Vušković, Recognizing Berge graphs, *Combinatorica* **25** (2005), 143-187.

[6] M. Chudnovsky and R. Kapadia, Detecting a theta or a prism, *in preparation*.

[7] M. Chudnovsky, K. Kawarabayashi, P. Seymour, Detecting even holes, *J. Graph Theory* **48** (2005), 85-111.

[8] M. Chudnovsky and A. Ovetsky, Colouring quasi-line graphs, *J. Graph Theory* to appear.

[9] M. Chudnovsky, N. Robertson, P. Seymour, R. Thomas, The strong perfect graph theorem, *Annals of Mathematics* **164** (2006), 51-229.

[10] M. Chudnovsky and S. Safra, The Erdős-Hajnal conjecture for bull-free graphs, *in preparation*.

[11] M. Chudnovsky and P. Seymour, Claw-free graphs I. Orientable prismatic graphs, *manuscript*, 2005.

[12] M. Chudnovsky and P. Seymour, Claw-free graphs II. Non-orientable prismatic graphs, *manuscript*, 2005.

[13] M. Chudnovsky and P. Seymour, Claw-free graphs III. Circular interval graphs, *manuscript*, 2005.

[14] M. Chudnovsky and P. Seymour, Claw-free graphs IV. Decomposition theorem, *manuscript*, 2005.

[15] M. Chudnovsky and P. Seymour, Claw-free graphs V. Global structure, *manuscript*, 2006.

[16] M. Chudnovsky and P. Seymour, Claw-free graphs VI. The structure of quasi-line graphs, *in preparation*.

[17] M. Chudnovsky and P. Seymour, Claw-free graphs VII. Colouring, *in preparation*.

[18] M. Chudnovsky and P. Seymour, Testing for a theta, *in preparation*.

[19] V. Chvátal, Star-cutsets and perfect graphs, *J. Combinatorial Theory, Ser. B* **39** (1985), 189-199.

[20] V. Chvátal and N. Sbihi, Bull-free Berge graphs are perfect, *Graphs and Combinatorics* **3** (1987), 127-139.

[21] M. Conforti, G. Cornuéjols, A. Kapoor, K. Vušković, Even-hole-free graphs, Part I: Decomposition theorem, *J. Graph Theory* **39** (2002), 6-49.

[22] M. Conforti, G. Cornuéjols, A. Kapoor, K. Vušković, Even-hole-free graphs, Part II: Recognition algorithm, *J. Graph Theory* **40** (2002), 238-266.

[23] M. Conforti, G. Cornuéjols, K. Vušković, Decomposition of odd-hole-free graphs by double star cutsets and 2-joins, *Discrete Applied Mathematics* **141** (2004), 41-91.

[24] M. Conforti and M.R. Rao, Testing balancedness and perfection of linear matrices, *Mathematical Programming* **61** (1993), 1-18.

[25] G. Cornuéjols and W.H. Cunningham, Compositions for perfect graphs, *Discrete Mathematics* **55** (1985), 245-254.

[26] J. Edmonds, Maximum matching and a polytope with $0, 1$-vertices, *J. Res. Nat. Bur. Standards* **69B** (1965), 125–130.

[27] P. Erdős and A. Hajnal, Ramsey-type theorems, *Discrete Applied Mathematics* **25** (1989), 37-52.

[28] D. König, Über Graphen und ihre Anwendung auf Determinantentheorie und Mengenlehre, *Math. Ann.* **77** (1916), 453-465.

[29] L. Lovász, A characterization of perfect graphs, *J. Combinatorial Theory, Ser. B* **13** (1972), 95-98.

[30] G.J. Minty, On maximal independent sets of vertices in claw-free graphs, *Journal of Combinatorial Theory, Ser. B* **28** (1980), 284–304.

[31] F. Maffray and N. Trotignon, Algorithms for perfectly contractile graphs, *SIAM J. Discrete Math* **19(3)** (2005), 553-574.

[32] J. Ramirez-Alfonsin and B. Reed (eds.), Perfect Graphs, *Wiley, Chichester* (2001), 130.

Maria Chudnovsky
Department of IEOR
Columbia University
New York, NY 10027
USA
mchudnov@columbia.edu

Paul Seymour
Department of Mathematics
Princeton University
Princeton, NJ 08544
USA
pds@math.princeton.edu

Designs and Topology

M.J. Grannell and T.S. Griggs

Abstract

An embedding of a graph in a surface gives rise to a combinatorial design whose blocks correspond to the faces of the embedding. Particularly interesting graphs include complete and complete multipartite graphs. Embeddings of these in which the faces are triangles, Hamiltonian cycles, or Eulerian cycles generate interesting designs. These designs include twofold, Mendelsohn and Steiner triple systems, and Latin squares. We examine some of these cases, looking at construction methods, structural properties and enumeration problems.

1 Context

Throughout this survey we will be predominantly concerned with triangular embeddings of graphs. These arise naturally in the context of the Heawood map-colouring conjecture. In its orientable form this asserts that the minimum number of colours required to colour a map on a surface S_g, the sphere with g handles, is given by

$$\chi(S_g) = \left\lfloor \frac{7 + \sqrt{1 + 48g}}{2} \right\rfloor, \quad g \geq 0.$$

For $g > 0$, the conjecture was finally established by Ringel, Youngs and others in 1968. The case $g = 0$ is the celebrated four colour theorem, finally established by Appel and Haken [7, 8] in 1976.

To see the connection between the Heawood conjecture and triangular embeddings, consider the dual problem obtained by placing a vertex in each region of the map and joining two vertices whenever the corresponding regions share a common border. We now require the minimum number of colours to vertex colour the resulting dual graph. The extremal case is the complete graph K_n requiring n colours. So it is natural to ask for the minimum genus g such that K_n may be embedded in S_g. Using Euler's formula $n + f - e = 2 - 2g$, where f denotes the number of faces and $e = \binom{n}{2}$ is the number of edges, we see that g is minimal when f is maximal and this will happen when the average number of edges per face is as small as possible. Euler's formula then gives $\lceil (n-3)(n-4)/12 \rceil$ as a lower bound for the genus. For $n \equiv 0$, 3, 4 or 7 (mod 12), this is achievable by taking all of the faces as triangles. When n does not lie in one of these congruence classes it is also achievable but a small number of non-triangular faces are required. The book by Ringel [78] gives the details and also deals with the nonorientable case of embedding K_n in N_γ, the sphere with γ crosscaps. In the nonorientable case Euler's formula is $n + f - e = 2 - \gamma$ and a lower bound for the minimum genus is $\lceil (n-3)(n-4)/6 \rceil$. In the cases $n \equiv 0$ or 1 (mod 3) except $n = 7$ this is achievable with all the faces as triangles. The surface of minimum nonorientable genus in which K_7 can be embedded is N_3.

The connection between graph embeddings and combinatorial designs arises from the observation that, when a graph is embedded in a surface, the faces that result can be regarded as the blocks of a design. This design may be thought of as being embedded in the surface. The first person to observe the connection between combinatorial designs and graph embeddings was Heffter. In a paper dated November 1890 [57] he presents a partition of the integers $1, 2, \ldots, 12s + 6$, $s \geq 0$ into $4s + 2$ triples so that for each triple $\{a, b, c\}$, $a + b + c \equiv 0 \pmod{12s + 7}$. He then shows how, if $4s+3$ is prime and the order of 2 modulo $4s+3$ is either $4s+2$ or $2s+1$, these triples can be used to construct a twofold triple system (for the formal definition see Section 2) of order $12s + 7$, the blocks of which are the triangular faces of an embedding of the complete graph K_{12s+7} in an orientable surface. As observed in both [52] and [78] it is still not known if there are infinitely many such values of s. But the method is applicable for $s = 0, 1, 2, 4, 5, 11$ and 14, numbers given explicitly in [57].

The only other paper published before 1970 which explores the relationship between combinatorial designs and graph embeddings appears to be that by Emch [36]. Although mainly combinatorial in nature, it does contain diagrams of the embeddings of the twofold triple system of order 6 in the projective plane, the embedding of a pair of Steiner triple systems of order 7 in the torus, as well as an interesting embedding of a pair of Steiner triple systems of order 9 in a pseudosurface formed from a torus by identifying three pairs of points. We will meet all of these embeddings later in the paper; see Figures 6.1, 6.2 and 12.1 respectively.

2 Preliminaries

In this section we review terminology taken from combinatorial design theory and topological graph theory, and we summarize some of the basic results. The principal item required from design theory is the following definition. A *Steiner triple system of order n* is a pair $(V, \mathcal{B})$ where V is an n-element set (the *points*) and $\mathcal{B}$ is a collection of 3-element subsets (the *blocks*) of V such that each 2-element subset of V is contained in exactly one block of $\mathcal{B}$. It is well known that a Steiner triple system of order n (briefly STS(n)) exists if and only if $n \equiv 1$ or 3 (mod 6) [62]. If, in the definition, the words "exactly one block" are replaced by "exactly two blocks", then we have a *twofold triple system of order n*, TTS(n) for short. If a TTS(n) has no repeated blocks, it is said to be *simple*. A simple twofold triple system of order n exists if and only if $n \equiv 0$ or 1 (mod 3) [28]. A (possibly non-simple) TTS(n) may be obtained by combining the block sets of two STS(n)s which have a common point set. An STS(n) can be considered as a decomposition of the complete graph K_n into triangles (copies of K_3); likewise a TTS(n) can be considered as a decomposition of the twofold complete graph $2K_n$ (in which there are two edges between each pair of vertices) into triangles.

Up to isomorphism, there is just one STS(n) for $n = 3, 7, 9$, while there are two for $n = 13$, precisely one of which is cyclic (that is, has an automorphism of order 13). There are 80 STS(15)s [27], of which two are cyclic, and there are 11,084,874,829 STS(19)s [60], of which four are cyclic. The number of nonisomorphic STS(n)s is $n^{n^2/6 - o(n^2)}$ [83] and, speaking asymptotically, almost all of these have only a trivial automorphism group [9].

A *Mendelsohn triple system of order n* is defined in a similar fashion to an STS(n) except that triples and pairs are taken to be ordered, so that the cyclically ordered triple (a, b, c) "contains" the ordered pairs $(a, b), (b, c)$ and (c, a). A Mendelsohn triple system of order n, MTS(n) for short, exists if and only if $n \equiv 0$ or 1 (mod 3) and $n \neq 6$ [73]. An MTS(n) may be considered as a decomposition of the complete directed graph on n vertices into directed 3-cycles. If the directions are ignored, then an MTS(n) gives a TTS(n).

A *transversal design of order n and block size* 3 is a triple $(V, \mathcal{G}, \mathcal{B})$ where V is a $3n$-element set (the *points*), $\mathcal{G}$ is a partition of V into 3 parts (the *groups*) each of cardinality n, and $\mathcal{B}$ is a collection of 3-element subsets (the *blocks*) of V such that each 2-element subset of V is either contained in exactly one block of $\mathcal{B}$ or in exactly one group of $\mathcal{G}$, but not both. A transversal design of order n and block size 3 is denoted by TD$(3, n)$; since we only consider block size 3, we will simply speak of a transversal design of order n. A TD$(3, n)$ may be considered as a decomposition of the complete tripartite graph $K_{n,n,n}$ into triangles with the tripartition defining the groups of the design. A TD$(3, n)$ is equivalent to a Latin square of side n in which the triples are given by (row, column, entry).

To see the connection between design theory and graph embeddings, consider the case of an embedding of the complete graph K_n in an orientable surface in which all the faces are triangles. Taking these triangles with a consistent orientation to form a set of blocks, the faces of the embedding yield a Mendelsohn triple system of order n. Similarly, a triangular embedding of K_n in a nonorientable surface gives a twofold triple system of order n.

We note here that all the surfaces we consider will be, unless otherwise stated, closed, connected 2-manifolds, without a boundary. That is, in the orientable case, S_g the sphere with g handles and, in the nonorientable case, N_γ the sphere with γ crosscaps. The surfaces S_1 and S_2 are the *torus* and *double torus* respectively and the surfaces N_1 and N_2 are the *projective plane* and *Klein bottle* respectively. Given a surface embedding of some simple graph G with vertex set $V(G)$, the *rotation* at a vertex $v \in V(G)$ is the cyclically ordered permutation of vertices adjacent to v, with the ordering determined by the embedding. The set of rotations at all the vertices of G is called the *rotation scheme* for the embedding. In the case of an embedding of G in an orientable surface, the rotation scheme provides a complete description of the embedding. This is not generally the case for a nonorientable surface because the rotation scheme does not enable the faces of the embedding to be unambiguously reconstructed: some additional information is required. However, in the cases we consider this will not be an issue, since sufficient extra information to determine the faces will be known.

Ringel [78] gives the following tests to determine if a rotation scheme represents a triangular embedding.

Rule Δ: A rotation scheme represents a triangular embedding of a simple graph G if, for each vertex $a \in V(G)$, whenever the rotation at a contains the sequence $\ldots bc \ldots$, then the rotation at b contains either the sequence $\ldots ac \ldots$ or the sequence $\ldots ca \ldots$.

Rule Δ^*:   If the rotations at each vertex can be directed in such a way that for each vertex $a \in V(G)$, whenever the rotation at a contains the sequence $\ldots bc \ldots$, then the rotation at b contains the sequence $\ldots ca \ldots$, then the embedding is in an orientable surface.

We refer the reader to [52, 78] for an explanation of current and voltage graphs which are used to construct graph embeddings. In Sections 3, 4, 5 and 10 we make extensive use of these methods. The origin of current graphs lies in the work of Gustin [53] who regarded these as combinatorial tools. Voltage graphs were introduced by Gross [51].

In a surface embedding of K_n, the rotation at each vertex will comprise a single cycle of length $n - 1$. As described in [31] this provides a test for an MTS(n), or a TTS(n), to be embeddable in an orientable, or a nonorientable surface, respectively. Let $(V, \mathcal{B})$ be a TTS(n). For each $x \in V$, define the *neighbourhood graph* G_x: its vertex set is $V \setminus \{x\}$, and two vertices y, z are joined by an edge if $\{x, y, z\} \in \mathcal{B}$. Clearly, G_x is a union of disjoint cycles. A TTS(n) occurs as a triangulation of a surface if and only if every neighbourhood graph consists of a single cycle. If the blocks of the TTS(n) can be ordered to form an MTS(n), then the surface is orientable, otherwise it is nonorientable.

Of much more interest is the relationship between embeddings of complete graphs and Steiner triple systems. Suppose that we have an embedding, not necessarily a triangular embedding, of the complete graph K_n with vertex set V in a surface S with the property that the faces can be properly 2-coloured, that is, no two faces with a common edge have the same colour. We will take the colour classes to be *black* and *white*. If either colour class consists entirely of triangles, then these triangles necessarily form the blocks of an STS(n) on the point set V. We will say that the STS(n) is *embedded* in the surface S. If both colour classes consist entirely of triangles, then we have two STS(n)s , black and white, *biembedded* in S. Slightly more generally, we will say that two STS(n)s, say B and W, are *biembeddable* in a surface S if there is a face 2-colourable triangular embedding of the complete graph K_n in S with the black (respectively white) faces forming a system isomorphic with B (respectively W).

The first obvious question is whether, given an STS(n), it has an embedding in an orientable and in a nonorientable surface. It turns out that this question has a positive answer, and the proof is not difficult. We will show in Section 8 how to construct a maximum genus embedding of an STS(n) where the faces comprise a set of black triangles representing the Steiner system, together with a single white face.

A sequence of deeper questions concerns biembeddings of STS(n)s, that is, face 2-colourable triangular embeddings of K_n. We list these in increasing order of difficulty.

1. For each $n \equiv 3$ or $7 \pmod{12}$ is there a biembedding of some pair of STS(n)s in an orientable surface? Similarly for each $n \equiv 1$ or $3 \pmod{6}$ is there such a biembedding in a nonorientable surface?

2. If such biembeddings exist, how many are there?

3. Given an STS(n), does it have a biembedding with some other STS(n) in an orientable and in a nonorientable surface?

4. Given a pair of STS(n)s, do they have a biembedding in an orientable and in a nonorientable surface?

Of course, a necessary condition for a positive answer to questions 3 and 4 in the orientable case is that $n \equiv 3$ or $7 \pmod{12}$. A complete answer to question 1 is given in Section 3. In subsequent sections, principally Sections 4, 5 and 6, we describe progress with questions 2, 3 and 4. The remaining sections are devoted to other related aspects such as Hamiltonian embeddings, biembeddings of Latin squares, and biembeddings of symmetric configurations.

3   Existence

In this section we establish the existence of biembeddings of STS(n)s. The orientable case $n \equiv 3 \pmod{12}$ and the nonorientable case $n \equiv 3 \pmod 6$ come from Ringel [78]. For the orientable case $n \equiv 7 \pmod{12}$ we turn to graphs first constructed by Youngs [86]. In each of these cases we present the general solution either by specifying appropriate current graphs or by giving the logs obtained from such graphs. For the nonorientable case $n \equiv 1 \pmod 6$ we refer the reader to [49] which gives explicit current graphs. In both the orientable case $n \equiv 3 \pmod{12}$ and the nonorientable case $n \equiv 3 \pmod 6$, we relate these solutions to the Bose construction for Steiner triple systems.

We first consider the orientable case $n \equiv 3 \pmod{12}$. The current graphs constructed by Ringel for this case are index 3 Möbius ladders, and the general form is shown in Figure 3.1. The ends labelled A should be identified, and likewise the ends labelled B. The graph is bipartite, which ensures that the resulting embedding is face 2-colourable. The vertex directions are indicated by solid and hollow circles, representing clockwise and anticlockwise respectively. Taking account of these directions we form the *logs* of the three circuits denoted by [0], [1] and [2] in the figure.

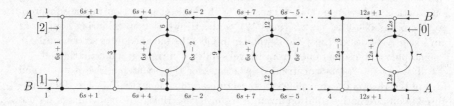

Figure 3.1: Orientable current graph for $n = 12s + 3$.

For the particular case $n = 15$, the logs are as follows.

[0] :	1	13	9	11	5	12	7	14	2	6	4	10	3	8
[1] :	14	7	8	5	9	4	10	6	11	2	3	1	13	12
[2] :	1	8	7	10	6	11	5	9	4	13	12	14	2	3

From an index 3 current graph with currents in $\mathbb{Z}_n$, we may obtain a rotation scheme for an embedding of K_n with vertex set $\mathbb{Z}_n$. The rotation at $i \in \mathbb{Z}_n$ is determined by adding i modulo n to each element of the log of $[a]$, where $i \equiv a \pmod 3$, $a \in \{0, 1, 2\}$.

An alternative approach to obtaining biembeddings of STS(n)s, where $n \equiv 3 \pmod{12}$ in orientable surfaces is given in [45] and uses the Bose construction.

Bose construction

Let $(G, +)$ be an Abelian group of odd order. Thus if $i, j \in G$ then $i * j = (i + j)/2$ is a well defined element of G. Let $V = G \times \mathbb{Z}_3$. On V form a collection $\mathcal{B}$ of triples as follows.

(1) $2s + 1$ triples of the form $\{(i, 0), (i, 1), (i, 2)\}$, $i \in G$,

(2) $3s(2s+1)$ triples of the form $\{(i, k), (j, k), (i*j, k+1)\}$, $i, j \in G$, $i \neq j$, $k \in \mathbb{Z}_3$.

Then it is easily verified that $(V, \mathcal{B})$ is a Steiner triple system of order $3|G|$.

A biembedding of STS(n)s where $n \equiv 3 \pmod{12}$ can now be obtained as follows. Build a Steiner triple system $(V, \mathcal{B})$, where $V = \mathbb{Z}_{4s+1} \times \mathbb{Z}_3$, by the Bose construction as above. Now define two Steiner triple systems $(\mathbb{Z}_n, \mathcal{B}_0)$ and $(\mathbb{Z}_n, \mathcal{B}_1)$, both isomorphic to $(V, \mathcal{B})$ using the bijections $f_m : V \mapsto \mathbb{Z}_n$, $m = 0, 1$ given by $f_m(i, k) = 3i + (-1)^m ks$ where $s = 6t + 1$. It is easy to prove that $\mathcal{B}_0 \cap \mathcal{B}_1 = \emptyset$, that is, the two STS($n$)s are disjoint. To show that the pair is biembeddable in an orientable surface, consider the triples in $\mathcal{B}_0$ (respectively $\mathcal{B}_1$) as the black (respectively white) triangles of a biembedding. For each pair of distinct points $u, v \in \mathbb{Z}_n$, we take the corresponding black and white triangles, both containing u and v as vertices, and glue these triangles together along the side uv. Let S be the resulting topological space; then S is certainly a generalized pseudosurface. We need to prove that, in fact, S is an orientable surface. This is done by exhibiting the rotation scheme and showing that it satisfies Ringel's Rule Δ^*. This is straightforward, though tedious, and details are given in [45]. Thus, use of the Bose construction provides a proof of the orientable case $n \equiv 3 \pmod{12}$ of the Heawood map-colouring conjecture by exclusively design-theoretic methods. In fact, the biembeddings so obtained are isomorphic to those obtained from Ringel's index 3 current graph construction.

The current graphs constructed by Ringel for the orientable case $n \equiv 7 \pmod{12}$ of the Heawood map-colouring conjecture are not bipartite. Nor are the graphs used in an alternative solution given by Youngs [85]. Hence the embeddings are not face 2-colourable and are consequently not biembeddings of Steiner triple systems. As recorded in Section 1, Heffter [57] had already in 1891 shown the existence of orientable biembeddings of STS(n)s for some $n \equiv 7 \pmod{12}$ but the case was not completed until nearly 80 years later. In [86] Youngs uses what he calls "zigzag diagrams" to construct index 1 bipartite current graphs, and hence biembeddings of Steiner triple systems for this case. In this context, index 1 means that there is a single circuit of the graph which traverses every edge precisely once in each direction and whose log contains every nonzero element precisely once. Hence for each $i \in \{1, 2, \ldots, (n-1)/2\}$ either i or $-i$ must appear as a current on one of the edges and each edge has exactly one of these $(n-1)/2$ currents. The biembeddings thus constructed are cyclic. The general forms of these current graphs are shown in Figures 3.2 and 3.3 for $n = 24m + 7$, $m \geq 2$, and $n = 24m + 19$, $m \geq 3$ respectively. In each case the ends labelled A should be identified, and likewise the ends labelled B. For the values $n = 7, 19, 31, 43$ and 67, Youngs gives specific diagrams.

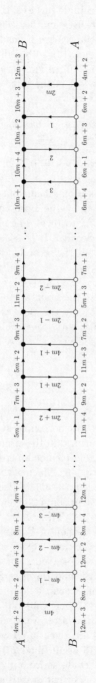

Figure 3.2: Orientable current graph for $n = 24m + 7$, $m \geq 2$.

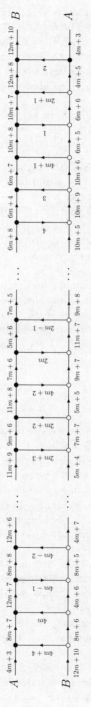

Figure 3.3: Orientable current graph for $n = 24m + 19$, $m \geq 3$.

Turning now to biembeddings of STS(n)s in nonorientable surfaces, the case $n \equiv$ 9 (mod 12) can also be found in [78]. The solution involves another class of index 3 current graphs which Ringel calls "cascades", and the remark is made that the method also works for the nonorientable case $n \equiv 3$ (mod 12), although no details are given. These were later worked out and are given in [10]. A simpler description is the following where, as above, [0], [1] and [2] are the logs of the three circuits.

$$
\begin{array}{llllllll}
[0]: & 1 & 2 & [24t+12 & 12t+8 & 24t+24 & 12t+14] & [-(6t+2) & 6t+4] \\
[1]: & 1 & -1 & [-(12t+6) & -(6t+4) & -(12t+12) & -(6t+7)] & [-(6t+2) & 6t+4] \\
[2]: & -2 & -1 & [-(12t+6) & -(6t+4) & -(12t+12) & -(6t+7)] & [12t+4 & -(12t+8)].
\end{array}
$$

Here the terms inside the square brackets are repeated for $t = 0, 1, \ldots, 2s - 1$ in the case of $n = 12s + 3$ and for $t = 0, 1, \ldots, 2s$ in the case of $n = 12s + 9$, with arithmetic in each case modulo n. In both cases the rotation scheme obtained gives two isomorphic Steiner triple systems again generated by the Bose construction with the group $G = \mathbb{Z}_n$. To see this, map each $i \in \mathbb{Z}_n$ to (a, b) where $a = \lfloor i/3 \rfloor$ and $b = i - 3a$. One of the two STS(n)s is then very clearly a Bose system and by applying the mapping $f((a, b)) = (a + b, b)$ it is seen that the second system is also a Bose system.

An alternative proof from the Bose construction for $n \equiv 3$ (mod 6) is given in [30] and is very similar to the construction given above for the orientable case. Build a Steiner triple system $(V, \mathcal{B})$, where $V = \mathbb{Z}_{2s+1} \times \mathbb{Z}_3$, by the Bose construction and define two Steiner triple systems $(V, \mathcal{B}_0)$ and $(V, \mathcal{B}_1)$, both isomorphic to $(V, \mathcal{B})$, using the bijections $f_m : V \mapsto V$, $m = 0, 1$, defined as follows.

$$
\begin{aligned}
f_m((i, 0)) &= (i, 0) \\
f_m((i, 1)) &= (i + m, 1) \\
f_m((i, 2)) &= (i - m + 2s, 2).
\end{aligned}
$$

Verification that this gives a biembedding of the two Steiner triple systems in a nonorientable surface follows the same procedure as outlined in the orientable case.

Perhaps surprisingly, the existence of a nonorientable biembedding of STS(n)s for $n \equiv 1$ (mod 6) was not established until fairly recently [49]. Much of the spectrum can be obtained from recursive constructions given in [19, 44, 46]. The cases $n \equiv 7$ or 25 (mod 36), $n \neq 7$, follow immediately from Construction 4.2 given in Section 4 and the known biembeddings for $n \equiv 3$ or 9 (mod 12). The case $n \equiv 13$ (mod 36) is more complex but comes from a nonorientable version of Construction 4 of [46] using a face 2-colourable triangular embedding of the complete tripartite graph $K_{6,6,6}$ having a parallel class in one of the colour classes, see [40], a nonorientable face 2-colourable triangular embedding of K_{13}, see [38], and the $n \equiv 3$ (mod 12) case. The case $n \equiv 1$ (mod 36) then follows from the Construction 4.2 using an inductive argument. This leaves the cases $n \equiv 19$ or 31 (mod 36) but the former would follow immediately if a method for dealing with the latter was known. But in [49] direct constructions using index 1 current graphs are given in all cases. There are four general subcases corresponding to $n \equiv 1, 7, 13$ or 19 (mod 24), as well as a number of particular cases. Limitations of space preclude us from giving details. We refer the reader to the original paper where all the current graphs are given in the same format as in this paper.

4 Growth estimates

We present two main recursive constructions. These have a degree of flexibility that enable us to obtain a lower bound on the number of biembeddings of STS(n)s for values of n lying in certain residue classes. Our first construction is new and produces biembeddings of STS($3n$)s from a biembedding of STS(n)s.

Construction 4.1

Take any biembedding of STS(n)s in either an orientable or a nonorientable surface. Pick a preferred point ∞ of these designs. Define *the cap at ∞* to comprise all the triangles, both black and white, incident with ∞ in the embedding. Next pick a preferred white triangle T incident with ∞. We distinguish three categories of white triangles:

  (i) those not on the cap at ∞,

 (ii) those on the cap at ∞ other than the preferred triangle T,

(iii) the triangle T.

Next take three copies of the given biembedding on three disjoint surfaces S^0, S^1 and S^2. We use superscripts in a similar way to identify corresponding points on these surfaces.

For each white triangular face (uvw) of type (i), we "bridge" S^0, S^1 and S^2 by gluing a torus to the three triangles $T^i = (u^i v^i w^i)$ in the following manner. Take a face 2-colourable triangular embedding in a torus of the complete tripartite graph $K_{3,3,3}$ having three vertex parts $\{u^i\}, \{v^i\}$ and $\{w^i\}$ and having black faces $(u^i w^i v^i)$ for $i = 0, 1, 2$ (see Figure 4.1). We use the same labels for the vertices of this graph as we do for the vertices of the three triangles T^i, but initially think of them as distinct points. (A similar gloss will be used on several occasions.) Now glue the black faces $(u^i w^i v^i)$ on the torus to the white faces $(u^i v^i w^i)$ on S^0, S^1 and S^2 respectively, so that points on the torus and on the surfaces S^i with the same label are identified.

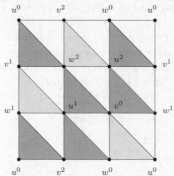

Figure 4.1: Toroidal embedding of $K_{3,3,3}$.

For each white triangular face $(uv\infty)$ of type (ii), we carry out a similar bridging operation but using a different type of bridge. For this we take a face 2-colourable triangular embedding of the graph $K_9 - K_3$ in the nonorientable surface N_4, defined by the following rotation scheme, where the colouration is determined by taking each $(u^i v^i \infty^i)$ as a black triangle.

$$
\begin{array}{llllllll}
\infty^0 & : & u^0 & v^0 & v^2 & u^1 & u^2 & v^1 \\
\infty^1 & : & u^1 & v^1 & v^0 & u^2 & u^0 & v^2 \\
\infty^2 & : & u^2 & v^2 & u^0 & v^1 & u^1 & v^0 \\
u^0 & : & u^1 & u^2 & \infty^1 & v^2 & \infty^2 & v^1 & \infty^0 & v^0 \\
u^1 & : & u^0 & u^2 & \infty^0 & v^2 & \infty^1 & v^1 & \infty^2 & v^0 \\
u^2 & : & u^0 & u^1 & \infty^0 & v^1 & v^2 & \infty^2 & v^0 & \infty^1 \\
v^0 & : & v^1 & v^2 & \infty^0 & u^0 & u^1 & \infty^2 & u^2 & \infty^1 \\
v^1 & : & v^0 & v^2 & u^2 & \infty^0 & u^0 & \infty^2 & u^1 & \infty^1 \\
v^2 & : & v^0 & v^1 & u^2 & \infty^2 & u^0 & \infty^1 & u^1 & \infty^0
\end{array}
$$

Table 4.1: N_4 embedding of $K_9 - K_3$.

We glue these bridges to S^0, S^1 and S^2 as before. Note that none of these bridges contain any edge $\infty^i \infty^j$.

To complete the construction, we construct a single bridge to join the three copies of the type (iii) triangle T. For this bridge we take a face 2-colourable triangular embedding of K_9 in the nonorientable surface N_5. Such an embedding, a biembedding of STS(9)s, is given in Section 3 and we can label the vertices so that the black faces include the triangles $(v^i u^i \infty^i)$ for $i = 0, 1, 2$. As before, we glue the white triangle $T^i = (u^i v^i \infty^i)$ on S^i to the black triangle $(v^i u^i \infty^i)$ on the bridge. Note that this bridge contains the three edges $\infty^i \infty^j$.

It is now routine to check that the resulting embedding represents a biembedding of two STS($3n$)s in a nonorientable surface. $\square$

We now make some observations about the construction that enable us to extend it. Firstly, the toroidal embedding of $K_{3,3,3}$ given in Figure 4.1 may be replaced by one in which the cyclic order of the three superscripts is reversed. The reversed embedding of $K_{3,3,3}$ is isomorphic to the original but is labelled differently (see Figure 4.2). For each white triangular face (uvw) of S we may carry out the bridging operation across S^0, S^1, S^2 using either the original $K_{3,3,3}$ embedding or the reversed embedding. The choice of which of the two to use can be made <u>independently</u> for each white triangle (uvw). Replacing one bridge by the reversed bridge is an example of a *surface trade*; these are discussed more generally in Section 7.

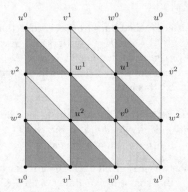

Figure 4.2: Reversed toroidal embedding of $K_{3,3,3}$.

As a consequence of this observation, we have the following result.

Theorem 4.1 *For $n \equiv 3$ or 9 (mod 18), there are at least $2^{n^2/54-o(n^2)}$ nonisomorphic face 2-colourable triangular embeddings of the complete graph K_n, and hence biembeddings of $STS(n)$s, in a nonorientable surface.*

Proof Take three fixed copies of the same face 2-colourable triangular embedding of K_m, $m \equiv 1$ or 3 (mod 6), and apply Construction 4.1 while varying the toroidal bridges. Since there are $(m-1)(m-3)/6$ toroidal bridges and two choices for each bridge, we may construct $2^{(m-1)(m-3)/6}$ differently labelled face 2-colourable embeddings of K_{3m}. The maximum possible size of an automorphism class of these is $(3m)!$. Hence there are at least $2^{m^2/6-o(m^2)}$ nonisomorphic face 2-colourable triangular embeddings of K_{3m}, and replacing $3m$ by n gives the result.   □

Our second observation about the construction is that it is not necessary for S^0, S^1 and S^2 to contain three copies of the same embedding of K_n. All that the construction requires is that the three embeddings have the "same" white triangular faces. To be more precise, by the term "same" we mean that there are mappings from the vertices of each surface onto the vertices of each of the other surfaces that preserve the white triangular faces. The sceptical reader may feel dubious that we can satisfy this requirement without in fact having three identically labelled copies of a single embedding. However, if we examine the black triangles of the embeddings generated as described in Theorem 4.1, we will see that it is indeed possible. We claim that in any two such embeddings, the black triangles are identical. To see this, note that the black triangles come from four sources, the original surfaces and the three types of bridges. Those lying on the surfaces S^0, S^1 and S^2 are unaltered during the construction and therefore are common to both embeddings. Those lying on the $K_{3,3,3}$ bridges are the same whether or not the bridges are reversed (see Figures 4.1 and 4.2). Those lying on N_4 bridges and on the N_5 bridge are common to both embeddings. It follows that the $2^{n^2/54-o(n^2)}$ nonisomorphic embeddings of K_n generated by Theorem 4.1 all contain identical black triangles. In each of these embeddings, by reversing the colours, we produce a plentiful supply of nonisomorphic embeddings in surfaces S^i on which to base a reapplication of the construction. All of these embeddings of K_n have the "same" white triangles.

We can select three surface embeddings from this collection to form S^0, S^1, S^2 in N^3 ways, where $N = 2^{\bar{n}^2/54 - o(n^2)}$. The $K_{3,3,3}$ bridges may be selected in $2^{(n-1)(n-3)/6}$ different ways. Any two of the resulting embeddings of K_{3n} (obtained by varying the surfaces S^0, S^1 and S^2, and the $K_{3,3,3}$ bridges) will be differently labelled. These results lead easily to the next theorem.

Theorem 4.2 *For* $n \equiv 9$ *or* 27 *(mod 54), there are at least* $2^{2n^2/81 - o(n^2)}$ *nonisomorphic face 2-colourable triangular embeddings of the complete graph* K_n, *and hence biembeddings of* $STS(n)s$, *in a nonorientable surface.*

Our second construction was first given by Širáň and ourselves in [44]. It produces biembeddings of $STS(3n-2)$s from a biembedding of $STS(n)$s. It uses many of the same ingredients as Construction 4.1, and we will be brief in our description of these common features. However, unlike Construction 4.1, this second construction can be used to produce both orientable and nonorientable biembeddings.

Construction 4.2

Take any biembedding of $STS(n)$s in either an orientable or a nonorientable surface S. Pick a preferred point ∞ and define the cap at ∞ as before. Delete this cap from S by removing the point ∞, all (open) edges incident with ∞ and all (open) triangular faces incident with ∞ to give an embedding of K_{n-1} in a surface S^* with a boundary $D = (u_1 u_2 \ldots u_{n-1})$. Each alternate edge of this Hamiltonian cycle is incident with a white triangle in S^*; suppose that these edges are $u_2 u_3, u_4 u_5, \ldots, u_{n-1} u_1$. Next take three copies of this embedding in three disjoint surfaces S^{*i}, $i = 0, 1, 2$, each with a boundary $D^i = (u_1^i u_2^i \ldots u_{n-1}^i)$. The white triangles on S^{*0}, S^{*1} and S^{*2} are bridged as before using toroidal face 2-colourable triangular embeddings of $K_{3,3,3}$.

After all the white triangles have been bridged we are left with a new connected triangulated surface with a boundary. We denote this surface by Σ. It has $3n-3$ vertices and the boundary comprises the three disjoint cycles D^i, each of length $n-1$. In order to complete the construction to obtain a face 2-colourable triangular embedding of K_{3n-2}, which gives a biembedding of two $STS(3n-2)$s, we must construct an auxiliary triangulated bordered surface T^* and paste it to Σ so that the three holes of Σ will be capped.

The bordered surface T^* is constructed from a surface T which has, as vertices, the points u_j^i for $i = 0, 1, 2$ and $j = 1, 2, \ldots, n-1$ together with one additional point which we call ∞^*. The construction of T uses voltage assignments. Suppose initially that $n \equiv 3 \pmod 6$.

Let ν be the plane embedding of the multigraph L with faces of length 1 and 3 coloured black and white, as depicted in Figure 4.3.

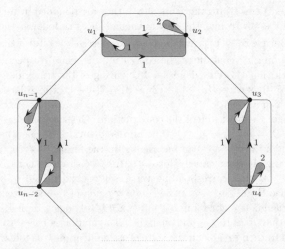

Figure 4.3: The plane embedding of the multigraph L.

Figure 4.3 also shows voltages α on directed edges of L, taken in the group $\mathbb{Z}_3 = \{0, 1, 2\}$. The edges with no direction assigned carry the zero voltage.

The lifted graph L^α has the vertex set $\{u^i_j;\ 1 \le j \le n-1,\ i \in \mathbb{Z}_3\}$. As before, we use the same letters for vertices of L^α as for vertices of our embedded graph in Σ, but we initially assume that these graphs are disjoint. The edge set of L^α can be described as follows. For each fixed $l = 1, 3, 5, \ldots, n-2$, the six vertices u^i_l, u^i_{l+1}, $i \in \mathbb{Z}_3$, induce a complete graph $J_l \simeq K_6$ in L^α. Moreover, two successive complete subgraphs J_l and J_{l+2} (indices mod $(n-1)$) are joined by three edges $u^i_{l+1}u^i_{l+2}$, $i \in \mathbb{Z}_3$. Thus we have a total of $15(n-1)/2 + 3(n-1)/2 = 9(n-1)$ edges in L^α, and there are neither loops nor multiple edges.

The lifted embedding $\nu^\alpha : L^\alpha \to T$ has $4(n-1)$ triangular faces: the white ones are bounded by the triangles $(u^0_l u^0_{l+1} u^2_{l+1})$, $(u^1_l u^1_{l+1} u^0_{l+1})$, $(u^2_l u^2_{l+1} u^1_{l+1})$ and $(u^0_l u^1_l u^2_l)$, where $l = 1, 3, 5, \ldots, n-2$, and the black ones are bounded by $(u^0_l u^1_{l-1} u^2_{l-1})$, $(u^1_l u^2_{l-1} u^0_{l-1})$, $(u^2_l u^0_{l-1} u^1_{l-1})$ and $(u^0_l u^2_l u^1_l)$, where $l = 2, 4, \ldots, n-1$. In addition, there are four more faces in the embedding ν^α; three faces, which we denote by F^i, bounded by $(n-1)$-gons of the form $(u^i_1 u^i_2 \ldots u^i_{n-1})$, $i \in \mathbb{Z}_3$, and one face F' bounded by the $(3n-3)$-gon $(u^0_1 u^1_2 u^1_3 u^1_4 \ldots u^2_{n-2} u^0_{n-1})$; here we use the fact that $n-1$ is coprime with 3. Thus the boundary of F' is a Hamiltonian cycle, say B, in L^α.

Now cut out from T the three (open) faces F^i, $i \in \mathbb{Z}_3$, bounded by the above three disjoint $(n-1)$-gons, obtaining thereby an orientable bordered surface T^*. Let L^* be the graph obtained from L^α by adding a new vertex ∞^* and joining it to each vertex of L^α, and keeping all edges in L^α unchanged. We construct an embedding $\nu^* : L^* \to T^*$ from ν^α in an obvious way: in the embedding ν^α (after the removal of the three open faces), we insert the vertex ∞^* in the centre of the face F' bounded by the $(3n-3)$-gon and join this point by open arcs within F' to every vertex on the boundary of F' (that is, with every vertex of the Hamiltonian cycle $\overline{B}$). Instead of F'

we now have $(3n-3)$ new triangular faces on T^*; they are bounded by 3-cycles of the form $\infty^* u_j^i u_{j+1}^{i'}$. We now colour the new triangular faces as follows: the face of ν^* bounded by the 3-cycle $\infty^* u_j^i u_{j+1}^{i'}$ will be black (respectively white) if the triangular face of the embedding ν^α containing the edge $u_j^i u_{j+1}^{i'}$ is white (respectively black). It is easy to check that this rule defines a 2-colouring of the triangular embedding $\nu^* : L^* \to T^*$. We thus have $4(n-1) + (3n-3) = 7(n-1)$ triangular faces on T^*, exactly half of which are black.

We are ready for the final step of the construction. The surface Σ has three holes with boundaries $D^i = (u_1^i u_2^i \ldots u_{n-1}^i)$. The bordered surface T^* has three holes as well, whose boundary cycles D^{*i} can be oriented in the form $D^{*i} = (u_{n-1}^i \ldots u_2^i u_1^i)$. It remains to do the obvious: namely, for $i = 0, 1, 2$ to paste together the boundary cycles D^i and D^{*i} so that corresponding vertices u_j^i get identified. As the result we obtain a surface $\Sigma \# T^*$, known as the *connected sum* of the bordered surfaces Σ and T^*, and a triangular embedding $\sigma : K \to \Sigma \# T^*$ of some graph K. It is then routine to check that $K \simeq K_{3n-2}$ and that the triangulation is face 2-colourable.

If $n \equiv 1 \pmod 6$ then we amend the voltage assignment on L as follows. We take one of the two-point subgraphs in Figure 4.3, say that containing u_1 and u_2, and replace the voltages 1 by 2 and vice versa, the remaining part of L being unaltered. The proof then proceeds on the same lines as before with the modified version of L. Note that this alteration ensures that the lifted embedding still has a $(3n-3)$-gon face even though $n-1$ is not coprime with 3. The order of the vertices around this face differs from that given previously, but it is still possible to insert a new vertex ∞^* and to complete a 2-colouring of the resulting triangular embedding. $\square$

In Construction 4.2, the surface T^* is orientable, as are the toroidal bridges. Hence, if the original biembedding of $STS(n)$s is orientable, then the resulting biembedding of $STS(3n-2)$s will be orientable. This is always possible for $n \equiv 3$ or $7 \pmod{12}$.

As with Construction 4.1, we may obtain growth estimates as given in [19] by Bonnington, Širáň and ourselves.

Theorem 4.3 *For $n \equiv 1$ or 7 (mod 18), there are at least $2^{n^2/54 - o(n^2)}$ nonisomorphic face 2-colourable triangular embeddings of the complete graph K_n, and hence biembeddings of $STS(n)$s, in a nonorientable surface.*

Theorem 4.4 *For $n \equiv 1$ or 19 (mod 54), there are at least $2^{2n^2/81 - o(n^2)}$ nonisomorphic face 2-colourable triangular embeddings of the complete graph K_n, and hence biembeddings of $STS(n)$s, in a nonorientable surface.*

By starting with orientable embeddings, we also obtain the following results.

Theorem 4.5 *For $n \equiv 7$ or 19 (mod 36), there are at least $2^{n^2/54 - o(n^2)}$ nonisomorphic face 2-colourable triangular embeddings of the complete graph K_n, and hence biembeddings of $STS(n)$s, in an orientable surface.*

Theorem 4.6 *For $n \equiv 19$ or 55 (mod 108), there are at least $2^{2n^2/81 - o(n^2)}$ nonisomorphic face 2-colourable triangular embeddings of the complete graph K_n, and hence biembeddings of $STS(n)$s, in an orientable surface.*

Not all residue classes that permit face 2-colourable triangular embeddings are covered by the theorems of this section. In particular, results are not given for $n \equiv 13$ or 15 (mod 18). We remark that further generalizations of Constructions 4.1 and 4.2 are possible. Some details of these and additional constructions are given in [46] where more than three copies of the initial embedding are used. These allow some inroads to be made into these two remaining residue classes modulo 18, but we do not have full coverage of these values.

An alternative approach is given by Korzhik and Voss [65, 66, 67, 63]. By starting with suitable current graphs and varying the vertex directions (see Section 5 for what this means), they construct for all suitably large n in each residue class modulo 12, $A2^{bn}$ nonisomorphic minimum genus embeddings of K_n in both orientable and nonorientable surfaces. The values of A and b vary with the residue class but in all cases $b > 1/12$. As observed in Section 1, in the nonorientable case, minimum genus embeddings of K_n are triangular embeddings when $n \equiv 0$ or 1 (mod 3), and in the orientable case when $n \equiv 0$, 3, 4 or 7 (mod 12). Since none of Korzhik and Voss' embeddings is face 2-colourable, they do not represent embeddings of Steiner triple systems but, rather, embeddings of twofold triple systems or, in the orientable case, Mendelsohn triple systems. Although these results cover all residue classes, the bound is a long way from 2^{an^2}. In a more recent development [64], Korzhik and Kwak combine the current graph approach with the cut-and-paste technique of Constructions 4.1 and 4.2 to prove that if $12s + 7$ is prime and if $n = (12s + 7)(6s + 7)$, then the number of nonorientable triangular embeddings of K_n is at least $2^{n^{3/2}(\sqrt{2}/72 + o(1))}$.

5   Orientable cyclic biembeddings

By a *cyclic biembedding* we mean a biembedding of two STS(n)s, each of which has the same cyclic automorphism, and such that this cyclic automorphism extends to an automorphism of the biembedding. We will assume that this cyclic automorphism is $z \mapsto z + 1$ (mod n). A cyclic STS(n) exists for every $n \equiv 1$ or 3 (mod 6) apart from $n = 9$ [76], see [25] for details. In the case where $n \equiv 3$ (mod 6), a cyclic STS(n) contains a unique short orbit and consequently there can be no cyclic biembeddings. As detailed in Section 3, Youngs [86] produced orientable cyclic biembeddings for all $n \equiv 7$ (mod 12) constructed from index 1 current graphs, and it is this case that we consider in this section. We take as our starting point the fact that every such biembedding can be obtained in this way from a current graph having the following properties.

(i) Each vertex has degree 3.

(ii) At each vertex, the sum of the directed currents is 0 (mod $12s + 7$) (*Kirchoff's current law*).

(iii) For each $i \in \{1, 2, \ldots, 6s + 3\}$, either i or $-i$ appears exactly once as a current on one of the edges and each edge has exactly one of these $6s + 3$ currents.

(iv) The directions (clockwise or anticlockwise) assigned to each vertex are such that a *complete circuit* is formed, that is, one in which every edge of the graph is traversed in each direction exactly once.

(v) The graph is bipartite.

Consideration of the degree and the currents shows that these current graphs have $4s + 2$ vertices. Furthermore, there can be no loops and, save for the exceptional case $s = 0$, no multiple edges. This last fact follows from consideration of the configuration shown in Figure 5.1.

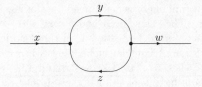

Figure 5.1: A possible multiple edge.

If this forms part of a current graph then $w \equiv x$ and so the whole current graph comprises two vertices with a triply repeated edge.

There is a close connection between current graphs and solutions of *Heffter's first difference problem* (HDP). In 1897, Heffter [58] posed the following question: can the integers $1, 2, \ldots, 3k$ be partitioned into k triples $\langle a, b, c \rangle$ such that, for each triple, $a + b \pm c \equiv 0 \pmod{6k+1}$? Examination of the triples formed by the directed currents at each vertex in either of the two vertex sets of a bipartite current graph shows that they form a solution to HDP for $k = 2s + 1$.

In view of the above observations, the problem of constructing orientable cyclic biembeddings of a pair of STS($12s + 7$)s, $s > 0$, may be reduced to three steps:

(a) Identifying simple connected cubic bipartite graphs having $4s + 2$ vertices.

(b) Assigning directions (clockwise or anticlockwise) at each of the vertices which then give rise to a complete circuit.

(c) Taking two solutions of HDP and labelling the edges of the graph in such a way that the triples arising from each of the vertex sets of the bipartition correspond to these two solutions.

These three steps have a large measure of independence from one another. However, we cannot exclude the possibility that for a particular graph it may be impossible to assign vertex directions to give a complete circuit, and, even if this is possible, it may not be possible to assign the HDP solutions to the edges. We note that a test for the existence of a complete circuit in a graph G is given by Xuong [84]. It asserts the existence of such a circuit, equivalent to a one-face orientable embedding of G, if and only if G has a spanning tree whose co-tree has no component with an odd number of edges.

Before proceeding further, it is appropriate to recall how Steiner triple systems arise from solutions to HDP. Given a difference triple $\langle a, b, c \rangle$ with $a + b \pm c \equiv 0$ (mod $6k+1$), we may form a cyclic orbit by developing the starter $\{0, a, a+b\}$ or the starter $\{0, b, a+b\}$. By taking all the difference triples from a solution of HDP and forming a cyclic orbit from each, a cyclic $STS(6k+1)$ is obtained. The converse is also true: given a cyclic $STS(6k+1)$, we may obtain a solution to HDP by taking from each orbit a block $\{0, \alpha, \beta\}$ and forming the difference triple $\langle \hat{\alpha}, \widehat{\beta - \alpha}, \hat{\beta} \rangle$, where

$$\hat{x} = \begin{cases} x & \text{if } 1 \le x \le 3k \\ 6k+1-x & \text{if } 3k+1 \le x \le 6k \end{cases}$$

Each solution to HDP produces 2^k different $STS(6k+1)$s; however, there may be isomorphisms between these systems. In addition, for a given value of k, there will generally be many distinct solutions to HDP. In this context, we say that two solutions to HDP for $k = 2s + 1$ are *multiplier equivalent* if one set of difference triples may be obtained from the other by first multiplying by a constant factor (mod $6k + 1$) and then reducing any residue x in the range $3k + 1 \le x \le 6k$ to $6k + 1 - x$. Further, we define a *Heffter class* to be a class of all solutions to HDP that are multiplier equivalent. The significance of this definition is that $STS(6k+1)$s obtained from multiplier equivalent solutions to HDP are themselves multiplier equivalent and hence isomorphic.

For $n = 19$, all the computations may be done by hand. The only cubic bipartite graph is $K_{3,3}$. Fixing the rotation at one vertex of $K_{3,3}$ there are twelve ways of assigning vertex directions to produce a complete circuit [15]. There are four solutions to HDP for $k = 3$ [22], but only two Heffter classes, namely:

$$
\begin{array}{llll}
\text{I}: & \langle 1,3,4 \rangle & \langle 2,7,9 \rangle & \langle 5,6,8 \rangle \\
& \langle 1,4,5 \rangle & \langle 2,6,8 \rangle & \langle 3,7,9 \rangle \\
& \langle 1,5,6 \rangle & \langle 2,8,9 \rangle & \langle 3,4,7 \rangle \\
\text{II}: & \langle 1,7,8 \rangle & \langle 2,3,5 \rangle & \langle 4,6,9 \rangle
\end{array}
$$

It is then easy to show that there is only one pair of solutions to HDP with which to label the edges of $K_{3,3}$ as described above; one solution coming from Heffter class I and the other from Heffter class II. The resulting orientable cyclic biembeddings of $STS(19)$s are then found to lie in just eight isomorphism classes. The rotations at 0 of these biembeddings together with an identification of the cyclic systems so biembedded are given in the Table below. They were first listed in [45]. The references to the cyclic $STS(19)$s, A1, A2, A3, A4, are as given in [72]. The rotation at i is obtained by adding i (mod 19) to the rotation at 0.

(1)	1	12	10	6	14	16	15	9	2	5	11	18	3	17	7	8	13	4	A1	A3
(2)	1	8	13	9	2	16	15	6	14	17	7	18	3	5	11	12	10	4	A1	A3
(3)	1	12	2	16	15	9	7	8	14	17	10	6	11	18	3	5	13	4	A2	A3
(4)	1	12	2	5	13	9	7	8	14	16	15	6	11	18	3	17	10	4	A2	A3
(5)	1	8	14	16	15	6	11	12	2	5	13	9	7	18	3	17	10	4	A2	A3
(6)	1	8	14	17	10	6	11	12	2	16	15	9	7	18	3	5	13	4	A2	A3
(7)	1	12	2	16	15	6	11	18	3	5	13	9	7	8	14	17	10	4	A2	A4
(8)	1	8	14	16	15	9	7	18	3	17	10	6	11	12	2	5	13	4	A2	A4

Table 5.1: Rotations at 0 of the eight $STS(19)$ cyclic biembeddings.

All four cyclic STS(19)s are cyclically biembeddable but none cyclically biembeds with itself. Only STS(19)s from Heffter class I (A1 and A2) may be cyclically embedded with STS(19)s from Heffter class II (A3 and A4). The first of these cyclic biembeddings was also previously given in [70] as well as two further cyclic embeddings of K_{19} corresponding to TTS(19)s.

For $n = 31$, the computations require a computer. There are two cubic bipartite graphs on 10 vertices and they may be obtained from $K_{5,5}$ by either removing a single 10-cycle, or a 6-cycle and a 4-cycle. Fixing the direction at one vertex gives a total of 160 sets of vertex directions in the former case and 128 sets of vertex directions in the latter case which result in complete circuits. There are 64 solutions to HDP for $k = 5$ [22], which lie in eight Heffter classes. Altogether there are 2,408 isomorphism classes of orientable cyclic biembeddings of STS(31)s, involving 76 of the 80 cyclic STS(31) [26]. Of these classes, 64 are cyclic biembeddings of a system with itself, representing 44 distinct systems. These were first given in [12] and further details of the argument again appear in [15]. As with $n = 19$, only systems from certain pairs of Heffter classes are cyclically biembeddable. The four STS(31)s which are not cyclically biembeddable all come from one particular Heffter class, represented by the difference triples $\langle 1, 5, 6 \rangle, \langle 2, 10, 12 \rangle, \langle 3, 13, 15 \rangle, \langle 4, 7, 11 \rangle, \langle 8, 9, 14 \rangle$. These STS(31)s are not cyclically biembeddable with any STS(31) from any Heffter class. Of course, this does not imply that these four systems have no biembeddings at all.

For $n = 43$, there are 13 cubic bipartite graphs on 14 vertices to consider [77]. Of these, two have edge-connectivity 2, and so cannot have currents assigned along their edges that are different as required by property (iii) above. This is because the current in one of the two edges of the cutset would have to be equal (but opposite in direction) to that in the other. The 11 remaining graphs admit direction and current assignments. Further details are given in [10, 15].

Before leaving this section we remark that [15] gives theoretical reasons, based on the above analysis, why certain pairs of cyclic STS(n)s cannot be cyclically biembedded together in an orientable surface. These are sufficient to give a complete explanation of cyclic biembeddability in orientable surfaces for $n = 19$ and $n = 31$, but not for all $n \equiv 7 \pmod{12}$.

6 Enumeration

Our purpose in this section is to briefly summarize the current state of knowledge about triangular embeddings of the complete graph K_n and hence of embeddings and biembeddings of designs, for small values of n. Specifically we will consider the cases $n = 3, 4, 7, 12$ and 15 for embeddings in an orientable surface and $n = 6, 7, 9, 10, 12$, 13 and 15 for embeddings in a nonorientable surface. We give enumeration results, by which we mean the number of nonisomorphic embeddings of the specified type. Automorphisms include those that, for face 2-colourable embeddings, exchange the colour classes and, in the orientable case, those that reverse the orientation. The first two cases are trivial. The STS(3) has a unique biembedding in the sphere which has automorphism group S_3 of order 6. There is a unique MTS(4), the embedding of which in the sphere is also unique. The automorphism group is S_4 of order 24 and odd permutations reverse the orientation.

The next two cases are less trivial but well-known. There is a unique TTS(6) and its unique embedding in the projective plane is shown below. The automorphism group is $PSL_2(5) \simeq A_5$ realized as $\langle z \mapsto (az+b)/(cz+d),\ a,b,c,d \in GF(5),\ ad-bc = 1\rangle$. This has order 60, the maximum possible, and acts transitively on *flags*, that is ordered triples (v, e, f) where e is an edge incident to vertex v and face f.

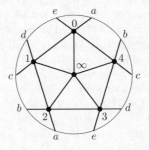

Figure 6.1: Embedding of TTS(6) in the projective plane.

The unique biembedding of the STS(7) in the torus has for its automorphism group the affine linear group AGL(1,7) of order 42. In the realization shown below, this is $\langle z \mapsto az + b,\ a,b \in GF(7),\ a \neq 0\rangle$. The automorphisms of even order exchange the colour classes but preserve the orientation. There is no embedding of the complete graph K_7 in the Klein bottle.

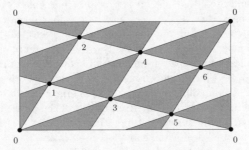

Figure 6.2: Biembedding of STS(7) in the torus.

Triangulations for $n = 9$ and $n = 10$ are necessarily nonorientable. In the former case there are precisely two embeddings. One of these is a biembedding of STS(9)s and has automorphism group $C_3 \times S_3$ of order 18. A realization is obtained by taking one system with block set {012, 345, 678, 036, 147, 258, 048, 156, 237, 057, 138, 246} and the other obtained from this by applying the permutation $\pi = (0\ 1)(2\ 6)(4\ 7)(3)(5)(8)$. In this realization, the permutations π and $(0\ 6\ 7)(1\ 8\ 4\ 3\ 2\ 5)$ generate the automorphism group. The automorphisms of even order exchange the colour classes. The other embedding is not face 2-colourable and is the TTS(9) having the following block set {BC0, CA1, AB2, BC3, CA4, AB5, A05, B10, C21, A32, B43, C54, A04, B15, C20, A31, B42, C53, 013, 124, 235, 340, 451, 502}. These embeddings were found by Altshuler and Brehm [6], and rediscovered by Bracho and Strausz [20], from which the given realization is taken. The automorphism group is C_6 of order 6 and is generated by the permutation (0 1 2 3 4 5)(A B C). The two embeddings of K_9 correspond to the twofold triple systems #36 and #35 respectively of the listing of the 36 nonisomorphic TTS(9)s as given in [25]. Using this listing it is not difficult for the reader to verify these results independently by examining the neighbourhood graphs of the systems as explained in Section 2. There are 394 nonisomorphic TTS(10)s without repeated blocks [23]. Of these, precisely 14 can be embedded. Four have trivial automorphism group, four have C_2 and there is one each with groups C_3, C_5, S_3, C_9, A_4 and A_5, [20].

The next two cases to consider are $n = 12$ and $n = 13$. There are 59 nonisomorphic embeddings of MTS(12)s in an orientable surface [5], and 182,200 nonisomorphic embeddings of TTS(12)s in a nonorientable surface [35]. There are two STS(13)s, one is cyclic and the other is not. We will refer to these here as C and N respectively. There are 615 biembeddings of C with C, of which 36 have an automorphism group of order 2 and four an automorphism group of order 3; the rest have only the trivial automorphism. There are 8,539 biembeddings of C with N, of which ten have an automorphism group of order 3 and the rest have only the trivial automorphism. Finally, there are 29,454 biembeddings of N with N, of which 238 have an automorphism group of order 2 and the rest have only the trivial automorphism. In each case, automorphisms of order 2 exchange the colour classes. These results come from [38] and were confirmed in [35] where all 243,088,286 nonorientable triangular embeddings of K_{13} were determined.

The final case which we consider is $n = 15$, and is of particular interest. To quote Ellingham and Stephens, [35], "it is probably infeasible to generate all triangular embeddings of K_{15} in N_{22}" and "it may be possible to generate all orientable embeddings of K_{15} in S_{11}" but "finding them may be a feasible (if still long-term) project on a large many-processor system". But the importance of this case is that $n = 15$ is the smallest value, apart from the well-known cases of the trivial STS(3) and unique STS(7), for which biembeddings of STS(n)s in an orientable surface can be investigated. There are 80 nonisomorphic STS(15)s; a standard numbering and some of their structural properties are given in [72]. They provide a laboratory for experimentation and for framing conjectures. However, before we consider orientable embeddings we first of all deal with the nonorientable case.

In [16], it was shown that every pair of the 80 isomorphism classes of STS(15) may be biembedded in a nonorientable surface. There are precisely three such biembeddings of system #1 with itself and five such biembeddings of system #1 with system #2 [11, 14]. System #1 is the point-line design of the projective geometry PG(3, 2) and system #2 is obtained from system #1 by making a Pasch trade, see Section 7. As a consequence of the results concerning the biembeddings of STS(n)s for $n = 9, 13$ and 15, we believe that there is reasonable evidence to support the following conjecture.

Conjecture 6.1 *Every pair of STS(n)s, $n \equiv 1$ or 3 (mod 6) and $n \geq 9$, can be biembedded in a nonorientable surface.*

Turning to orientable biembeddings of the STS(15)s, we firstly observe that there are precisely three systems having an automorphism of order 5. Each of these systems has a biembedding with itself having an automorphism group of order 10. One of these is the one originally given by Ringel [78], and which can also be obtained from the Bose construction, see Section 3 for details. The other two may be obtained by Ringel's method from index 3 current graphs [13]. In each case an automorphism of order 2 with a single fixed point, exchanges the colour classes but preserves the orientation. In [17] a computer search for biembeddings of the 80 systems, each with itself, was based on examining all possible automorphisms of order 2 having a single fixed point and exchanging the colour classes. As a result, it was shown that 78 of the 80 systems have orientable biembeddings of this type. The exceptions are systems numbered #2 and #79 in the standard listing. In the case of #2, it is further shown in [17] <u>not</u> to have an orientable biembedding with itself. It was also shown that, in the case of #79, any such biembedding can only have the trivial automorphism group. However more recent and, at the time of writing, unpublished work by the present authors and Martin Knor has disposed of this possibility. Hence we can state the following theorem.

Theorem 6.1 *Of the* 80 *nonisomorphic STS(15)s,* 78 *have a biembedding with themselves in an orientable surface. The two exceptions which have no such biembedding are #2 and #79 in the standard listing.*

An orientable biembedding of system #79 with system #77 having an automorphism of order 3 is also given in [17] and is the first known example of a biembedding of a pair of nonisomorphic STS(15)s, though of course, as described in Section 5, there are already many known biembeddings of pairs of nonisomorphic STS(n)s for $n = 19$ and $n = 31$.

Again, with Martin Knor, we have established a programme to find further such biembeddings. Of particular interest is whether there exists a biembedding of system #2 with some other system. In fact we have discovered such a biembedding and hence can state another theorem.

Theorem 6.2 *Each of the* 80 *nonisomorphic STS(15)s has a biembedding with some STS(15) in an orientable surface.*

7   Trades

The concept of a trade is well established in combinatorial design theory. Below we give definitions sufficient for our purposes. A good overview is given in [29] and the listings we make use of appear in [61]. In this section we describe <u>surface</u> trades in triangular embeddings. By this we mean replacing one set of triangular faces with another set that covers the same edges. By applying such trades one may generally move between nonisomorphic embeddings of the same graph. Referring to the constructions presented in Section 4, the replacement of one $K_{n,n,n}$ toroidal bridge by the reversed bridge provides an example of a surface trade. Underlying any such surface trade there is a combinatorial trade on some (possibly partial) twofold triple system. However, the existence of a combinatorial trade amongst the triples formed by a set of triangular faces does not ensure the existence of a corresponding surface trade since applying the trade may transform the surface into a generalized pseudosurface. The geometrical arrangement of the faces is important both for the feasibility of the trade and for questions of orientability.

One may also consider surface trades in the context of the "distance" between different triangular embeddings of a graph G, where distance is defined as the minimum number of faces in which two triangular embeddings of G can differ. We describe below various surface trades which were used in [43] to show that the minimum distance between two different nonorientable triangular embeddings of K_n is at least 4, a number that increases to 6 if one or both of the embeddings is orientable. Moreover, these distances are achievable for some values of n.

A triangular embedding of a graph G, with vertex set V of cardinality n, determines a *partial twofold triple system*, $\text{PTTS}(n) = (V, \mathcal{B})$, where $\mathcal{B}$ is the collection of triples of points of V formed by the vertices of the triangular faces; this has the property that every pair of points corresponding to an edge of G appears in precisely two triples (triangular faces of the embedding), but those corresponding to the edges of the complementary graph do not appear in any triple. When G is a complete graph K_n, the resulting $\text{PTTS}(n)$ is a $\text{TTS}(n)$. A combinatorial trade on a $\text{PTTS}(n)$ may be defined as follows.

Suppose that T_1 and T_2 are disjoint sets of triples taken from a finite base set U. If every pair of points of U occurs in the triples of T_1 with precisely the same multiplicity (0, 1 or 2) with which it appears in the triples of T_2, then the pair $\mathcal{T} = \{T_1, T_2\}$ is called a *combinatorial trade*. The *volume* of the trade $\mathcal{T}$, $\text{vol}(\mathcal{T})$, is the common cardinality of T_1 and T_2, and the *foundation* of the trade $\mathcal{T}$, $\text{found}(\mathcal{T})$, is the set of points of U which appear amongst the triples of T_1 (or T_2).

The rationale for the above definition is that if $P_1 = (V, \mathcal{B}_1)$ is a $\text{PTTS}(n)$ whose triples include those of T_1, then by replacing these triples with those of T_2, we form another $\text{PTTS}(n)$, $P_2 = (V, \mathcal{B}_2)$ say, and the triples of $\mathcal{B}_1$ and $\mathcal{B}_2$ cover exactly the same pairs of points from V with the same multiplicities.

Now consider the effect of making a trade on an embedding. Suppose that M_1 is a triangular embedding of the simple connected graph G in some surface S and that $P_1 = (V, \mathcal{B}_1)$ is the associated $\text{PTTS}(n)$. Further suppose that $\mathcal{T} = \{T_1, T_2\}$ is a trade with $\text{found}(\mathcal{T}) \subseteq V$ and that $T_1 \subseteq \mathcal{B}_1$. Put $\mathcal{B}_2 = (\mathcal{B}_1 \setminus T_1) \cup T_2$, so that $P_2 = (V, \mathcal{B}_2)$ is a $\text{PTTS}(n)$ covering all the edges of G precisely twice and no other pairs from V. If we now regard the triples from $\mathcal{B}_2$ as triangular faces and sew these faces together along the common edges, then this operation may or may not result in a surface embedding M_2 of G; the reason that the process may fail to yield such an embedding is that the sewing operation may yield a generalized pseudosurface. However, when the operation succeeds in producing a surface embedding, then we say that $\mathcal{T}$ forms a *surface trade* between the embeddings M_1 and M_2 of the graph G.

A variety of interesting questions may be posed concerning trades and embeddings. For example, does every combinatorial trade on a $\text{PTTS}(n)$ yield at least one surface trade? Is it possible to characterize those combinatorial trades which, no matter how they lie on the surface, always transform a surface embedding into a surface embedding (rather than into a generalized pseudosurface embedding)? Which surface trades are guaranteed to preserve orientability? How many different surface trades with foundation size less than n must a triangular embedding of K_n possess?

And if $b = b(n)$ denotes the minimum integer such that any two triangular embeddings of K_n may be transformed into one another by a trade of volume at most b, how does b vary with n? In order to make progress with such questions it is helpful to have a catalogue of small surface trades.

Apart from the trivial case $G = K_3$, no triangular embedding of a simple connected graph G can give rise to a PTTS(n) with a repeated triple. Furthermore, in this trivial case, it is clear that no trade exists. We may therefore assume that $G \neq K_3$, and that the associated PTTS(n) does not contain any repeated triples. We consider here the case of trades $\mathcal{T}$ on PTTS(n)s with $\mathrm{vol}(\mathcal{T}) \leq 6$. Up to isomorphism, there are precisely five such combinatorial trades, one having $\mathrm{vol}(\mathcal{T}) = 4$ and the other four having $\mathrm{vol}(\mathcal{T}) = 6$. These five trades are all given in [61], and it is shown in [18] that there are no further trades $\mathcal{T} = \{T_1, T_2\}$ having $\mathrm{vol}(\mathcal{T}) \leq 6$.

The five trades are listed below. The first three have common names as given. In each case T_1 is isomorphic with T_2.

1. (Pasch or quadrilateral trade) $T_1 = \{123, 145, 624, 635\}$,
   $T_2 = \{124, 135, 623, 645\}$.

2. (6-cycle trade) $T_1 = \{123, 145, 167, 834, 856, 872\}$,
   $T_2 = \{134, 156, 172, 823, 845, 867\}$.

3. (Semihead trade) $T_1 = \{127, 136, 145, 235, 246, 347\}$,
   $T_2 = \{126, 135, 147, 237, 245, 346\}$.

4. (Trade-X) $T_1 = \{123, 124, 156, 256, 345, 346\}$,
   $T_2 = \{125, 126, 134, 234, 356, 456\}$.

5. (Trade-Y) $T_1 = \{124, 125, 136, 137, 267, 345\}$,
   $T_2 = \{126, 127, 134, 135, 245, 367\}$.

Surface trades are not new. For example, in Figure 1 of [21], which relates to triangulations of the projective plane, the pair $\{a, b\}$ gives a geometrical realization of trade-X, the pair $\{c, d\}$ a realization of a Pasch trade, and the pair $\{e, f\}$ a realization of a semihead trade. Trade-X represents a sequence of diagonal flips. However, our purpose in this section is to show how one can determine the precise geometrical circumstances in which a surface trade results from a combinatorial trade. We give the details for Pasch trades and we summarize the other cases, leaving the interested reader to consult our joint paper with Bennett, Korzhik and Širáň [18] for further information.

So, consider the possibility of the triangular faces (defined by their vertex triples) 123, 145, 624, 635 of an embedding M being traded with the triangular faces 124, 135, 623, 645 to form an embedding M'. Initially we ignore the question of orientability. At the point 1, and up to reversal, there are two possibilities for the rotation in M, namely

(a) $1: \ 23 \cdots 45 \cdots$ or

(b) $1: \ 23 \cdots 54 \cdots$,

where $\cdots$ denotes undetermined sections of the rotation.

In M' there are faces 124 and 135, but in case (b) the partial rotations $4\cdots 2$ and $3\cdots 5$ preclude these unless the undetermined sections of these partial rotations are empty, that is, case (b) has the form $1 : 2354$. In this case M also contains the faces 124 and 135, and so M' would have two copies of each of these faces. So we may exclude case (b). Returning to case (a) and applying similar reasoning at the other vertices shows that the partial rotations in M and in M' at the points $1, 2, \ldots, 6$ are, up to reversals, as shown in Table 7.1. Note also that these partial rotations in M and M' are isomorphic; for example the permutation (3 4) takes one to the other.

M	M'
1 : $23\cdots 45\cdots$	1 : $24\cdots 35\cdots$
2 : $31\cdots 64\cdots$	2 : $36\cdots 14\cdots$
3 : $12\cdots 56\cdots$	3 : $15\cdots 26\cdots$
4 : $51\cdots 62\cdots$	4 : $56\cdots 12\cdots$
5 : $14\cdots 36\cdots$	5 : $13\cdots 46\cdots$
6 : $24\cdots 35\cdots$	6 : $23\cdots 45\cdots$

Table 7.1: Partial Pasch surface trade.

Next consider the question of orientability. Assuming a consistent orientation of M and starting with $1 : 23\cdots 45\cdots$, we require $2 : 31\cdots 64\cdots$ and $4 : 51\cdots 62\cdots$. However, these give respectively $6 : 42\cdots$ and $6 : 24\cdots$, contradicting orientability. Therefore a consistent orientation of M, and similarly M', is not possible. Thus a surface trade based on the combinatorial Pasch trade is necessarily between nonorientable embeddings.

We have shown the necessity of Table 7.1 for the existence of a Pasch surface trade, but we have not demonstrated that such a trade exists. In order to do this, take the rows of the partial rotation scheme for M with the undetermined sections eliminated and then determine any resulting non-triangular faces. From each such face, eliminate multiple vertices, if any, by the insertion of additional triangles involving new faces as illustrated below in Figure 7.1, where the twice repeated vertex x is eliminated from the face F by the insertion of new vertices x_1 and x_2.

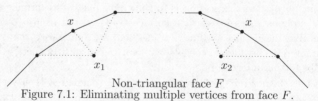

Non-triangular face F
Figure 7.1: Eliminating multiple vertices from face F.

Having completed this elimination, for a non-triangular face without multiple vertices, insert a new vertex into the interior of that face and join it by non-intersecting edges to all the vertices on the boundary, thereby forming a triangular embedding of some simple connected graph.

Application of this algorithm to the case of the Pasch trade given in Table 7.1 give the rotations M and M' as shown below in Table 7.2

$$
\begin{array}{ll}
M & M' \\
1: \ 23x45y & 1: \ 24x35y \\
2: \ 31y64z & 2: \ 36y14z \\
3: \ 12z56x & 3: \ 15z26x \\
4: \ 51x62z & 4: \ 56x12z \\
5: \ 14z36y & 5: \ 13z46y \\
6: \ 24x35y & 6: \ 23x45y \\
x: \ 1364 & x: \ 1364 \\
y: \ 1265 & y: \ 1265 \\
z: \ 2354 & z: \ 2354 \\
\end{array}
$$

Table 7.2: Example of a Pasch surface trade.

The same algorithm may be applied to produce examples of other surface trades from partial rotation schemes; it preserves orientability in the sense that if a partial rotation scheme is potentially orientable, then the resulting triangular embedding M will be orientable. This does not, however, ensure that the traded embedding M' is orientable. Also note that it is always possible to render both M and M' nonorientable by gluing on a nonorientable triangular embedding which shares a common face with M and M'.

The results of [18] for all five surface trades having volume at most 6, are summarized in Table 7.3. In the case of Trade-X, every possible geometric realization permits a surface trade. In the case of a face 2-colourable embedding M both Trade-X and Trade-Y necessarily involve both colour classes. The entry "28" against the semihead trade reduces to 19 if we allow M and M' to be exchanged. This only arises for semihead trades because the geometric realizations of the partial rotations in M and M' can be nonisomorphic in this case.

Name	Number of nonisomorphic geometric realizations	Comments
Pasch	1	M and M' are necessarily nonorientable.
6-cycle	4	In one case it is possible for one or both of M and M' to be orientable.
Semihead	28	In one case it is possible for one or both of M and M' to be orientable.
Trade-X	7	In one case it is possible for both M and M' to be orientable, but not to have one orientable and the other nonorientable.
Trade-Y	3	In one case it is possible for one or both of M and M' to be orientable.

Table 7.3: Small surface trades.

Perhaps the most compelling reason for considering surface trades is the possibility of using them to obtain lower bounds of the form 2^{an^2} on the numbers of triangular embeddings of K_n for residue classes not covered by the methods described in Section 4. Such potential use depends on constructing embeddings having a large number of independent trades, possibly using current graphs. So far, at least, we have not been able to implement this strategy.

8   Maximum genus embeddings

Whenever a biembedding of two $STS(n)$s exists, it represents a minimum genus face 2-colourable embedding of K_n in a surface and hence may be considered to be a minimum genus embedding of each of the two $STS(n)$s involved. From Euler's formula, in the orientable case the minimum genus is $(n-3)(n-4)/12$ and in the nonorientable case it is $(n-3)(n-4)/6$.

Our focus in this section lies at the opposite extreme, namely on cellular embeddings of Steiner triple systems of maximum genus. To be precise, we seek a face 2-colourable embedding of a complete graph K_n in a surface in which the black faces are triangles and so determine an $STS(n)$, while there is just one white face and the interior of that face is homeomorphic to an open disc. This latter condition ensures that the embedding is cellular and it precludes artificial inflation of the genus by the addition of unnecessary handles or crosscaps. In the orientable case, the corresponding genus is $(n-1)(n-3)/6$, and in the nonorientable case, $(n-1)(n-3)/3$. To avoid trivialities, we shall assume that $n > 3$ and then the single white face, which has $n(n-1)/2$ edges, may be referred to unambiguously as the *large face*. In topological graph theory, graphs which are cellularly embeddable with precisely one face are called "upper-embeddable". By analogy with this usage, we use the term *upper-embedding* for embeddings of $STS(n)$s of the type just described, appending the qualifier "orientable" or "nonorientable" as appropriate.

By contrast with biembeddings, it is easy to prove that for $n > 3$ every $STS(n)$ has both an orientable and a nonorientable upper-embedding. It is also possible to give detailed results about the possible automorphisms of such embeddings. We represent handles and crosscaps in diagrams as shown in Figure 8.1. The results of this section are taken from our joint paper with Širáň [47].

Figure 8.1: Representation of handles and crosscaps.

Theorem 8.1 *Every $STS(n)$ has an orientable upper-embedding.*

Proof The triples of the $STS(n)$ will be represented as black triangles of the embedding. The initial step is to take all of the black triangles containing a fixed point ∞ of the $STS(n)$. From these, one constructs a face 2-coloured planar embedding of a connected simple graph G on n vertices, having for its faces the $(n-1)/2$ black triangles incident with ∞, and one white face. The graph G and its embedding are illustrated in Figure 8.2.

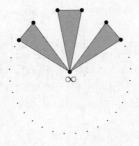

Figure 8.2: The planar embedding of G.

We proceed to add the remaining $(n-1)(n-3)/6$ triples of the STS(n), one at a time, increasing the genus by 1 at each step. Consider at any stage the boundary of the white face. We assume that every point of the STS(n) appears on this boundary at least once. This assumption is certainly true for the initial embedding illustrated in Figure 8.2. If the next triple to be added is $\{u, v, w\}$ then we locate one occurrence of each of these points on the boundary of the white face, add a handle to the white face, and paste on the triangle (u, v, w) (or (u, w, v), depending on the order of the selected points around the white face). This is illustrated in Figure 8.3 which shows the location of the triangle relative to the handle.

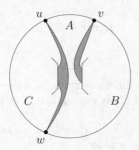

Figure 8.3: Adding a black triangle.

If the points u, v, w originally divided the boundary of the white face into three sections A, B and C, then it is easy to see that after the addition of the black triangle (u, v, w) as shown in Figure 8.3, there still remains just one white face with boundary $A(vw)C(uv)B(wu)$. This face has three more edges than at the previous stage and every point of the STS(n) still appears on the boundary. It is also clear that if the interior of the white face was homeomorphic to an open disc prior to the addition of the black triangle, then it remains so after this addition. □

We remark that it is not necessary to start with the planar embedding specified in the proof. All that is required is a planar embedding of some graph G containing

only black triangles from the STS(n) and a single white face, the interior of which
is homeomorphic to an open disc, incident with all the points of the STS(n).

Theorem 8.2 *Every STS(n) with $n > 3$ has a nonorientable*
upper-embedding.

Proof The proof is identical with that of Theorem 8.1 up to the addition of the
final black triangle. This is added to the white face using two crosscaps rather than
one handle. Figure 8.4 illustrates this step. For clarity, the edges uv, vw and wu are
labelled a, b and c respectively.

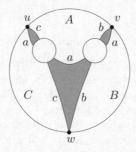

Figure 8.4: Adding the final black triangle.

Using the same notation as in the proof of Theorem 8.1, the boundary of the
white face after the addition of the black triangle (u, v, w) is $A(vw)B(vu)$ $C(wu)$.
The resulting surface has $((n - 1)(n - 3)/6) - 1$ handles and 2 crosscaps, giving
nonorientable genus $(n - 1)(n - 3)/3$.   □

It follows from Theorems 8.1 and 8.2 that for each admissible n, the number of
orientable (or nonorientable) upper-embeddings of STS(n)s is at least as great as
the number of STS(n)s.

We next give some results about the possible automorphisms of upper-embeddings
of STS(n)s. We repeat the assumption that $n > 3$.

Theorem 8.3 *If ϕ is an automorphism of an orientable or nonorientable upper-*
embedding of an STS(n) then ϕ, represented as a permutation of the points, has one
of two forms:
(a) *ϕ comprises a product of disjoint cycles of equal length, or*
(b) *ϕ comprises a single fixed point together with a product of disjoint cycles of equal*
length.
Furthermore, ϕ preserves the direction around the large face, and the common cycle
length is odd.

Proof Suppose that ϕ has two fixed points, a and b. Since ϕ must preserve the large
face and the edge ab appears somewhere on the boundary of this face, it must fix
the points adjacent to the edge ab on this boundary. By repetition of this argument,

ϕ fixes every point of the STS(n). Thus ϕ is the identity mapping and so is both of type (a) and type (b). It follows that if ϕ is not the identity mapping then it can have at most one fixed point.

Next suppose that ϕ contains two disjoint cycles of lengths p and q, where $1 < p < q$. Then ϕ^p is an automorphism with p fixed points and a cycle of length at least 2. By the previous paragraph, this is not possible. Hence ϕ must take one of the forms (a) or (b) defined in the statement of the theorem.

Now assume that ϕ has the form (a) and that it reverses the direction around the large face. Clearly ϕ is not the identity. Consider any edge ab which is mapped by ϕ to an edge $a'b'$ appearing on the boundary of the large face as shown in Figure 8.5.

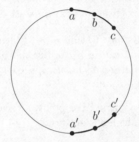

Figure 8.5: Points around the large face.

If c is adjacent to b on this boundary then it must be mapped to c' adjacent to b' as shown in Figure 8.5. Proceeding in this fashion we deduce that $\phi(a') = a$ and, further, that $\phi^2(x) = x$ for every point x of the STS(n). Since ϕ is not the identity and has the form (a), we see that ϕ must be the product of disjoint transpositions, contradicting the fact that n is odd.

Next, assume that ϕ has the form (b) and that it reverses the direction around the large face. Again, ϕ is clearly not the identity. Suppose that ϕ fixes the point ∞ (and no other point). Arguing as before we see that ϕ fixes ∞ and contains $(n-1)/2$ disjoint transpositions. Suppose that three of these are $(a_1 \; b_1)$, $(a_2 \; b_2)$ and $(a_3 \; b_3)$. Consider the edge a_1b_1. Since this edge is stabilized by ϕ, it must appear midway between two successive occurrences of ∞ on the boundary of the large face. But the edge a_2b_2 must also appear midway between the same two successive occurrences of ∞, and the same is true of the edge a_3b_3. Since there are only two midway positions, we have a contradiction. We conclude that ϕ preserves the direction around the large face.

Finally, consider the cycle length. If ϕ has the form (a), then the cycle length is necessarily odd. If ϕ has the form (b) and the cycle length is k, suppose that k is even. Then $\psi = \phi^{k/2}$ is an automorphism which comprises a fixed point and $(n-1)/2$ transpositions. If $(a_1 \; b_1)$ is one of these transpositions, then ψ will reverse the direction of the edge a_1b_1 and so fails to preserve the direction around the large face, a contradiction. Thus k must be odd. □

By using arguments based on voltage graphs, more can be said in case (a) of Theorem 8.3. The following result is given in [47].

Theorem 8.4 *If ϕ is an automorphism of an orientable upper-embedding of an STS(n), and if ϕ comprises a product of disjoint cycles of equal length k, then either $k = 1$ (in which case ϕ is the identity permutation) or $k = 3$ (in which case $n \equiv 3$ (mod 6)).*

Direct constructions using voltage graphs are then used in [47] to show that the restrictions described in Theorems 8.3 and 8.4 are, in a sense, best possible. For automorphisms without a fixed point, the following results are obtained.

Theorem 8.5 *If $n \equiv 3$ (mod 6), then there exists an orientable upper-embedding of an STS(n) having an automorphism that is a product of disjoint 3-cycles.*

Theorem 8.6 *If $n \equiv 1$ or 3 (mod 6) and $n > 3$, then every cyclic STS(n) has a nonorientable upper-embedding with a cyclic automorphism. Consequently, if $k|n$, then every cyclic STS(n) has a nonorientable upper-embedding having an automorphism which is the product of disjoint k-cycles.*

For automorphisms with a single fixed point, i.e. case (b) of Theorem 8.3, constructions given in [47] yield the following result.

Theorem 8.7 *Let S be an STS(n) with an automorphism ϕ having a single fixed point and l cycles each of length k, where k is odd and $n = kl + 1$. Then there exist both an orientable and a nonorientable upper-embedding of S having ϕ as an automorphism.*

9 Hamiltonian embeddings

A *Hamiltonian embedding* of K_n is an embedding of K_n in a surface, which may be orientable or nonorientable, in such a way that each face is a Hamiltonian cycle. In a triangular embedding of a complete graph, each face is as small as possible. At the opposite extreme, for every n there exists an embedding of K_n having a single face [32]. Around this single face every vertex appears $n - 1$ times. The problem of constructing Hamiltonian embeddings of K_n is intermediate between the two extremes: the face lengths are as large as possible subject to the restriction that no vertex is repeated on the boundary of any face. In design theory terminology, if the embedding is face 2-colourable then the faces in each colour class form an n-cycle system, in other words a decomposition of the edge set of K_n into Hamiltonian cycles. Whether or not the embedding is face 2-colourable, the complete set of faces forms a twofold n-cycle system.

In a Hamiltonian embedding of K_n the number of faces is $n - 1$. In the nonorientable case Euler's formula gives the genus as $\gamma = (n-2)(n-3)/2$. In the orientable case the genus is $g = (n - 2)(n - 3)/4$, which implies that $n \equiv 2$ or 3 (mod 4) is a necessary condition for the embedding. Face 2-colourability requires n to be odd, so that $n \equiv 1$ or 3 (mod 4). The recent paper by Ellingham and Stephens [33] established the existence of Hamiltonian embeddings in nonorientable surfaces for $n = 4$ and $n \geq 6$. We summarize their results in sufficient detail to give the flavour, giving a somewhat simpler construction in the case $n \equiv 3 \pmod 4$. We also present a novel construction given by Širáň and ourselves in [48] which produces Hamiltonian embeddings of K_n from triangular embeddings of K_n.

Theorem 9.1 (Ellingham and Stephens) *For $n = 4$ or $n \geq 6$, K_n has a Hamiltonian embedding in a nonorientable surface. Moreover, when n is odd, there is such an embedding that is face 2-colourable. There is no orientable or nonorientable Hamiltonian embedding of K_5.*

Proof First consider the case n even and write $n = 2k + 2$. Take K_n to have vertex set $\mathbb{Z}_{2k+1} \cup \{\infty\}$. Let C_i be the Hamiltonian cycle $(\infty, i, i+1, i-1, i+2, i-2, \ldots, i+k, i-k)$. The cycle C_0 is illustrated in Figure 9.1 and C_i is obtained from it by rotating i places clockwise.

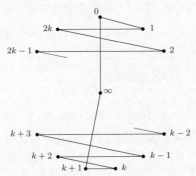

Figure 9.1: The cycle C_0.

The set of $2k + 1$ Hamiltonian cycles $\{C_i : i = 0, 1, \ldots, 2k\}$ may be sewn together along common edges to produce a Hamiltonian embedding of K_{2k+2}. To verify this, we compute the rotations at ∞ and i. These are as follows.

$$
\begin{array}{cccccccccccc}
\infty & : & 0 & k & 2k & k-1 & 2k-1 & k-2 & \ldots & 2 & k+2 & 1 & k+1 \\
i & : & \infty & i+1 & i+2 & i+3 & i+4 & i+5 & \ldots & i-4 & i-3 & i-2 & i-1
\end{array}
$$

Since each of these is a single cycle of length $2k + 1$, it follows that the construction produces a Hamiltonian embedding of K_{2k+2}. To see that the embedding is nonorientable for $k \geq 1$, delete the point ∞ and the edges incident with ∞, and examine the boundary of the resulting single face embedding of K_{2k+1}. This has the form

$$
(\overbrace{0, 1, 2k, 2, 2k-1, \ldots, k-1, k+2, k}^{C_0}, \underbrace{k+1, k+2, k, k+3, \ldots, 0, 1}_{C_{k+1}} \overbrace{\ldots \ldots \ldots \ldots}^{C_1}),
$$

where the bracings indicate the Hamiltonian cycles from which the sections are derived. Since the edge 01 is encountered twice in the same direction, the embedding cannot be orientable.

Next consider the case $n = 4s + 3$, $s \geq 1$. Take K_{4s+3} to have vertex set $\{\infty, a_0, a_1, \ldots, a_{2s}, b_0, b_1, \ldots, b_{2s}\}$. With subscript arithmetic modulo $2s + 1$, let H_i be the Hamiltonian cycle

$$H_i = (\infty a_i b_i b_{2s+i} a_{1+i} a_{2s+i} b_{1+i} b_{2s-1+i} a_{2+i} a_{2s-1+i} b_{2+i} b_{2s-2+i} \cdots$$
$$\cdots a_{s+2+i} b_{s-1+i} b_{s+1+i} a_{s+i} a_{s+1+i} b_{s+i}).$$

The cycle H_0 is illustrated in Figure 9.2 and H_i is obtained from it by rotating $2i$ places clockwise.

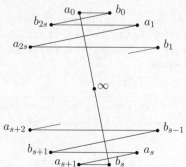

Figure 9.2: The cycle H_0.

A second Hamiltonian decomposition may be formed from this one by applying the mapping $a_j \to a_{j+1}$ ($j = 0, 1, \ldots, 2s$). This produces Hamiltonian cycles G_i which may be written most conveniently with the cyclic order reversed as

$$G_i = (\infty b_{s+i} a_{s+2+i} a_{s+1+i} b_{s+1+i} b_{s-1+i} a_{s+3+i} \cdots$$
$$\cdots b_{1+i} a_i a_{2+i} b_{2s+i} b_i a_{1+i}).$$

The set of $4s + 2$ Hamiltonian cycles $\{H_i, G_i : i = 0, 1, \ldots, 2s\}$ described above may be sewn together along common edges to produce a Hamiltonian embedding of K_{4s+3}. To verify this, we compute the rotations at ∞, a_i and b_i. These are as follows.

∞ :	a_0	b_s	a_1	b_{s+1}	a_2	b_{s+2}	$\cdots$	a_{2s}	b_{s-1}
a_i :	∞	b_i	a_{1+i}	b_{1+i}	a_{2+i}	b_{2+i}	$\cdots$	a_{2s+i}	b_{2s+i}
b_i :	∞	a_{1+i}	b_{2s+i}	a_i	b_{2s-1+i}	a_{2s+i}	$\cdots$	b_{1+i}	a_{2+i}

Since each of these is a single cycle of length $4s + 2$, it follows that the construction produces a Hamiltonian embedding of K_{4s+3}. Because each of $\{H_i : i = 0, 1, \ldots, 2s\}$ and $\{G_i : i = 0, 1, \ldots, 2s\}$ is a Hamiltonian decomposition of K_{4s+3}, it also follows that the Hamiltonian embedding is face 2-colourable. To see that the embedding is nonorientable for $s \geq 1$, delete the point ∞ and the edges incident with ∞, and examine the boundary of the resulting single face embedding of K_{4s+2}. This has the form

$$\overbrace{(a_0 b_0 \ldots b_s}^{H_0} a_{s+2} a_{s+1} b_{s+1} \ldots a_1}^{H_1} \ldots \ldots \ldots \ldots \ldots a_s b_s \ldots b_{2s}}^{H_s} a_1 a_0 b_0 \ldots a_{s+1}}^{H_{s+1}} \ldots \ldots),$$

with $\underbrace{}_{G_0}$ and $\underbrace{}_{G_s}$ indicated below.

where the bracings again indicate the Hamiltonian cycles from which the sections are derived. Since the edge $a_0 b_0$ is encountered twice in the same direction, the embedding cannot be orientable.

For $n = 4s + 1$, Ellingham and Stephens take a similar Hamiltonian decomposition of K_n into cycles H_i, again apply a permutation of the vertices to give a second Hamiltonian decomposition into cycles G_i, and then combine the two decompositions to produce the embedding. The permutation required is somewhat more complicated than that given above for $n = 4s + 3$. By this method, the embedding is certainly face 2-colourable, and it is again easily shown to be nonorientable. For the details, we refer the reader to the original paper [33].

To see that K_5 does not have a Hamiltonian embedding, suppose the contrary. Take the vertices as $0, 1, 2, 3, 4$, and delete the vertex 0 together with edges incident with it to obtain a single-face embedding of K_4 whose face boundary may be taken, without loss of generality, as $(1, a, b, 2, c, d, 3, e, f, 4, g, h)$, where $\{a, b\} = \{3, 4\}, \{c, d\} = \{1, 4\}, \{e, f\} = \{1, 2\}$ and $\{g, h\} = \{2, 3\}$. Since every edge of K_4 must appear twice, it is easy to check that there are precisely four possibilities, all of which lie in one isomorphism class. One of the possibilities for the face boundary is $(1, 3, 4, 2, 4, 1, 3, 2, 1, 4, 3, 2)$. Consideration of the rotation at the vertex 2 shows that this does not produce a surface embedding. □

We next show how Hamiltonian embeddings of K_n may be derived by surface surgery from triangular embeddings of K_n. Such triangular embeddings exist for $n \equiv 0$ or $1 \pmod 3$; whether the triangular embedding is in an orientable or nonorientable surface is immaterial. To avoid trivial cases we assume that $n \geq 4$. This work comes from our joint paper with Širáň [48].

Construction 9.1

Take a triangular embedding of K_n on the vertex set $\{\infty, a_1, a_2, \ldots, a_{n-1}\}$ and, without loss of generality, take the rotation scheme to have the following form.

$$
\begin{array}{ccccccccc}
\infty & : & a_1 & a_2 & a_3 & a_4 & \cdots & a_{n-2} & a_{n-1} \\
a_1 & : & \infty & a_2 & b_{1,1} & b_{1,2} & \cdots & b_{1,n-4} & a_{n-1} \\
a_2 & : & \infty & a_3 & b_{2,1} & b_{2,2} & \cdots & b_{2,n-4} & a_1 \\
& \vdots & & & & & \vdots & & \vdots \\
a_i & : & \infty & a_{i+1} & b_{i,1} & b_{i,2} & \cdots & b_{i,n-4} & a_{i-1} \\
& \vdots & & & & & \vdots & & \vdots \\
a_{n-1} & : & \infty & a_1 & b_{n-1,1} & b_{n-1,2} & \cdots & b_{n-1,n-4} & a_{n-2}
\end{array}
$$

where, for each $i = 1, 2, \ldots, n - 1$, $(b_{i,1} \ b_{i,2} \ \cdots \ b_{i,n-4})$ is some permutation of $\{a_1, a_2, \ldots, a_{n-1}\} \setminus \{a_{i-1}, a_i, a_{i+1}\}$, with subscript arithmetic modulo $n - 1$.

From the n lines of the rotation scheme, create $n - 1$ Hamiltonian cycles by discarding the first line and, for each i, replacing the line corresponding to a_i by the cycle $A_i = (\infty a_i a_{i+1} b_{i,1} b_{i,2} \ldots b_{i,n-4} a_{i-1})$. It is easy to see that these cycles form a Hamiltonian decomposition of $2K_n$. The Hamiltonian face corresponding to A_i is formed from the triangular faces that comprise the rotation at a_i in the original triangular embedding, with the triangle $(\infty \ a_i \ a_{i+1})$ removed. It remains to show that these Hamiltonian faces may be sewn together along common edges to produce a Hamiltonian embedding of K_n. In order to prove this, it is only necessary to prove that the resulting rotation about any vertex comprises a single cycle of length $n - 1$, rather than a set of shorter cycles with total length $n - 1$. This may be done as in the proof of Theorem 9.1, and the details are given in [48]. To consider the question of orientability, delete the point ∞ and the edges incident with ∞ from the embedding to obtain a single face embedding of K_{n-1}. It is then easy to show that an orientable triangular embedding of K_n will, by this construction, produce a nonorientable Hamiltonian embedding of K_n. Although it appears conceivable that a nonorientable triangular embedding might produce an orientable Hamiltonian embedding of K_n for $n \equiv 3, 6, 7$ or $10 \pmod{12}$, we have no examples of this and examination of the boundary of the single face suggests that such situations are likely to be rare.   □

An advantage of Construction 9.1 is that it produces a large number of nonisomorphic Hamiltonian embeddings. The following result is proved in [48].

Theorem 9.2 *If there exist M nonisomorphic triangular embeddings of K_n, $n \equiv 0$ or $1 \pmod 3$, then there exist at least $M/4n^2(n - 1)$ nonisomorphic Hamiltonian embeddings of K_n.*

Some easy consequences that follow from this and the results given in Section 4 are as follows.

Corollary 9.3 *For $n \equiv 0$ or $1 \pmod 3$ there are at least $2^{n/6-o(n)}$ nonisomorphic Hamiltonian embeddings of K_n.*

Proof For $n \equiv 0$ or $1 \pmod 3$, Korzhik and Voss [67] established that there are at least $2^{n/6-o(n)}$ nonisomorphic triangular embeddings of K_n. The result follows immediately from this and Theorem 9.2.   □

Corollary 9.4 *For $n \equiv 1, 3, 7$ or $9 \pmod{18}$ there are at least $2^{n^2/54-o(n^2)}$ nonisomorphic Hamiltonian embeddings of K_n.*

Corollary 9.5 *The constant $1/54$ that appears in the exponent in the preceding corollary may be improved to $2/81$ for $n \equiv 1, 3, 7, 9, 19, 21, 25$ or $27 \pmod{54}$.*

Finally in this section, we mention a further result of Ellingham and Stephens [34] that gives a recursive construction for <u>orientable</u> Hamiltonian embeddings of K_n.

Theorem 9.6 *Suppose that $s \geq 1$ and that K_{4s+2} has an orientable Hamiltonian embedding. Then K_{8s+2} also has an orientable Hamiltonian embedding.*

With the aid of an orientable Hamiltonian embedding of K_{10} found by a computer search, this facilitates the construction of an infinite family of such embeddings. Apart from rumours of an orientable Hamiltonian embedding of K_{30}, and the resulting infinite series, we know of no other orientable cases.

10   Latin squares

The constructions of Section 4 and their generalizations rely on face 2-colourable triangular embeddings of complete tripartite graphs $K_{n,n,n}$. It is therefore of interest to investigate these. Note that the faces in each colour class form a decomposition of $K_{n,n,n}$ into triples and hence a TD$(3, n)$ transversal design or, equivalently, a Latin square of side n. If we adopt a similar definition of biembeddability for Latin squares to that given for Steiner triple systems in Section 2, then a face 2-colourable triangular embedding of $K_{n,n,n}$ may be regarded as a biembedding of two Latin squares of side n. We may reasonably enquire about existence of these for each n, the number of biembeddings for each n, whether every Latin square is biembeddable, and whether every pair of Latin squares of the same size is biembeddable. Much of the material in this Section is taken from our joint papers with Knor and Širáň [46, 39, 40, 42].

The first result, taken from [40], is the equivalence of face 2-colourability and orientability.

Theorem 10.1 *A triangular embedding of $K_{n,n,n}$ is orientable if and only if it is face 2-colourable.*

Proof   Suppose that $K_{n,n,n}$ has tripartition $\{A, B, C\}$. If an orientable embedding is given, then triangles with clockwise orientation (A, B, C) may be coloured black and those with clockwise orientation (A, C, B) may be coloured white. Conversely, suppose that a face 2-colourable triangular embedding is given. If a black triangle of the embedding with vertices $a \in A$, $b \in B$, $c \in C$ is oriented, say clockwise, as (A, B, C), then all black triangles incident with a also have clockwise orientation (A, B, C), while the white triangles incident with a have orientation (A, C, B). Since the vertices of these triangles span $B \cup C$, all remaining black triangles have clockwise orientation (A, B, C) and all remaining white triangles have clockwise orientation (A, C, B). It follows that the rotation scheme for the embedding satisfies Ringel's Rule Δ^* (see Section 2) and therefore represents an orientable embedding. $\quad\square$

The existence of orientable triangular embeddings of $K_{n,n,n}$ for every n was established by Ringel and Youngs in [79], and a proof using a voltage graph based on a dipole with n parallel edges embedded in a sphere is indicated by Stahl and White [80]. Generalizing this voltage graph slightly to the one shown in Figure 10.1 gives Construction 10.1.

Construction 10.1

Suppose that $\{a_0, a_1, \ldots, a_{n-1}\} = \{0, 1, \ldots, n-1\}$ and that for $0 \le i \le n-1$, the differences $a_i - a_{i-1}$ are coprime with n, where subscripts are taken modulo n. Then the lift of the embedding M shown in Figure 10.1, with voltages as shown in the group $\mathbb{Z}_n$, gives an embedding of the complete bipartite graph $K_{n,n}$ in an orientable surface in which every face is bounded by a Hamiltonian cycle. If, for each i, a new vertex w_i is placed into that face obtained by lifting the 2-gon with voltages a_i and $-a_{i-1}$, and this new vertex is joined by non-intersecting edges to all the vertices lying on the boundary of that cycle, then a triangular embedding of $K_{n,n,n}$ in an orientable surface is formed. $\quad\square$

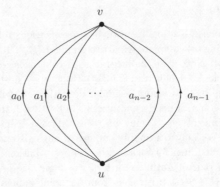

Figure 10.1: Dipole embedded in a sphere.

A careful analysis of possible isomorphisms between embeddings obtained from this construction yields the following growth estimate.

Theorem 10.2 *If n is prime then there are at least $\frac{(n-2)!}{6n}$ nonisomorphic orientable triangular embeddings of the complete tripartite graph $K_{n,n,n}$.*

For a proof see [42] where results are also given for the case when n is not prime.

The particular case of Construction 10.1 when $a_i = i$ for $0 \leq i \leq n-1$ results in one colour class of triangular faces containing all triangles of the form $(u_j v_{j+k} w_k)$ and the other containing all triangles of the form $(u_j v_{j-k+1} w_k)$ for $0 \leq j, k \leq n-1$. The corresponding Latin squares are both copies of the cyclic square

$$
C_n = \begin{array}{|ccccc|}
\hline
0 & 1 & 2 & \dots & n-1 \\
1 & 2 & 3 & \dots & 0 \\
2 & 3 & 4 & \dots & 1 \\
\vdots & \vdots & \vdots & \ddots & \vdots \\
n-1 & 0 & 1 & \dots & n-2 \\
\hline
\end{array}
$$

Thus, Construction 10.1 asserts, *inter alia*, that for each n the cyclic Latin square C_n is biembeddable with a copy of itself. In fact, as is shown in [42], this embedding is the unique regular triangular embedding of $K_{n,n,n}$ in an orientable surface. By saying that an orientable embedding M of a graph G is *regular*, we mean that for every two *flags*, that is ordered triples (v_1, e_1, f_1) and (v_2, e_2, f_2), where e_i is an edge incident to vertex v_i and face f_i, $1 \leq i \leq 2$, there exists an automorphism of M which maps v_1 to v_2, e_1 to e_2, and f_1 to f_2. Note that this definition requires automorphisms which reverse the global orientation of the surface. A regular embedding has the greatest possible number of automorphisms because the image of any one flag under an automorphism is sufficient to determine the automorphism completely. Thus the total number of automorphisms in a regular orientable triangular embedding of $K_{n,n,n}$ is just the number of flags, which is easily seen to be $12n^2$. Conversely, an orientable triangular embedding M of $K_{n,n,n}$ having $12n^2$ automorphisms must be regular.

This regular embedding may be constructed directly from the Latin square C_n and an isomorphic copy C'_n. Index rows and columns of these squares by the group $\mathbb{Z}_n$ so that the entry in row i, column j of C_n is $C_n(i,j) = i + j$, and then define C'_n by $C'_n(i,j) = i + j - 1$.

To see how these squares are combined to produce the embedding, consider the case $n = 3$, so that

$$C_n = \begin{vmatrix} 0 & 1 & 2 \\ 1 & 2 & 0 \\ 2 & 0 & 1 \end{vmatrix} \qquad C'_n = \begin{vmatrix} 2 & 0 & 1 \\ 0 & 1 & 2 \\ 1 & 2 & 0 \end{vmatrix}$$

Then take the nine points of $K_{3,3,3}$ to be $0_r, 1_r, 2_r, 0_c, 1_c, 2_c, 0_e, 1_e, 2_e$. Black triangles with clockwise orientation (r, c, e), are read from the first square so that, for example, the $(0, 2)$ entry 2 gives the triangle $(0_r 2_c 2_e)$. White triangles with clockwise orientation (r, e, c) are read from the second. The resulting rotation scheme is

$$
\begin{array}{llllllll}
0_r : & 0_c & 0_e & 1_c & 1_e & 2_c & 2_e \\
1_r : & 0_c & 1_e & 1_c & 2_e & 2_c & 0_e \\
2_r : & 0_c & 2_e & 1_c & 0_e & 2_c & 1_e \\
0_c : & 0_e & 0_r & 2_e & 2_r & 1_e & 1_r \\
1_c : & 0_e & 2_r & 2_e & 1_r & 1_e & 0_r \\
2_c : & 0_e & 1_r & 2_e & 0_r & 1_e & 2_r \\
0_e : & 0_r & 0_c & 1_r & 2_c & 2_r & 1_c \\
1_e : & 0_r & 1_c & 1_r & 0_c & 2_r & 2_c \\
2_e : & 0_r & 2_c & 1_r & 1_c & 2_r & 0_c \\
\end{array}
$$

Returning to the general case, this biembedding has n^2 automorphisms of the form $\phi_{\alpha,\beta} : (i_r, j_c, k_e) \rightarrow ((i + \alpha)_r, (j + \beta)_c, (k + \alpha + \beta)_e)$, and these all preserve the colour classes, the orientation, and the rows, columns and entries. In addition, the mapping $\chi : (i_r, j_c, k_e) \rightarrow (i_c, -j_e, -k_r)$ gives an automorphism of order 3 which permutes rows, columns and entries, but preserves the colour classes and the orientation. The mapping $\mu : (i_r, j_c, k_e) \rightarrow (i_c, j_r, k_e)$ gives an automorphism of order 2 which preserves the colour classes but reverses orientation, and the mapping $\nu : (i_r, j_c, k_e) \rightarrow (-i_c, -j_r, (-k - 1)_e)$ gives an automorphism of order 2 which reverses the colour classes but preserves the orientation. It follows that the group of automorphisms generated by these mappings has order at least $12n^2$. Since this is the maximum possible order, we deduce that this group is the full automorphism group of the biembedding and that the biembedding is regular.

A useful feature of the cyclic Latin square is that for odd values of n it contains a transversal and hence any associated biembedding contains a parallel class of triangles in the corresponding colour class. In particular, for odd n, the regular biembedding has a parallel class of triangles in each colour. A parallel class in one colour is required for the $K_{3,3,3}$ bridges used in the recursive constructions for biembeddings of Steiner triple systems in Section 4, and for the $K_{n,n,n}$ bridges used in generalizations of these constructions. There is a similar recursive construction for Latin squares first given in [46] which we now present and which enables us to give lower bounds on the numbers of biembeddings of Latin squares in certain cases.

Construction 10.2

Take any biembedding of two Latin squares of side n in a (necessarily orientable) surface S. Next take m copies of the given biembedding on m disjoint surfaces $S^0, S^1, \ldots, S^{m-1}$. We use superscripts in a similar way to identify corresponding points on these surfaces. We attempt to join these surfaces together to produce a biembedding of Latin squares of side mn. To do this we will use as bridges biembeddings of Latin squares of side m. So let T denote the bridging surface supporting such an embedding, say M, and assume that the graph $K_{m,m,m}$ embedded in T has vertex parts $\{a^i\}, \{b^i\}$ and $\{c^i\}$ and that the embedding has black faces $(a^i c^i b^i)$ for $i = 0, 1, \ldots, m-1$. *Note this requires M to have a parallel class of black triangles.*

For each white triangular face (uvw) in S we bridge $S^0, S^1, \ldots, S^{m-1}$ using a copy of M, obtained by renaming a^i, b^i and c^i as u^i, v^i and w^i respectively. The black face $(u^i w^i v^i)$ from the copy of M is glued to the white face $(u^i v^i w^i)$ in S^i.

It is now routine to check that the resulting embedding represents a biembedding of two Latin squares of side mn, that is a triangular embedding of $K_{mn,mn,mn}$ in an orientable surface. $\qquad\square$

As with the constructions of Section 4, certain generalizations are possible. We may use alternative bridges provided they all have a common parallel class of black triangles having the same orientation. Likewise, we may vary the embeddings in the surfaces S^i provided that they all have the same white triangles with the same orientations. Reapplication of the construction may also be possible in certain circumstances. For reasons of space we cannot present all the ramifications here. However the following points are worthy of remark as they produce large lower bounds for the number of biembeddings in many cases. For further details see [46].

Remark Take Construction 10.2 with $m = 3$, and use as bridges the two differently labelled $K_{3,3,3}$ embeddings given in Section 4. Since a face 2-colourable triangular embedding of $K_{n,n,n}$ has n^2 white faces, varying the bridges gives 2^{n^2} differently labelled embeddings of $K_{3n,3n,3n}$. Replacing $3n$ by n, we may express this by saying that there are at least $2^{n^2/9}$ differently labelled orientable triangular embeddings of $K_{n,n,n}$ for $n \equiv 0 \pmod 3$. Since the maximum possible size of an isomorphism class is $6(n!)^3$, this gives a lower bound of $2^{n^2/9-o(n^2)}$ for the number of nonisomorphic biembeddings of Latin squares when $n \equiv 0 \pmod 3$.

Remark In view of the previous remark, it is clearly useful to have a large supply of differently labelled orientable triangular embeddings of $K_{m,m,m}$, all having a common oriented parallel class of triangular faces in one of the two colour classes. So, on the assumption that one such embedding, say M, exists, apply to it all permutations which fix this parallel class, including its orientation, and which preserve the tripartition. There are $3(m!)$ such permutations. Suppose that π is one of these permutations and that π is also an automorphism of M. Since π preserves the orientation, the parallel class and the tripartition, π is determined by the image of any single vertex. Consequently, there are at most $3m$ such permutations π. It follows that, provided one such embedding exists, there are at least $3(m!)/3m = (m-1)!$

differently labelled orientable triangular embeddings of $K_{m,m,m}$ all having a common oriented parallel class of triangular faces in one of the two colour classes. Hence for m odd there are at least $((m-1)!)^{n^2}$ differently labelled orientable triangular embeddings of $K_{mn,mn,mn}$.

The same bound also holds for those even values of m for which there exists a biembedding of two Latin squares of side m, at least one of which has a transversal. Such biembeddings do not exist for $m = 2$ and $m = 4$, but they do exist for $m = 6$ and $m = 8$ and, in the light of the computational results described below, it would be surprising if they did not exist for all even $m \geq 10$.

The failure of the construction method for $m = 2$ and $m = 4$ is not quite the end of the story. We have one more construction which is new but similar to Construction 10.2. It takes a biembedding of Latin squares of side n and produces a biembedding of Latin squares of side $2n$. The notation is similar to the previous case.

Construction 10.3

Take any biembedding of two Latin squares of side n in a surface S. Next take two copies of the given biembedding on disjoint surfaces S^0 and S^1 with the colour classes on S^1 <u>reversed</u> so that a white triangle $(u^0 v^0 w^0)$ in S^0 corresponds to a black triangle $(u^1 v^1 w^1)$ in S^1. The bridges are formed from copies of a face 2-colourable embedding M of $K_{2,2,2}$ in a sphere having vertex parts $\{a^0, a^1\}, \{b^0, b^1\}, \{c^0, c^1\}$, a black face $(a^0 c^0 b^0)$ and a white face $(a^1 c^1 b^1)$. For each white triangular face (uvw) in S we bridge S^0 and S^1 using a copy of M, obtained by renaming a^i, b^i and c^i as u^i, v^i and w^i respectively. The black (respectively white) face $(u^i w^i v^i)$ from the copy of M is glued to the white (respectively black) face $(u^i v^i w^i)$ in S^i.

Again it is now routine to check that the resulting embedding represents a biembedding of two Latin squares of side $2n$.   □

We next turn our attention to some computational results. Again for reasons of space, we must merely summarize these, pointing out what appear to be interesting features. Fuller details are given in [40]. When we speak of the number of Latin squares of side n, we refer to the number of *main classes*, that is the number of nonisomorphic $TD(3, n)$ designs. A representative of each main class for $n = 4, 5, 6$ and 7 is given in [24].

Firstly, for each of $n = 1, 2$ and 3 there is only one Latin square of side n and one biembedding. For $n = 4$ there are two Latin squares of side n, but only one biembedding which, from above, is the regular biembedding of the cyclic square with a copy of itself. The other Latin square of side 4 is the Cayley table of the Klein 4-group. This is not biembeddable, either with itself or the cyclic square as can be easily shown. Let the Latin square be given by

$$
L_1 =
\begin{array}{c||cccc}
 & 4 & 5 & 6 & 7 \\
\hline\hline
0 & 8 & 9 & X & Y \\
1 & 9 & 8 & Y & X \\
2 & X & Y & 8 & 9 \\
3 & Y & X & 9 & 8 \\
\end{array}
$$

For clarity we represent the rows, columns and entries by different symbols. Without loss of generality it can be assumed that the rotation about the point 8 is

$$8: \quad 0 \ 4 \ 1 \ 5 \ 2 \ 6 \ 3 \ 7$$

This determines the coordinates of the entry 8 in the Latin square, say L_2, with which we are attempting to biembed L_1, namely (row, column) = (0, 7), (1, 4), (2, 5) and (3, 6). Now the only way of completing row 0 and column 4 of the Latin square L_2 without the rotation about either the point 0 or the point 4 not being a complete cycle is as follows.

$$
L_2 =
\begin{array}{c|cccc}
  & 4 & 5 & 6 & 7 \\
\hline
0 & X & Y & 9 & 8 \\
1 & 8 &   &   &   \\
2 & Y & 8 &   &   \\
3 & 9 &   & 8 &   \\
\end{array}
$$

But now it is impossible to place any entry in the (3, 5) position.

There are two Latin squares of side 5 and three biembeddings, but these biembeddings all involve two copies of the cyclic square, and the other square is not biembeddable. For $n = 6$ there are 12 Latin squares and 29 biembeddings. The Latin squares of side 6 numbered 3, 4, 7 and 10 in the listing of [24] do not feature in any of the 29 biembeddings, but the remaining eight squares each have a biembedding with a copy of themselves. For $n = 7$ there are 147 Latin squares and 23,664 biembeddings of which 4,761 are biembeddings of a Latin square with itself. However, although every Latin square of side 7 features in some biembedding, several do not biembed with themselves. But perhaps the most interesting feature of these biembeddings is that it is possible to partition the set of 147 squares into 16 subsets, of cardinalities 1, 1, 1, 2, 3, 3, 3, 6, 6, 8, 8, 9, 18, 19, 26 and 33 respectively, so that within each subset most squares biembed with most squares, and no two squares from different subsets biembed. More details of this partition appear in [40]. This bizarre partitioning, which also occurs for the Latin squares of side 6 although in not such a startling fashion, is wholly unexplained. It may just be a feature for small values of n but it may be more general and have a deeper significance. It also suggests that some form of surface trade may be involved.

From the previous paragraph it will be seen that there are six Latin squares, one each of sides 4 and 5, and four of side 6, that do not feature in any biembeddings. These include, as well as the Cayley table of the Klein 4-group, that of the non-Abelian group of order 6, #7 in the listing of [24]. It is an interesting question whether these squares are the only ones with this property. In an attempt to answer this question, with Martin Knor we have looked at those Latin squares of side 8 that come from the Cayley tables of the five groups of order 8. One of these groups is $\mathbb{Z}_2 \times \mathbb{Z}_2 \times \mathbb{Z}_2$ and another is $\mathbb{Z}_4 \times \mathbb{Z}_2$, both of which might be considered as close relatives of the Klein 4-group ($= \mathbb{Z}_2 \times \mathbb{Z}_2$). The two non-Abelian groups of order 8, namely the dihedral group $\mathcal{D}_4$ and the quaternion group $\mathcal{Q}$ are also of interest. However, we have found that each of the resulting five Latin squares biembeds and we know of no further cases of non-biembeddable Latin squares. In examining the biembeddings of these five squares of side 8, we find that, apart from the cyclic square, these biembeddings never contain two copies of the same square.

11 Symmetric configurations

The term "configuration" is nowadays used rather loosely; it has come to refer to any fixed small number of blocks which form part of a design. In this section we revert to the original meaning and define an (n_r, b_k) *configuration* to be an incidence structure of n points and b lines such that

1. each line contains k points,

2. each point lies on r lines,

3. two different points are connected by at most one line.

If $b = n$, and therefore $r = k$, the configuration is said to be *symmetric* and denoted by n_k. Our interest, in the case where $k = 3$, is in the problem of biembedding a pair of symmetric configurations of triples in a closed surface. The embedded graph is the incidence graph of each of the two configurations, where two vertices are joined by an edge if they occur together in some triple. This graph is 6-regular and, by Euler's formula, the supporting surface must be either the torus or the Klein bottle. Examples of symmetric configurations are the Fano plane or STS(7), which is the unique 7_3 configuration, and the Pappus and Desargues configurations which are 9_3 and 10_3 configurations respectively. Already in the nineteenth century enumeration results of n_3 configurations were available for small values of n. Kantor [59] showed that there is one 8_3, three 9_3 and ten 10_3 configurations and Martinetti [71] extended this catalogue by enumerating all 31 11_3 configurations.

We now have a sequence of questions concerning biembeddings of n_3 configurations which are analogous to those asked at the end of Section 2 in relation to Steiner triple systems.

1. Given an n_3 configuration, does it have a biembedding with some other n_3 configuration in the torus, the Klein bottle or both? In particular for each $n \geq 7$, is there an n_3 configuration which has such a biembedding in one or the other or both of the surfaces?

2. Given a pair of n_3 configurations do they have a biembedding in the torus, the Klein bottle or both?

3. If such biembeddings exist, how many are there?

An answer to these questions in the case of the torus was provided by Altshuler [4] and then for both the torus and the Klein bottle by Negami [74, 75]. But all three papers are written from a different viewpoint; the term "configuration" is not mentioned at all. In each case, the problem of biembedding symmetric configurations is related to the classification of which 6-regular graphs have a triangulation in the torus or the Klein bottle (or both). Negami refers to these as 6-regular *toroidal graphs* and 6-regular *Klein bottlal graphs* respectively, but for the latter we use the term *Klein bottleable graphs*. The simpler terms "torus graph" and "Klein bottle graph" might be thought preferable, but these are used by Negami to describe embeddings rather than graphs and it would be confusing for us to use them for a different purpose. In the main, our account and notation follows that given in [74].

Considering first triangulations of the torus, we define *the standard 6-regular triangulation* $T(p, q, r)$ of the torus. To do this consider the triangulation, shown in Figure 11.1, of the domain

$$\{(x, y) \in R^2 : \ 0 \leq x \leq r, \ \ 0 \leq y \leq p\},$$

where p and r are positive integers.

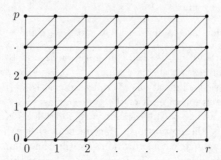

Figure 11.1: Triangulation of $\{(x, y) \in R^2 : \ 0 \leq x \leq r, \ \ 0 \leq y \leq p\}$.

In order to convert this into a triangulation of the torus, first identify the upper and lower sides of the rectangle in the usual way to form an open-ended cylinder. The embedded graph of this triangulation we denote by H_r^p and we make use of this again when considering embeddability in the Klein bottle. Now glue one of the boundaries of the cylinder to the other so that the point $(0, y)$, $0 \leq y \leq p$ coincides with the point (r, y'), $0 \leq y' \leq p$ if $y - y' \equiv q$ (mod p), where q is an integer satisfying $0 \leq q < p$. Informally we make a "twist" in the cylinder before gluing the two boundaries. This procedure defines the standard triangulation $T(p, q, r)$. Note that $T(p, q, r)$ is face 2-colourable and that a rotation of the diagram by π gives an isomorphism between the face sets of the two colour classes. For our purposes, the main result in both [4] and [74] is the following theorem.

Theorem 11.1 *If G is a 6-regular toroidal graph and M is an embedding of G in the torus, then M is isomorphic to some standard triangulation $T(p, q, r)$.*

We remark that different ordered triples (p, q, r) and (p', q', r') can lead to isomorphic triangulations. For example, as shown in [74], $T(p, q, r)$ is isomorphic to $T(p, q', r)$ if $q' \equiv -(q + r)$ (mod p). Also, the embedded graph of $T(p, q, r)$ need not be simple, although Negami identifies those which are not. He also goes on to prove that if G is a simple 6-regular toroidal graph, then the embedding is unique up to isomorphism.

To determine if an n_3 configuration has a biembedding in the torus, it therefore suffices to decide if its incidence graph is isomorphic to the embedded graph of some $T(p, q, r)$. If this is the case, then the biembedding exists. If it is not the case, then the configuration has no biembedding in the torus. When the biembedding exists, it is unique and the two biembedded configurations are isomorphic.

Also in [74], Negami lists the isomorphism classes for standard triangulations $T(p, q, r)$ on fewer than 15 vertices. For 11 vertices or less, those with simple embedded graphs comprise $T(n, 2, 1)$, $7 \leq n \leq 11$ together with $T(3, 0, 3)$. In general, $T(n, 2, 1)$ is the biembedding of the cyclic symmetric configuration on the base set $\{0, 1, 2, \ldots, n-1\}$ generated from the triple $\{0, 2, 3\}$ under the action of the mapping $z \mapsto z + 1 \pmod{n}$, and the two colour classes that result are isomorphic under $z \mapsto -z \pmod{n}$. The particular case $T(3, 0, 3)$ is the biembedding of the Pappus configuration with a copy of itself. It follows that the unique 7_3 and 8_3 configurations, two of the three 9_3 configurations and one of each of the ten 10_3 and 31 11_3 configurations are biembeddable in the torus, and that the remaining configurations on 11 vertices or less are not. Further analysis shows that for the 12_3, 13_3 and 14_3 configurations respectively, four of 229, two of 2,036 and two of 21,399 are biembeddable in the torus, and the remainder are not. The classification also implies that any connected cyclic symmetric configuration n_3 has a unique biembedding with an isomorphic copy of itself in the torus. (Here "connected" means that the incidence graph is connected.) This is because the incidence graph of such a configuration is isomorphic to the embedded graph of $T(p, q, r)$ for some values of p, q, r. An alternative and purely combinatorial proof of this result appears in [41].

Turning now to biembeddings of symmetric configurations n_3 in the Klein bottle, the classification of which 6-regular graphs have triangulations in this surface is given in [75]. This paper is a preprint and seems not to have been published in a journal. But the results are both important and interesting and deserve to be better known. We describe the relevant graphs beginning with H_r^p defined above. This has $p(r+1)$ vertices, those vertices with coordinates $(0, j)$ or (r, j) for $0 \leq j \leq p-1$ have degree 4, but all other vertices have degree 6. From the graph H_r^p and its cylindrical embedding, two families of triangulations of the Klein bottle may be constructed.

The first of these is achieved by identifying, for each y, $0 \leq y \leq p$, the points with coordinates $(0, y)$ and $(r, p - y)$. These embeddings are called *Klein bottle triangulations of handle type* and denoted by $Kh(p, r)$.

The construction of the second family of triangulations depends on the parity of p. Again referring to H_r^p, if $p = 2m$ is even, identify the point $(0, y)$ with $(0, y + m)$ and the point (r, y) with $(r, y + m)$, $0 \leq y \leq m$. If $p = 2m + 1$ is odd, use the graph H_{r-1}^p and join the point $(0, y)$ to $(0, y + m)$ and the point $(r - 1, y)$ to $(r - 1, y + m)$, $0 \leq y \leq p$, with arithmetic on the second coordinate modulo p. In this second case, when p is odd, the Klein bottle is formed by placing the additional joins across two crosscaps. The resulting triangulations, for p even or odd, are called *Klein bottle triangulations of crosscap type* and denoted by $Kc(p, r)$.

In both $Kh(p, r)$ and $Kc(p, r)$ the number of vertices is pr and the two families of triangulations are distinct. The classification theorem given in [75] is now as follows.

Theorem 11.2 *If G is a 6-regular Klein bottleable graph and M is an embedding of G in the Klein bottle, then M is isomorphic to precisely one of $Kh(p, r)$, $p \geq 3, r \geq 3$ or $Kc(p, r)$, $p \geq 5, r \geq 2$.*

As with the toroidal graphs, Negami proves that the triangular embedding of any 6-regular Klein bottleable graph is unique. In fact, the triangulations $Kh(p, r)$ are face 2-colourable while the triangulations $Kc(p, r)$ are not. So, in seeking the answer to the question of biembeddability of an n_3 configuration in the Klein bottle, it is only necessary to determine whether or not its incidence graph is isomorphic to the embedded graph of some $Kh(p,r)$. As in the toroidal case, there is an isomorphism between the face sets of the two colour classes of $Kh(p,r)$. It remains to consider the question of whether any symmetric configuration can be biembedded in both the torus and the Klein bottle. This is not so and follows from the fact that none of the embedded graphs of $T(p,q,r)$ triangulations are isomorphic to any of those of $Kh(p,r)$ triangulations. An alternative and perhaps simpler proof, which does not rely on the above classification, is given in [69].

Combining the results for the torus and the Klein bottle, we have the following theorem.

Theorem 11.3 *A symmetric configuration n_3 is biembeddable in the torus if and only if its incidence graph is isomorphic to the embedded graph of some $T(p,q,r)$. It is biembeddable in the Klein bottle if and only if its incidence graph is isomorphic to the embedded graph of some $Kh(p,r)$, $p \geq 3, r \geq 3$. Any such biembedding is unique and the two n_3 configurations that appear in the biembedding are isomorphic. No n_3 configuration has a biembedding in both the torus and the Klein bottle.*

The third 9_3 configuration which is not biembeddable in the torus corresponds to $Kh(3,3)$ and is therefore biembeddable in the Klein bottle.

Perhaps some readers may feel it is somewhat unsatisfactory that the answer to the question of the biembeddability of symmetric configurations is given in terms of whether their incidence graphs are isomorphic to any of the embedded graphs from $T(p,q,r)$ or $Kh(p,r)$. But this is a situation in which a design-theoretic problem can be successfully attacked by methods of topological graph theory. This is in contrast to Section 3, where the existence of an orientable triangulation of the complete graph K_n, $n \equiv 3 \pmod{12}$, was determined by exclusively design-theoretic methods and shows the interplay between the two areas.

Finally in this section we mention the work of White and in particular the papers [37, 82]. As the titles imply the emphasis here is on finding topological models of configurations on appropriate surfaces. The biembedding of the Pappus configuration with itself in the torus appears explicitly in these papers as well as an embedding of the Desargues configuration in the double torus.

12   Concluding remarks

In this final section, we review some open problems and briefly discuss other work in this area. We begin with the questions 2 to 4 posed at the end of Section 2, which we consider in reverse order.

The results given in Section 6 show that not every pair of Steiner triple systems of order $n = 15$ has an orientable biembedding, and it seems possible that similar nonexistence results may apply to all $n \equiv 3$ or $7 \pmod{12}$ with $n \geq 15$. However, for $n = 15$, the situation regarding nonorientable biembeddings is, as we described, quite different, with every pair of STS(15)s having at least one biembedding. This led us to make Conjecture 6.1 that every pair of STS(n)s, $n \equiv 1$ or $3 \pmod 6$ and $n \geq 9$, has at least one nonorientable biembedding. A proof of this conjecture would represent a major step forward.

Confining our attention to the orientable case, we know that the STS(7) and all 80 STS(15)s have minimum genus embeddings. Does every STS(n), $n \equiv 3$ or 7 (mod 12) have such an embedding, necessarily a biembedding? We think that the answer is likely to be in the affirmative though it may be a very difficult result to prove. But we did show in Section 8 that every STS(n) has a maximum genus embedding in which the black faces are triangles corresponding to the triples of the STS(n) and there is just one white face. An intermediate result where the black faces are triangles and there are $(n-1)/2$ white faces, all of which are Hamiltonian cycles, might be of interest.

The theorems of Section 4 give, for n lying in certain residue classes, a lower bound of the form 2^{an^2} for the number of biembeddings of STS(n)s in both orientable and nonorientable surfaces. What is the true order of magnitude of this number? We can obtain a crude upper estimate by using the known upper bound for the number of labelled Steiner triple systems of order n, namely $(e^{-1/2}n)^{n^2/6}$ [83]. It follows easily from this fact that, in both the orientable and nonorientable cases, the number of nonisomorphic biembeddings is less than $n^{n^2/3}$.

If it were the case that each pair of STS(n)s has a biembedding, then we could obtain a lower bound for the number of nonisomorphic biembeddings in a similar fashion, since the number of such pairs is at least $n^{n^2/3-o(n^2)}$. So, if the rate of growth of the number of nonisomorphic biembeddings were really of the order 2^{an^2} then this would imply that almost all STS(n)s are not biembeddable either orientably or nonorientably. Conjecture 6.1, based on the STS(15) data, therefore constrains us to the view that the correct rate of growth in the number of biembeddings is $n^{n^2/3-o(n^2)}$, at least in the nonorientable case.

Whatever the true rate of growth for biembeddings (that is, face 2-colourable triangulations of K_n), one would expect to see similar and related growth estimates for the number of minimum genus embeddings of K_n for all residue classes.

Turning now to other problems associated with biembeddings of pairs of STS(n)s, we showed in Section 3 how certain Steiner triple systems obtained from the Bose construction can be biembedded. Specifically, the groups used are cyclic. In the orientable (respectively nonorientable) cases can the result be generalized to any Abelian group of order $4s+1$ (respectively $2s+1$)? The Bose construction itself has a number of generalizations. In the version given in Section 3, the group G is used to construct a commutative idempotent quasigroup with operation $*$ defined by $i * j = (i+j)/2$. But there are many other such quasigroups. Some of these generalizations may have topological implications.

With regard to the cyclic biembeddings described in Section 5, it seems likely that infinitely many pairs of cyclic STS($12s+7$)s do not biembed cyclically in an orientable surface. Indeed, there may be infinitely many cyclic STS($12s+7$)s that do not appear in any orientable cyclic biembedding. It seems somewhat more likely that, possibly with finitely many exceptions, each such pair biembeds cyclically in a nonorientable surface.

Most of the work surveyed in this paper has been concerned with embeddings of various kinds of triple system. An exception is Section 9 where embeddings of the complete graph K_n in which each face is a Hamiltonian cycle are considered. Theorem 9.1 gives a complete solution to the existence of such embeddings in the case of a nonorientable surface. However, the existence question for orientable surfaces is far from settled. But more generally, one could consider embeddings of K_n in which all the faces are cycles of any constant length. The logical place to begin would be with quadrangulations. The necessary condition for a quadrangulation of the complete graph K_n in a nonorientable surface is $n \equiv 0$ or 1 (mod 4) and in an orientable surface is $n \equiv 0$ or 5 (mod 8). In two papers [55, 56], Hartsfield and Ringel construct such embeddings for $n \equiv 1$ (mod 4) in the former case and $n \equiv 5$ (mod 8) in the latter. The necessary and sufficient condition for a 4-cycle system, that is a decomposition of K_n into 4-cycles, is $n \equiv 1$ (mod 8). Thus any biembedding of a pair of 4-cycle systems would necessarily be in a nonorientable surface.

In the case of Latin square biembeddings, face 2-colourability is equivalent to orientability. The results given in Section 10 show that not every pair has a biembedding, and it seems likely that there are infinitely many such pairs. However, it may be the case that all but a finite number of Latin squares appear in some biembedding. In fact, we may already have identified all the exceptional non-biembeddable Latin squares; one each of side 4 and side 5 and four of side 6. But again it may be difficult to prove that every Latin square, apart from these six exceptions, has a biembedding. However we do know that every Latin square which is the Cayley table of a cyclic group is biembeddable. Does this result extend to the Cayley table of any group, apart from $\mathcal{K}_4$ and $\mathcal{D}_3$? Our computational results concerning groups of order 8 suggest that it might, though these Latin squares do not have biembeddings with isomorphic copies of themselves, unlike the situation with the cyclic groups. Other classes of Latin square which would be of particular interest are the composition tables of Steiner quasigroups and Steiner loops, defined respectively as follows. Let $(V, \mathcal{B})$ be an STS(n). Define on V an operation $*$ by $x * x = x$, $x \in V$ and $x * y = z$ if $\{x, y, z\} \in \mathcal{B}$. Then $(V, *)$ is a *Steiner quasigroup* or *squag*. Alternatively define on $V \cup \{e\}$ an operation $\circ$ by $x \circ x = e$, $e \circ x = x \circ e = x$, $x \in V \cup \{e\}$ and $x \circ y = z$ if $\{x, y, z\} \in \mathcal{B}$. Then $(V, \circ)$ is a *Steiner loop* or *sloop*. Does the Latin square composition table of every squag or sloop have a biembedding? Finally one can make estimates and conjectures concerning the growth rate for the number of biembeddings of Latin squares and these have similar forms to those described above for Steiner triple systems.

Concerning symmetric configurations, we know that an n_3 configuration can only biembed with itself and that if it does then the biembedding is unique. But relatively few symmetric configurations seem to have such minimum genus embeddings in the torus or the Klein bottle. Possibly other higher genus embeddings such as the one mentioned of the Desargues configuration in the double torus would be interesting.

Our survey has been concerned with embeddings, usually triangulations, of graphs in surfaces. But some of the ideas can be extended to pseudosurfaces. We follow [81] in making the definitions. A *pseudosurface* is the topological space which results when finitely many identifications of finitely many points each, are made on a given surface. More precisely, distinct points $\{p_{i,j} : i = 1, 2, \ldots, k, \ j = 0, 1, \ldots, m_i\}$ on a given surface are identified to form points $p_i = \{p_{i,j} : j = 0, 1, \ldots, m_i\}$, $i = 1, 2, \ldots, k$ called *singular points* or *pinch points*. The number m_i is the *multiplicity* of the pinch point p_i. It is at these pinch points that a pseudosurface fails to be a 2-manifold. A *generalized pseudosurface* is the connected topological space which results when finitely many identifications of finitely many points each, are made on a topological space of finitely many components each of which is a pseudosurface. The points subject to such identifications are also called pinch points and their multiplicities are defined in the obvious way.

The relationship between twofold triple systems and generalized pseudosurfaces is given in [3]; there is a one-to-one correspondence between $TTS(n)$s and triangular embeddings of the complete graph K_n in generalized pseudosurfaces. The correspondence is explored in greater depth in [68], where details of the generalized pseudosurfaces associated with twofold triple systems on 10 or less points can be found. Many of the generalized pseudosurfaces have an irregular structure but certain twofold triple systems correspond to more regular generalized pseudosurfaces. The simplest of these, for $n \equiv 1$ or 3 (mod 6), is a $TTS(n)$ obtained by combining the block sets of two identical $STS(n)$s. Each pair of repeated blocks gives a triangle embedded in a sphere. By identifying points which have the same label, a generalized pseudosurface is obtained which is the union of $s = n(n-1)/6$ spheres and has n pinch points all of the same multiplicity $m = (n-1)/2$. Other generalized pseudosurfaces having a similar structure are obtained as follows. A Steiner system $S(2, 4, n)$ is a pair $(V, \mathcal{B})$ where V is an n-element and $\mathcal{B}$ is a collection of 4-element subsets (the *blocks*) of V such that each 2-element subset of V is contained in exactly one block of $\mathcal{B}$. Such systems exist if and only if $n \equiv 1$ or 4 (mod 12) [54]. Each block corresponds to an embedding of a tetrahedron in the sphere. Again by identifying points which have the same label, a generalized pseudosurface is obtained which is the union of $s = n(n-1)/12$ spheres and has n pinch points all of multiplicity $m = (n-1)/3$. A generalized pseudosurface which is the union of $s = n(n-1)/24$ (respectively $n(n-1)/60$) spheres and has n pinch points all of multiplicity $m = (n-1)/4$ (respectively $(n-1)/5$) arises from the decomposition of the complete graph K_n into octahedra (respectively icosahedra). The former problem is solved, the spectrum is $n \equiv 1$ or 9 (mod 24) [50, 1] and is equivalent to the exact decomposition of the blocks of an $STS(n)$ into Pasch configurations, see Section 7. The necessary condition for the latter problem is $n \equiv 1$, 16, 21 or 36 (mod 60) but only the case $n \equiv 1$ (mod 60) is resolved [2].

But probably a more interesting problem concerning pseudosurfaces is the following. The necessary and sufficient condition for the biembedding of two $STS(n)$s in an orientable surface is $n \equiv 3$ or 7 (mod 12). But as Emch's example given below shows, there does exist a face 2-colourable triangular embedding of the complete graph K_9 in a pseudosurface formed from an orientable surface, in fact the torus, with three pinch points of multiplicity 1.

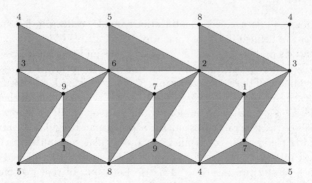

Figure 12.1 Pseudosurface biembedding of STS(9)s

In [78], a rotation scheme is given for an embedding of the complete graph K_8 in the double torus having 16 triangular faces and 2 quadrangular faces, the vertices of the quadrangular faces comprising all 8 points of the embedding. By placing two new points, say x and y, one in each quadrangle, inserting edges joining each point to the vertices of the corresponding quadrangle, and then identifying the two points x and y, we obtain a triangular embedding of the complete graph K_9 in a pseudosurface having just one pinch point of multiplicity 1. But this embedding is not face 2-colourable. These two examples naturally lead to the question of determining the pseudosurface having the least number of pinch points and/or pinch points having the least multiplicities obtained from an orientable surface for a biembedding of two STS(n)s when $n \equiv 1$ or 9 (mod 12).

More research work is needed!

Acknowledgements

We thank the following people who have been our co-authors for many of the papers on which we have drawn in preparing this survey: Geoff Bennett, Paul Bonnington, Martin Knor, Vladimir Korzhik and Jozef Širáň. We also thank the Leverhulme Trust for supporting our work with colleagues in Slovakia under the grant F/00269/E.

References

[1] P. Adams, E. J. Billington and C. A. Rodger, Pasch decompositions of lambda-fold triple systems, *J. Combin. Math. Combin. Comput.* **15** (1994), 53–63.

[2] P. Adams and D. E. Bryant, Decomposing the complete graph into Platonic graphs, *Bull. Inst. Combin. Appl.* **17** (1996), 19–26.

[3] S. R. Alpert, Twofold triple systems and graph imbeddings, *J. Combin. Theory Ser. A* **18** (1975), 101–107.

[4] A. Altshuler, Construction and enumeration of regular maps on the torus, *Discrete Math.* **4** (1973), 201–217.

[5] A. Altshuler, J. Bokowski and P. Schuchert, Neighborly 2-manifolds with 12 vertices, *J. Combin. Theory Ser. A* **75** (1996), 148–162.

[6] A. Altshuler and U. Brehm, Neighborly maps with a few vertices, *Discrete Comput. Geom.* **8** (1992), 93–104.

[7] K. Appel and W. Haken, Every planar map is four colourable, Part I: discharging, *Illinois J. Math.* **21** (1977), 429–490.

[8] K. Appel, W. Haken and J. Koch, Every planar map is four colourable, Part II: reducibility, *Illinois J. Math.* **21** (1977), 491–567.

[9] L. Babai, Almost all Steiner triple systems are asymmetric, in *Topics on Steiner systems, Ann. Discrete Math.*, **7** (1980), 37–39.

[10] G. K. Bennett, *Topological embeddings of Steiner triple systems and associated problems in design theory*, Ph.D. thesis, The Open University (2004).

[11] G. K. Bennett, M. J. Grannell and T. S. Griggs, Bi-embeddings of the projective space PG(3,2), *J. Statist. Plann. Inference* **86** (2000), 321–329.

[12] G. K. Bennett, M. J. Grannell and T. S. Griggs, Cyclic bi-embeddings of Steiner triple systems on 31 points, *Glasgow Math. J.* **43** (2001), 145–151.

[13] G. K. Bennett, M. J. Grannell and T. S. Griggs, Bi-embeddings of Steiner triple systems of order 15, *Graphs Combin.* **17** (2001), 193–197.

[14] G. K. Bennett, M. J. Grannell and T. S. Griggs, On the bi-embeddability of certain Steiner triple systems of order 15, *European J. Combin.* **23** (2002), 499–505.

[15] G. K. Bennett, M. J. Grannell and T. S. Griggs, On cyclic bi-embeddings of Steiner triple systems of order 12s+7, *J. Combin. Des.* **10** (2002), 92–110.

[16] G. K. Bennett, M. J. Grannell and T. S. Griggs, Non-orientable biembeddings of Steiner triple systems of order 15, *Acta Math. Univ. Comenianae* **73** (2004), 101–106.

[17] G. K. Bennett, M. J. Grannell and T. S. Griggs, Orientable self-embeddings of Steiner triple systems of order 15, *Acta Math. Univ. Comenianae* **75** (2006), 163–172.

[18] G. K. Bennett, M. J. Grannell, T. S. Griggs, V. P. Korzhik and J. Širáň, Small surface trades in triangular embeddings, *Discrete Math.* **306** (2006), 2637–2646.

[19] C. P. Bonnington, M. J. Grannell, T. S. Griggs and J. Širáň, Exponential families of nonisomorphic triangulations of complete graphs, *J. Combin. Theory Ser. B* **78** (2000), 169–184.

[20] J. Bracho and R. Strausz, Nonisomorphic complete triangulations of a surface, *Discrete Math.* **232** (2001), 11–18.

[21] B. Chen and S. Lawrencenko, Structural characterization of projective flexibility, *Discrete Math.* **188** (1998), 233–238.

[22] C. J. Colbourn, The construction of disjoint cyclic Steiner triple systems, *University of Saskatchewan, Department of Computational Science Research report* **81-4** (1981).

[23] C. J. Colbourn, M. J. Colbourn, J. J. Harms and A. Rosa, A complete census of (10,3,2) block designs and of Mendelsohn triple systems of order ten. III. (10,3,2) block designs without repeated blocks, *Congr. Numer.* **37** (1983), 211–234.

[24] C. J. Colbourn and J. H. Dinitz, (eds.), *The CRC Handbook of Combinatorial Designs*, CRC Press, Boca Raton (1996).

[25] C. J. Colbourn and A. Rosa, *Triple Systems*, Clarendon Press, New York (1999).

[26] M. J. Colbourn and R. A. Mathon, On cyclic Steiner 2-designs, in *Topics on Steiner systems, Ann. Discrete Math.*, **7** (1980), 215–253.

[27] F. N. Cole, L. D. Cummings and H. S. White, The complete enumeration of triad systems in 15 elements, *Proc. Nat. Acad. Sci. U.S.A.* **3** (1917), 197–199.

[28] M. Dehon, On the existence of 2-designs $S_\lambda(2,3,v)$ without repeated blocks, *Discrete Math.* **43** (1983), 155–171.

[29] D. Donovan, E. S. Mahmoodian, C. Ramsay and A. P. Street, Defining sets in combinatorics: a survey, in *Surveys in Combinatorics 2003, London Math. Soc. Lecture Notes*, **307** (2003), 115–174.

[30] P. M. Ducrocq and F. Sterboul, On G-triple systems, *Publications du Laboratoire de Calcul de l'Université des Sciences et Techniques de Lille* **103** (1978), 18pp.

[31] P. M. Ducrocq and F. Sterboul, Les *G*-systemes triples, *Ann. Discrete Math.* **9** (1980), 141–145.

[32] J. Edmonds, On the surface duality of linear graphs, *J. Res. Nat. Bur. Standards Sect. B* **69B** (1965), 121–123.

[33] M. N. Ellingham and C. Stephens, The nonorientable genus of joins of complete graphs with large edgeless graphs, *J. Combin. Theory Ser. B*, in press.

[34] M. N. Ellingham and C. Stephens, The orientable genus of some joins of complete graphs with large edgeless graphs, submitted, 2005.

[35] M. N. Ellingham and C. Stephens, Triangular embeddings of complete graphs (neighborly maps) with 12 and 13 vertices, *J. Combin. Des.* **13** (2005), 336–344.

[36] A. Emch, Triple and multiple systems, their geometric configurations and groups, *Trans. Amer. Math. Soc.* **31** (1929), 25–42.

[37] R. M. Figueroa-Centeno and A. T. White, Topological models for classical conifgurations, *J. Statist. Plann. Inference* **86** (2000), 421–434.

[38] M. J. Grannell, T. S. Griggs and M. Knor, Face two-colourable triangulations of K_{13}, *J. Combin. Math. Combin. Comput.* **47** (2003), 75–81.

[39] M. J. Grannell, T. S. Griggs and M. Knor, Regular Hamiltonian embeddings of the complete bipartite graph $K_{n,n}$ in an orientable surface, *Congr. Numer.* **163** (2003), 197–205.

[40] M. J. Grannell, T. S. Griggs and M. Knor, Biembeddings of Latin squares and Hamiltonian decompositions, *Glasgow Math. J.* **46** (2004), 443–457.

[41] M. J. Grannell, T. S. Griggs and M. Knor, Biembeddings of symmetric configurations of triples, *Proceedings of MaGiA conference, Kočovce 2004, Slovak University of Technology* (2004), 106–112.

[42] M. J. Grannell, T. S. Griggs, M. Knor and J. Širáň, Triangulations of orientable surfaces by complete tripartite graphs, *Discrete Math.* **306** (2006), 600–606.

[43] M. J. Grannell, T. S. Griggs, V. P. Korzhik and J. Širáň, On the minimal non-zero distance between triangular embeddings of a complete graph, *Discrete Math.* **269** (2003), 149–160.

[44] M. J. Grannell, T. S. Griggs and J. Širáň, Face 2-colourable triangular embeddings of complete graphs, *J. Combin. Theory Ser. B* **74** (1998), 8–19.

[45] M. J. Grannell, T. S. Griggs and J. Širáň, Surface embeddings of Steiner triple systems, *J. Combin. Des.* **6** (1998), 325–336.

[46] M. J. Grannell, T. S. Griggs and J. Širáň, Recursive constructions for triangulations, *J. Graph Theory* **39** (2002), 87–107.

[47] M. J. Grannell, T. S. Griggs and J. Širáň, Maximum genus embeddings of Steiner triple systems, *European J. Combin.* **26** (2005), 401–416.

[48] M. J. Grannell, T. S. Griggs and J. Širáň, Hamiltonian embeddings from triangulations, *Bull. London Math. Soc.*, in press.

[49] M. J. Grannell and V. P. Korzhik, Nonorientable biembeddings of Steiner triple systems, *Discrete Math.* **285** (2004), 121–126.

[50] T. S. Griggs, M. J. de Resmini and A. Rosa, Decomposing Steiner triple systems into four-line configurations, *Ann. Discrete Math.* **52** (1992), 215–226.

[51] J. L. Gross, Voltage graphs, *Discrete Math.* **9** (1974), 239–246.

[52] J. L. Gross and T. W. Tucker, *Topological Graph Theory*, John Wiley, New York (1987).

[53] W. Gustin, Orientable embeddings of Cayley graphs, *Bull. Amer. Math. Soc.* **69** (1963), 272–275.

[54] H. Hanani, The existence and construction of balanced incomplete block designs, *Ann. Math. Statist.* **32** (1961), 361–386.

[55] N. Hartsfield and G. Ringel, Minimal quadrangulations of orientable surfaces, *J. Combin. Theory Ser. B* **46** (1989), 84–95.

[56] N. Hartsfield and G. Ringel, Minimal quadrangulations of nonorientable surfaces, *J. Combin. Theory Ser. B* **50** (1989), 186–195.

[57] L. Heffter, Über das Problem der Nachbargebiete, *Math. Ann.* **38** (1891), 477–508.

[58] L. Heffter, Über Tripelsysteme, *Math. Ann.* **49** (1897), 101–112.

[59] S. Kantor, Die Configurationen $(3,3)_{10}$, *Sitzungsber. Akad. Wiss. Wien, Math.-Nat. Kl.* **84** (1881), 1291–1314.

[60] P. Kaski and P. R. J. Östergård, The Steiner triple systems of order 19, *Math. Comp.* **73, no. 248** (2004), 2075–2092.

[61] G. B. Khosrovshahi, H. R. Maimani and R. Torabi, On trades: an update, *Discrete Appl. Math.* **95** (1999), 361–376.

[62] T. P. Kirkman, On a problem in combinations, *Cambridge and Dublin Math. J.* **2** (1847), 191–204.

[63] V. P. Korzhik, Exponentially many nonisomorphic orientable triangular embeddings of K_{12s} and K_{12s+3}, in preparation.

[64] V. P. Korzhik and J. H. Kwak, A new approach to constructing exponentially many nonisomorphic nonorientable triangular embeddings of complete graphs, submitted (2006).

[65] V. P. Korzhik and H-J. Voss, On the number of nonisomorphic orientable regular embeddings of complete graphs, *J. Combin. Theory Ser. B* **81** (2001), 58–76.

[66] V. P. Korzhik and H-J. Voss, Exponential families of nonisomorphic nontriangular orientable genus embeddings of complete graphs, *J. Combin. Theory Ser. B* **86** (2002), 186–211.

[67] V. P. Korzhik and H-J. Voss, Exponential families of nonisomorphic nonorientable genus embeddings of complete graphs, *J. Combin. Theory Ser. B* **91** (2004), 253–287.

[68] W. Kühnel, Topological aspects of twofold triple systems, *Expo. Math.* **16** (1998), 289–332.

[69] S. Lawrencenko and S. Negami, Constructing the graphs that triangulate both the torus and the Klein bottle, *J. Combin. Theory Ser. B* **77** (1999), 211–218.

[70] S. Lawrencenko, S. Negami and A. T. White, Three nonisomorphic triangulations of an orientable surface with the same complete graph, *Discrete Math.* **135** (1994), 367–369.

[71] V. Martinetti, Sulle configurazioni piane μ_3, *Ann. Mat. Pura Appl.* **15** (1887), 1–26.

[72] R. A. Mathon, K. T. Phelps and A. Rosa, Small Steiner triple systems and their properties, *Ars Combin.* **15** (1983), 3–110.

[73] N. S. Mendelsohn, A natural generalization of Steiner triple systems, in *Computers in Number Theory*, Academic Press, New York (1971), 323–338.

[74] S. Negami, Uniqueness and faithfulness of embedding of toroidal graphs, *Discrete Math.* **44** (1983), 161–180.

[75] S. Negami, Classification of 6-regular Klein-bottleable graphs, *Research Reports on Information Sciences, Department of Information Sciences, Tokyo Institute of Technology* **A-96** (1984), 16pp.

[76] R. Peltesohn, Eine Lösung der beiden Heffterschen Differenzenprobleme, *Composito Math.* **6** (1939), 251–257.

[77] R. C. Read and R. J. Wilson, *An Atlas of Graphs*, Clarendon Press, Oxford (1998).

[78] G. Ringel, *Map Color Theorem*, Springer-Verlag, New York (1974).

[79] G. Ringel and J. W. T. Youngs, Das Geschlecht des vollständige dreifarbaren Graphen, *Comment. Math. Helv.* **45** (1970), 152–158.

[80] S. Stahl and A. T. White, Genus embeddings for some complete tripartite graphs, *Discrete Math.* **14** (1976), 279–296.

[81] A. T. White, Block designs and graph imbeddings, *J. Combin. Theory Ser. B* **25** (1978), 166–183.

[82] A. T. White, Modelling finite geometries on surfaces, *Discrete Math.* **244** (2002), 479–493.

[83] R. M. Wilson, Nonisomorphic Steiner triple systems, *Math. Z.* **135** (1974), 303–313.

[84] N. H. Xuong, How to determine the maximum genus of a graph, *J. Combin. Theory Ser. B* **26** (1979), 217–225.

[85] J. W. T. Youngs, The Heawood map-colouring problem: cases 1, 7 and 10, *J. Combin. Theory Ser. A* **8** (1970), 220–231.

[86] J. W. T. Youngs, The mystery of the Heawood conjecture, in *Graph Theory and its Applications*, Academic Press, New York (1970), 17–50.

M.J. Grannell and T.S. Griggs
Department of Mathematics
The Open University
Walton Hall
Milton Keynes MK7 6AA
United Kingdom
m.j.grannell@open.ac.uk
t.s.griggs@open.ac.uk

The number of points on an algebraic curve over a finite field

J.W.P. Hirschfeld, G. Korchmáros and F. Torres

Abstract

How many points are there on a curve with coordinates in a given finite field when the curve has (a) no singular points or (b) singular points counted once or (c) singular points counted with multiplicity?
What is the maximum number of points on a curve of given genus?
Can curves attaining this maximum number be characterised?

1 Introduction

Problems in combinatorics, especially in finite geometry, often require a count of the number of solutions of an equation in one or more unknowns defined over a finite field $\mathbf{F}_q$. When two unknowns, say X, Y, occur, the equation is of type $f(X, Y) = 0$ with $f \in \mathbf{F}_q[X, Y]$, and the geometric approach for solving it depends on the theory of algebraic curves over finite fields.

Curves over a finite field have applications in the theory of linear error-correcting codes in two areas: (a) the construction of Goppa or algebraic-geometry codes; (b) obtaining bounds for the maximum length of codes when given the dimension and minimum distance.

In cryptography, ciphers are constructed from both elliptic and hyperelliptic curves

It is natural to think about a plane algebraic curve $\mathcal{F}$ of equation $f(X, Y) = 0$ as the set of the points $P = (x, y)$ in the affine plane over the coordinate field K such that $f(x, y) = 0$. But important numerical results on curves and their intersections, such as Bézout's theorem, have an easier formulation when the following are taken into consideration.

(i) $\mathcal{F}$ is enhanced with its infinite points, that is, when $\mathcal{F}$ is viewed as a curve in the projective plane over K;

(ii) K is algebraically closed, that is, every equation $g(X) = 0$ with $g \in K[X]$ has at least one solution.

Every field is a subfield of an algebraically closed field. The algebraic closure $\overline{\mathbf{F}}_q$ of $\mathbf{F}_q$ contains a unique finite field of order q^n. These finite fields cover $\overline{\mathbf{F}}_q$. So, the idea is to work with plane projective curves over $\overline{\mathbf{F}}_q$ but state the results in the projective subplane over $\mathbf{F}_q$.

The deepest results on the number of points of an algebraic curve over $\mathbf{F}_q$, such as the Hasse–Weil theorem, the Serre bound and the Stöhr–Voloch theorem, are formulated for irreducible, non-singular algebraic curves. Nevertheless, these results can be applied to singular curves, since every irreducible algebraic curve $\mathcal{F}$ defined over $\mathbf{F}_q$ has a non-singular model Γ over $\mathbf{F}_q$, that is, $\mathcal{F}$ is birationally equivalent over $\mathbf{F}_q$ to an irreducible non-singular curve Γ. It should be noted that a birational map does not ensure that every singular point of $\mathcal{F}$ has an image point in Γ. Also, a singular point of $\mathcal{F}$ may have more than one image point in Γ. These phenomena can cause difficulties but do not significantly worsen the bounds, since an irreducible curve can have only a few singular points.

The paper starts with basic facts on algebraic curves. Then, the questions posed in the abstract are addressed for irreducible non-singular curves. Particular results for plane singular curves are discussed in the later sections.

For more details on all the topics covered, see [15].

2   Background

2.1   Planes

Definition 2.1 Let K be a field.

(i) The *affine plane* $\mathrm{AG}(2, K) = \mathbf{A}^2(K)$ is a pair $(\mathcal{P}, \mathcal{L})$ where

$$\mathcal{P} = \{P = (x, y) \mid x, y \in K\}, \quad \mathcal{L} = \{\ell = aX + bY + c \mid a, b, c \in K\},$$

and a *point* $P = (x, y)$ lies on a *line* $\ell = aX + bY + c$ if $ax + by + c = 0$. When $K = \mathbf{F}_q$, write $\mathrm{AG}(2, q)$.

(ii) The *projective plane* $\mathrm{PG}(2, K) = \mathbf{P}^2(K)$ is a pair $(\mathcal{P}, \mathcal{L})$ where

$$\begin{aligned}
\mathcal{P} &= \{P = (x, y, z) = (\lambda x, \lambda y, \lambda z) \mid x, y, z, \lambda \in K;\ \lambda \neq 0\}, \\
\mathcal{L} &= \{\ell = aX + bY + cZ \mid a, b, c \in K\},
\end{aligned}$$

and a *point* $P = (x, y, z)$ lies on a *line* $\ell = aX + bY + cZ$ if $ax + by + cz = 0$. When $K = \mathbf{F}_q$, write $\mathrm{PG}(2, q)$.

2.2   Plane curves

Definition 2.2   (i) The *plane affine curve*

$$\mathcal{F} = \mathbf{v}_a(F) = \{P = (x, y) \in \mathrm{AG}(2, K) \mid F(x, y) = 0\}.$$

(ii) The *degree of* $\mathcal{F}$, written $\deg \mathcal{F}$, is $\deg F$.

Any affine transformation sends an affine curve to another having the same degree. Therefore, $\deg \mathcal{F}$ of an affine curve $\mathcal{F}$ is an affine invariant.

Definition 2.3   (i) A *component* of the affine curve $\mathcal{F} = \mathbf{v}_a(F)$ is an affine curve $\mathcal{G} = \mathbf{v}_a(G)$ such that G divides F.

(ii) The affine curve $\mathcal{F} = \mathbf{v}_a(F)$ is *irreducible* when it has no proper component, that is, when F is irreducible.

Any line containing at least $n + 1$ points from an affine curve $\mathcal{F}$ of degree n is a component of $\mathcal{F}$. To show this, $\ell = \mathbf{v}_a(Y)$ may be assumed by covariance. Let $\mathcal{F} = \mathbf{v}_a(F(X, Y))$. Then $|\ell \cap \mathbf{v}_a(F)| \geq n + 1$ implies that $F(X, 0)$ has more than n roots. Therefore, $F(X, 0) = 0$, and hence X divides $F(X, Y)$.

Let $\mathcal{F} = \mathbf{v}_a(F)$ be an affine curve with $\deg F = d$, and let $\ell = -bX + aY + c$ be a line containing the point $P_0 = (x_0, y_0)$ on $\mathcal{F}$. Then, for any point $P = (x, y) \in \ell$,

$$-bx + ay = -bx_0 + ay_0,$$
$$b(x - x_0) = a(y - y_0) = abt,$$
$$x = x_0 + at, \quad y = y_0 + bt$$

for some $t \in K$. Then

$$F(x, y) = F(x_0 + at, y_0 + bt) = G(t) \quad = \quad G_0 + G_1 t + G_2 t^2 + \ldots + G_d t^d$$
$$= \quad G_m t^m + \ldots + G_d t^d, \qquad (2.1)$$

with $G_m \neq 0$, $G_d \neq 0$.

Definition 2.4 The affine curve $\mathcal{F} = \mathbf{v}_a(F)$ is *irreducible* when F is irreducible.

Lemma 2.5 *The two irreducible curves $\mathcal{F}_1 = \mathbf{v}_a(F_1)$ and $\mathcal{F}_2 = \mathbf{v}_a(F_2)$ are the same if and only if $F_2 = \lambda F_1$ for some $\lambda \in K \backslash \{0\}$.*

Definition 2.6 If $F \in K[X, Y]$ satisfies

$$F = F_1^{n_1} F_1^{n_2} \ldots F_s^{n_s}$$

with each F_i irreducible, then $\mathcal{F} = \mathbf{v}_a(F)$ has *components* $\mathcal{F}_i = \mathbf{v}_a(F_i)$ with *multiplicity* n_i for $i = 1, \ldots, s$.

The multiplicity of a component is an affine invariant.

Definition 2.7 Let ℓ be a line which is not a component of $\mathcal{F}$.

(i) The integer m of (2.1) is the *intersection number of ℓ and $\mathcal{F}$ at P_0*: write

$$m = I(P_0, \ell \cap \mathcal{F});$$

(ii) if $m = 1$ for some line ℓ through P_0, then P_0 is a *simple* or *non-singular* point of $\mathcal{F}$;

(iii) if $m \geq 2$ for all lines ℓ through P_0, then P_0 is a *singular* or *multiple* point of $\mathcal{F}$;

(iv) if $m_0 = \min\{m \mid \ell \text{ a line through } P_0\}$, then m_0 is the *multiplicity of P_0 on $\mathcal{F}$*, or P_0 is an *m_0-fold point* of $\mathcal{F}$, and write

$$m_0 = m_{P_0}(\mathcal{F}) = m_{P_0}(F);$$

(v) if $m > m_0$ for any line ℓ, then ℓ is a *tangent* to $\mathcal{F}$ at P_0.

The intersection number and the multiplicity of a point are affine invariants.

Definition 2.8 If $m_P(\mathcal{F}) = 2$, then P is a *double* point of $\mathcal{F}$. A double point P with two distinct tangents to $\mathcal{F}$ at P is a *node*, and with only one tangent to $\mathcal{F}$ at P is a *cusp*. If $m_P(\mathcal{F}) = 3$, then P is a *triple* point of $\mathcal{F}$.

Remark Let M be a subfield of K and suppose that $\mathcal{F}$ is defined over M, that is, $\mathcal{F} = \mathbf{v}(f(X,Y))$ with $f(X,Y) \in M[X,Y]$. If P is a double point with two distinct tangents, neither of them is defined over M, then P is an *isolated point over M*.

Lemma 2.9 *If P_0 is a simple point of $\mathcal{F}$, then, in (2.1),*

$$G_1 = \left.\frac{\partial F}{\partial X}\right|_{P_0} b - \left.\frac{\partial F}{\partial Y}\right|_{P_0} a.$$

Corollary 2.10 *The tangent to $\mathcal{F}$ at a simple point $P = (x,y)$ is*

$$\ell_P = \left.\frac{\partial F}{\partial X}\right|_{P} (X - x) + \left.\frac{\partial F}{\partial Y}\right|_{P} (Y - y).$$

Note the meaning of this corollary: the line ℓ_P has intersection multiplicity at least 2 with $\mathcal{F}$ at P.

Definition 2.11 A non-singular point P of $\mathcal{F}$ is a *point of inflexion* of $\mathcal{F}$ if

$$I(P, \ell_P \cap \mathcal{F}) \geq 3.$$

Here, P is also called an *inflexion* or, in some sources, a *flex*; the tangent ℓ_P at P is the *inflexional tangent*. Tangents and inflexional tangents are covariant.

Remark The behaviour of $P = (0,0)$ for an affine curve $\mathcal{F} = \mathbf{v}_a(F)$ follows simply from the form of F. Write

$$F(X,Y) = F_m + F_{m+1} + \ldots + F_d,$$

where F_i is homogeneous of degree i in X and Y, and $F_m \neq 0$. Then

(1) if $m > 0$, the point P lies on $\mathcal{F}$;

(2) if $m = 1$, the point P is simple and F_1 is the tangent at P;

(3) if $m \geq 2$, the term $F_m = \prod \ell_i$, where $\ell_1, \ldots, \ell_m$ are the tangents at P;

(4) if $\ell_1, \ldots, \ell_m$ are distinct, then P is an *ordinary* multiple point.

Definition 2.12 If the plane projective curve $\mathcal{F}$ has degree n and singular points $P_1, \ldots, P_r$ of multiplicities $m_1, \ldots, m_r$, then its *genus* is

$$g = \tfrac{1}{2}(d - 1)(d - 2) - \sum_i \tfrac{1}{2} m_i(m_i - 1).$$

Remark The genus of a curve is a non-negative birational invariant.

2.3   Affine and projective curves

Consider the zeros of
$$F = Y - X^3$$
in the affine plane $AG(2,3)$ defined over $\mathbf{F}_3 = \{0, 1, -1 \mid 3 = 0\}$; they are
$$(0,0), \quad (1,1), \quad (-1,-1).$$
Similarly, the zeros of $F' = Y^2 - X^3$ in $AG(2,3)$ are
$$(0,0), \quad (1,1), \quad (1,-1).$$

Now consider the homogeneous versions $F^* = YZ^2 - X^3$ and $F'^* = Y^2Z - X^3$ of these polynomials. Then, in the projective plane $PG(2,3)$, the zeros of F^* are
$$(0,0,1), \quad (1,1,1), \quad (-1,-1,1), \quad (0,1,0),$$
and the zeros of F'^* are
$$(0,0,1), \quad (1,1,1), \quad (1,-1,1), \quad (0,1,0).$$

From the affine viewpoint, the curves given by F and F' are different, since the curve given by F' has a singularity, a cusp, at the origin, whereas the curve given by F has no singularity, even though the origin is an inflexion.

From the projective viewpoint, the two curves given by F^* and F'^* are equivalent, as the interchange of Y and Z interchanges the polynomials. What is happening is now revealed. The curve given by F^* has a cusp at $(0,1,0)$ and an inflexion at $(0,0,1)$; the curve given by F'^* has an inflexion at $(0,1,0)$ and a cusp at $(0,0,1)$. The point $(0,1,0)$ is not seen in the affine version since it lies 'at infinity'.

Thus, when considering plane curves, it is necessary to encompass the projective plane so that no singularities get lost.

For any polynomial $F \in \mathbf{F}_q[X,Y]$ of degree d, let
$$F^*(X,Y,Z) = Z^d F(X/Z, Y/Z);$$
then F^* is homogeneous.

A polynomial $F \in K[X_1, X_2, \ldots, X_n]$ is *irreducible* if it has no non-constant factors over any extension of K.

Definition 2.13   (i) Given $F \in \mathbf{F}_q[X,Y]$ or $F^* \in \mathbf{F}_q[X,Y,Z]$, let
$$V_i = V_i(F) = V_i(F^*) = \{(x,y,z) \in PG(2,q^i) \mid F^*(x,y,z) = 0\};$$
that is, V_i is the set of zeros of F^* over $\mathbf{F}_{q^i}$.

(ii) The *curve* $\mathcal{F} = \mathbf{v}(F) = \mathbf{v}(F^*) = V_1 \cup V_2 \cup \ldots$; that is, the curve consists of points over the ground field $\mathbf{F}_q$ and all algebraic extensions.

(iii) A point P is $\mathbf{F}_{q^i}$-*rational* if its coordinates lie in $\mathbf{F}_{q^i}$; it has *degree* i if it is $\mathbf{F}_{q^i}$-rational but not $\mathbf{F}_{q^j}$-rational for $j < i$. For $\mathbf{F}_q$-rational, the term *rational* is also used.

(iv) The set of K-rational points of $\mathcal{F}$ is denoted $\mathcal{F}(K)$.

3 The zeta function

The definition of a curve can be generalised to higher-dimensional space as an algebraic variety of dimension one. For simplicity, in this article the account is mainly restricted to the plane case, although the results hold in general unless otherwise specified.

Now, let the curve $\mathcal{F}$ be non-singular and let

$$N_i = |V_i|.$$

Define the *zeta function* of $\mathcal{F}$ to be the following formal power series:

$$\zeta_{\mathcal{F}}(T) = \exp(\sum N_i T^i / i).$$

Theorem 3.1 (Hasse–Weil)

$$\zeta_{\mathcal{F}}(T) = f(T)/\{(1-T)(1-qT)\},$$

where

(i) $f(T) = (1 - \alpha_1 T) \ldots (1 - \alpha_{2g} T) \in \mathbf{Z}[T]$;

(ii) $\alpha_j \alpha_{g+j} = q$, $j = 1, \ldots, g$;

(iii) $|\alpha_j| = \sqrt{q}$, $j = 1, \ldots, 2g$.

Corollary 3.2 (i) $N_i = 1 + q^i - (\alpha_1^i + \ldots + \alpha_{2g}^i)$.

(ii) $|N_i - (1 + q^i)| \leq 2g\sqrt{q^i}$.

Let $f(T) = 1 + c_1 T + \ldots + c_{2g} T^{2g}$. For rational points of $\mathcal{F}$, this gives the following result.

Corollary 3.3 (i) $N_1 = 1 + q - (\alpha_1 + \ldots + \alpha_{2g}) = 1 + q + c_1$.

(ii) $|N_1 - (1 + q)| \leq 2g\sqrt{q}$.

Corollary 3.4 *For a plane non-singular curve of degree d,*

$$|N_1 - (1 + q)| \leq (d-1)(d-2)\sqrt{q}.$$

Corollary 3.5 *For $g = 0$,*

(i) $\zeta_{\mathcal{F}}(T) = 1/\{(1-T)(1-qT)\}$;

(ii) $N_1 = q + 1$.

Corollary 3.6 *For $g = 1$,*

(i) $\zeta_{\mathcal{F}}(T) = (1 + cT + qT^2)/\{(1-T)(1-qT)\}$;

(ii) $q + 1 - 2\sqrt{q} \leq N_1 \leq q + 1 + 2\sqrt{q}$.

An improvement of Corollary 3.3(ii) due to Serre is the following result, where $\lfloor x \rfloor$ is the integer part of x.

Theorem 3.7 $|N_1 - (1 + q)| \leq g\lfloor 2\sqrt{q} \rfloor$.

Example 3.8 Let $q = 2$ and $F = X^3 + Y^3 + Z^3$; then

$$V_1 = \{(0, 1, 1), \quad (1, 0, 1), \quad (1, 1, 0)\}).$$

So $N_1 = 3, c = 0$. Hence

$$\zeta_{\mathcal{F}}(T) = \frac{1 + 2T^2}{(1 - T)(1 - 2T)}$$

and

$$\sum N_i T^i = \log \zeta_{\mathcal{F}}(T) = \sum T^i/i + \sum (2T)^i/i + \sum (-1)^{j-1}(2T^2)^j/j.$$

Therefore,

$$N_h = \begin{cases} 1 + 2^h & \text{for } h \text{ odd}, \\ 1 + 2^h + 2.2^{h/2} & \text{for } h \equiv 2 \pmod{4}, \\ 1 + 2^h - 2.2^{h/2} & \text{for } h \equiv 0 \pmod{4}. \end{cases}$$

4 Equality in the Hasse–Weil bound

The *Hermitian* curve $\mathcal{U}_{2,q}$ is the case that

$$F = X^{\sqrt{q}+1} + Y^{\sqrt{q}+1} + Z^{\sqrt{q}+1}. \tag{4.1}$$

This gives an example of a curve $\mathcal{F}$ in which the upper bound in Corollary 3.3(ii) is achieved. Here, $g = \frac{1}{2}(q - \sqrt{q})$, whence

$$q + 1 + 2g\sqrt{q} = q + 1 + (q - \sqrt{q})\sqrt{q} = q\sqrt{q} + 1 = N_1.$$

In fact,

$$N_2 = q\sqrt{q} + 1 = q^2 + 1 - (q - \sqrt{q})q = q^2 + 1 - 2gq,$$

showing that, over $\mathbf{F}_{q^2}$, the curve achieves the lower bound.

Definition 4.1 A curve $\mathcal{F}$ over $\mathbf{F}_q$ is *maximal* if $N_1 = q + 1 + 2g\sqrt{q}$.

Thus $\mathcal{U}_{2,q}$ is one example. Note that q is necessarily a square.
So it is natural to ask the following.

Question 4.2   (i) *Which are the maximal curves?*

(ii) *For which genera does a maximal curve exist?*

(iii) *Classify the maximal curves for a given genus.*

Theorem 4.3 *If $\mathcal{F}$ is maximal curve of genus g defined over $\mathbf{F}_q$, then*

$$g \leq \frac{1}{2}(q - \sqrt{q}).$$

Proof From Theorem 3.1 (iii) and Corollary 3.3 (ii), it follows that $\alpha_i = -\sqrt{q}$ for all i. Hence, also using Corollary 3.3 (iii),

$$q + 1 + 2g\sqrt{q} = N_1 \le N_2 = q^2 + 1 - 2gq.$$

The result follows. $\square$

It also follows that the zeta function of a maximal curve $\mathcal{F}$ is

$$\zeta_{\mathcal{F}}(T) = \frac{(1 + \sqrt{q}T)^{2g}}{(1 - T)(1 - qT)},$$

and, of the Hermitian curve, is

$$\zeta_{\mathcal{U}_2}(T) = \frac{(1 + \sqrt{q}T)^{q-\sqrt{q}}}{(1 - T)(1 - qT)}.$$

Also, if $\mathcal{F}$ is $\mathbf{F}_q$-maximal and N_m is the number of its $\mathbf{F}_{q^m}$-rational points, then

$$N_m = q^m + 1 + (-1)^{m-1}2gq^{m/2}, \qquad \text{for } m = 1, 2, \ldots. \tag{4.2}$$

Given one maximal curve such as the Hermitian curve, other examples flow from the following result, which is ascribed to Serre in Lachaud [20].

Theorem 4.4 *A curve $\mathcal{F}'$ whose function field is a subfield of the function field of a maximal curve $\mathcal{F}$, is also maximal.*

Proof (Outline) The inverse roots α_i' appearing in the zeta function of $\mathcal{F}'$ are a subset of the inverse roots α_i for $\mathcal{F}$. $\square$

Theorem 4.5 (Rück–Stichtenoth [21]) *If a curve $\mathcal{F}$, defined over $\mathbf{F}_q$, is maximal and has genus $g = \frac{1}{2}(q - \sqrt{q})$, then $\mathcal{F}$ is isomorphic to the Hermitian curve $\mathcal{U}_{2,q}$.*

A related result that is both weaker and stronger than this last one is the following. Here it is not assumed that $\mathcal{F}$ is absolutely irreducible.

Theorem 4.6 ([16]) *If $\mathcal{F}$ is a plane curve defined over $\mathbf{F}_q$ with $q > 4$, of degree $\sqrt{q} + 1$, with no linear component, and with at least $q\sqrt{q} + 1$ rational points, then $\mathcal{F}$ is projectively equivalent to the Hermitian curve $\mathcal{U}_{2,q}$.*

It may happen that for a given genus g there is no curve over $\mathbf{F}_q$ that attains the Serre bound.

Definition 4.7 (i) Let $N_q(g) = \max N_1$, taken over all non-singular curves of genus g.

(ii) A curve for which $N_1 = N_q(g)$ is *optimal*.

The simplest case of the Stöhr–Voloch theorem, [22], is the following:

Theorem 4.8 *Let $\mathcal{F}$ be a plane irreducible curve of degree d over $\mathbf{F}_q$ with q odd such that not all points are inflexions. Then*

$$N_1 \le \tfrac{1}{2}d(q + d - 1).$$

Example 4.9 From Theorem 3.7 it follows that $N_7(3) \leq 23$, but Theorem 4.8 implies that $N_7(3) \leq 20$, as a curve of genus 3 can be considered as a plane quartic and it cannot have an infinite number of inflexions. Now, the curve $\mathcal{F} = \mathbf{v}(F)$ over $\mathbf{F}_7$, with

$$F = X^4 + Y^4 + Z^4 + 3(X^2Y^2 + X^2Z^2 + Y^2Z^2),$$

has 20 rational points, namely,

$$(\pm 1, \pm 3, 1), \ (\pm 3, \pm 1, 1), \ (\pm 3, \pm 2, 1), \ (\pm 2, \pm 3, 1), \ (\pm 2, \pm 2, 1)$$

Hence $N_7(3) = 20$, and so $\mathcal{F}$ is optimal. See Top [23].

5   Examples of maximal curves

Now, consider Theorem 4.4 in order to obtain more maximal curves. Beginning from the Hermitian curve $\mathcal{U}_{2,q}$, to find curves that it covers, consider the following curves written in affine form:

$$
\begin{aligned}
\mathcal{D}_t &= \mathbf{v}(X^{(\sqrt{q}+1)/t} + Y^{(\sqrt{q}+1)/t} + 1), \\
\mathcal{A}_t &= \mathbf{v}(Y^{\sqrt{q}} + Y - X^{(\sqrt{q}+1)/t}).
\end{aligned}
\tag{5.1}
$$

The curve $\mathcal{D}_t$ is a *Fermat* curve and the curve $\mathcal{A}_t$ is an *Artin–Schreier* curve; here, $\sqrt{q} \equiv -1 \pmod{t}$. Both $\mathcal{D}_1$ and $\mathcal{A}_1$ are affine forms of $\mathcal{U}_{2,q}$.

Lemma 5.1 *With $m = (\sqrt{q} + 1)/t$, the genus and number of rational points for each of $\mathcal{D}_t$ and $\mathcal{A}_t$ is as follows:*

(a) $\mathcal{D}_t:$ $\begin{aligned} g &= \tfrac{1}{2}(m-1)(m-2), \\ N_1 &= 1 + q + (m-1)(m-2)\sqrt{q}; \end{aligned}$

(b) $\mathcal{A}_t:$ $\begin{aligned} g &= \tfrac{1}{2}(m-1)(\sqrt{q}-1), \\ N_1 &= 1 + q + (m-1)(q-\sqrt{q}). \end{aligned}$

Corollary 5.2   (a) *For $\mathcal{A}_2$, the genus $g = \tfrac{1}{4}(\sqrt{q}-1)^2$;*

(b) *for $\mathcal{A}_4$, the genus $g = \tfrac{1}{8}(\sqrt{q}-1)(\sqrt{q}-3)$;*

(c) *for $\mathcal{D}_2$, the genus $g = \tfrac{1}{8}(\sqrt{q}-1)(\sqrt{q}-3)$.*

Theorem 5.3 (Fuhrmann–Torres [7]) *If a curve $\mathcal{F}$, defined over $\mathbf{F}_q$, is maximal and has genus $g < \tfrac{1}{2}(q - \sqrt{q})$, then $g \leq \tfrac{1}{4}(\sqrt{q}-1)^2$.*

This leads to the following characterisations for q odd and even.

Theorem 5.4 (Fuhrmann–Garcia–Torres [6]) *If a curve $\mathcal{F}$, defined over $\mathbf{F}_q$ with q odd, is $\mathbf{F}_{q^2}$-maximal and has genus $g = \tfrac{1}{4}(\sqrt{q}-1)^2$, then $\mathcal{F}$ is isomorphic to the Artin–Schreier curve $\mathcal{A}_2$.*

Theorem 5.5 (Abdón–Torres [1]) *If a curve $\mathcal{F}$, defined over $\mathbf{F}_q$ with q even and $q \geq 16$, is $\mathbf{F}_{q^2}$-maximal and has genus $g = \tfrac{1}{4}q(q-2)$, then $\mathcal{F}$ is isomorphic to the curve $\mathcal{T}_2 = \mathbf{v}(T_2)$, with*

$$T_2(X, Y) = Y^{q/2} + Y^{q/4} + \ldots + Y^2 + Y + X^{q+1}.$$

Corollary 5.2 then raises two questions:

(1) Are $\mathcal{A}_4$ and $\mathcal{D}_2$ isomorphic?

(2) If so, is a maximal curve of genus $g = \frac{1}{8}(\sqrt{q}-1)(\sqrt{q}-3)$ isomorphic to these curves?

For (1), the following result gives the answer.

Theorem 5.6 *The curves,*

$$\begin{aligned} \mathcal{A}_4 &= \mathbf{v}(Y^{\sqrt{q}} + Y - X^{(\sqrt{q}+1)/4}), \\ \mathcal{D}_2 &= \mathbf{v}(X^{(\sqrt{q}+1)/2} + Y^{(\sqrt{q}+1)/2} + 1), \end{aligned}$$

have the same genus but are not isomorphic.

Theorem 5.7 *([3]) If $\mathcal{F}$ is a non-singular, plane, maximal curve of degree $(\sqrt{q}+1)/2$, then $\mathcal{F}$ is isomorphic to the Fermat curve $\mathcal{D}_2$.*

This theorem is equivalent to saying that, if $\mathcal{F}$ is a maximal curve, which has genus $\frac{1}{8}(\sqrt{q}-1)(\sqrt{q}-3)$ and a plane non-singular model, then $\mathcal{F}$ is isomorphic to $\mathcal{D}_2$.

6   Theoretical background

Definition 6.1   (i) For a plane curve $\mathcal{F} = \mathbf{v}(F)$ with $F \in K[X, Y]$, its *function field* is $\Sigma = K[X, Y]/(F)$.

(ii) The *automorphism group* $\mathrm{Aut}_K(\Sigma)$ of the curve is the group of all K-automorphisms of Σ.

(iii) For any subgroup G of $\mathrm{Aut}_K(\Sigma)$, the set,

$$\Sigma^G = \{z \in \Sigma \mid \sigma(z) = z \text{ for all } \sigma \in G\},$$

is a subfield of Σ, the *fixed subfield of G*.

(iv) The curve $\mathcal{F}'$ whose function field is Σ^G is the *quotient curve of $\mathcal{F}$ with respect to G* and denoted by $\mathcal{F}/G$.

(v) Let Σ' be any subfield of Σ properly containing K. Then the extension Σ/Σ' is algebraic of degree $n = [\Sigma : \Sigma']$. If Σ has a finite automorphism group G of order n such that $\Sigma' = \Sigma^G$, then the extension Σ/Σ' is a *Galois cover of degree n*, and $G = \mathrm{Gal}(\Sigma/\Sigma')$.

Let $\mathcal{F}$ be an absolutely irreducible plane curve of degree d, which is a (possibly singular) plane model of a projective, geometrically irreducible, non-singular, algebraic curve $\mathcal{X}$ defined over $\mathbf{F}_q$. To each point of $\mathcal{X}$ there corresponds a *place* or a *branch* of $\mathcal{F}$; associated to each place is a unique tangent. If P is a place of $\mathcal{F}$ and $\alpha = m_P(\mathcal{F})$ is the minimum of the intersection numbers $I(P, l \cap \mathcal{F})$ for all lines l through P and so the multiplicity of P on $\mathcal{F}$, then α is the *order* of P. The *tangent* l_P at P is the unique line for which $I(P, l_P \cap \mathcal{F}) > \alpha$ and $\beta = I(P, l_P \cap \mathcal{F}) - \alpha$ is the *class* of P. With respect to the linear system $\mathcal{L}_2$ of lines of $\mathrm{PG}(2, \overline{\mathbf{F}}_q)$, a point of order $\alpha = r$ and class $\beta = s - r$ is said to have *order sequence* $(0, r, s)$. This definition of order sequence can be generalised to curves in higher-dimensional spaces; see [15, Chapter 7].

A *point of inflexion* is a point with order sequence $(0, 1, s)$ and $s \geq 3$. If $\mathcal{F}$ has only a finite number of points of inflexion, the order sequence of a generic point is $(0, 1, 2)$ and $\mathcal{F}$ is said to be *classical* for $\mathcal{L}_2$.

If $\mathcal{F}$ is non-classical, then the order sequence at a generic point is $(0, 1, p^v)$, with $p^v > 2$, or, equivalently, the order sequence of $\mathcal{X}$ with respect to γ_n^2, the linear series cut out by lines.

For any curve $\mathcal{F}$, whether classical or non-classical, only a finite number of points have a different order sequence from the generic one. In the case that $\mathcal{F} = \mathcal{U}_{2,q}$ with degree $\sqrt{q} + 1$,

$$(0, r, s) = \begin{cases} (0, 1, \sqrt{q} + 1) & \text{for } P \text{ rational,} \\ (0, 1, \sqrt{q}) & \text{for } P \text{ generic.} \end{cases}$$

The curve $\mathcal{F}$ is *Frobenius classical* if $P^q \notin l_P$, apart from a finite number of places; so it is *Frobenius non-classical* if $P^q \in l_P$. If the order sequence at P is $(0, 1, p^v)$, then the *Frobenius order sequence* at P is

$$(0, \nu) \quad \text{with } \nu = 1 \text{ or } p^v.$$

Then $\mathcal{F}$ is Frobenius classical if $\nu = 1$ and Frobenius non-classical if $\nu = p^v$.

Theorem 4.8 is generalised as follows.

Theorem 6.2 (Stöhr–Voloch [22]) *Let $\mathcal{F}$ be a plane irreducible curve of degree d over $\mathbf{F}_q$. Then*

$$N_1 \leq \tfrac{1}{2}\{d(d - 3)\nu + (q + 2)d\},$$

where $(0, \nu)$ is the Frobenius order sequence.

The most general form of this theorem is the following.

Theorem 6.3 (Stöhr–Voloch [22]) *Suppose that*

(a) *$\mathcal{X}$ is an irreducible curve of genus g;*

(b) *γ_d^n is a linear series on $\mathcal{X}$ of dimension n and order d;*

(c) *the order sequence on $\mathcal{X}$ is $(\epsilon_0, \ldots, \epsilon_n)$;*

(d) *the Frobenius order sequence on $\mathcal{X}$ is $(\nu_0, \ldots, \nu_{n-1})$.*

Then

$$N_1 \leq \tfrac{1}{n}\{(2g - 2)(\nu_0 + \ldots + \nu_{n-1}) + (q + n)d\}.$$

Theorem 6.4 (Hefez–Voloch [12],[13]) *Suppose that*

(a) *$\mathcal{F}$ is a plane non-singular curve of degree d;*

(b) *$\mathcal{F}$ is Frobenius non-classical.*

Then

$$N_1 = d(q - d + 2).$$

An example of this is the Hermitian curve $\mathcal{U}_{2,q}$.

7   Some optimal curves

For $g = 0$, every curve is $\mathbf{F}_q$-optimal, as the number of $\mathbf{F}_q$-rational points of the projective line over $\mathbf{F}_q$ is $q + 1$, while $N_q(0) \leq q + 1$ by the Hasse–Weil bound, Corollary 3.3.

For $g = 1$, the situation is known and is now described.

With $S_q = N_1$, the number of rational points on a curve $\mathcal{F}$, consider the case that $\mathcal{F}$ is an elliptic curve; equivalently, $\mathcal{F}$ is a non-singular plane cubic. For more details, see [14, Chapter 11].

From Theorem 3.6,

$$(\sqrt{q} - 1)^2 \leq N_1 \leq (\sqrt{q} + 1)^2.$$

In fact, the precise values that N_1 can take are given by the next result.

Theorem 7.1 (Waterhouse) *There exists an elliptic cubic over* $\mathbf{F}_q$, $q = p^h$, *with precisely* $N_1 = q + 1 - t$ *rational points, where* $|t| \leq 2\sqrt{q}$, *for precisely the values of* t *in Table 1.*

Table 1: Values of t

	t	p	h
(1)	$t \not\equiv 0 \pmod{p}$		
(2)	$t = 0$		*odd*
(3)	$t = 0$	$p \not\equiv 1 \pmod{4}$	*even*
(4)	$t = \pm\sqrt{q}$	$p \not\equiv 1 \pmod{3}$	*even*
(5)	$t = \pm 2\sqrt{q}$		*even*
(6)	$t = \pm\sqrt{2q}$	$p = 2$	*odd*
(7)	$t = \pm\sqrt{3q}$	$p = 3$	*odd*

Let $N_q(1)$ denote the maximum number of rational points on any non-singular cubic over $\mathbf{F}_q$ and $L_q(1)$ the minimum number. The prime power $q = p^h$ is *exceptional* if h is odd, $h \geq 3$, and p divides $\lfloor 2\sqrt{q} \rfloor$.

Remark   The only exceptional $q < 1000$ is $q = 128$.

Corollary 7.2 *The bounds* $N_q(1)$ *and* $L_q(1)$ *are as follows:*

(i) $N_q(1) = \begin{cases} q + \lfloor 2\sqrt{q} \rfloor, & \text{if } q \text{ is exceptional} \\ q + 1 + \lfloor 2\sqrt{q} \rfloor, & \text{if } q \text{ is non-exceptional;} \end{cases}$

(ii) $L_q(1) = \begin{cases} q + 2 - \lfloor 2\sqrt{q} \rfloor, & \text{if } q \text{ is exceptional} \\ q + 1 - \lfloor 2\sqrt{q} \rfloor, & \text{if } q \text{ is non-exceptional.} \end{cases}$

Proof   This is an immediate consequence of Theorem 7.1.   □

Corollary 7.3 *The number N_1 takes every integer value between $q + 1 - \lfloor 2\sqrt{q} \rfloor$ and $q + 1 + \lfloor 2\sqrt{q} \rfloor$ if and only if* (a) $q = p$ *or* (b) $q = p^2$ *with* $p = 2$ *or* $p = 3$ *or* $p \equiv 11 \pmod{12}$.

Corollary 7.4 *For $q \leq 128$, the values that N_1 cannot take between $L_q(1)$ and $N_q(1)$ are given in Table 2.*

Table 2: The values that N_1 cannot take between $L_q(1)$ and $N_q(1)$ for $q \leq 128$

q	Forbidden values
8	$7, 11$
16	$11, 15, 19, 23$
25	26
27	$22, 25, 31, 34$
32	$23, 27, 29, 31, 35, 37, 39, 43$
49	$43, 57$
64	$51, 53, 55, 59, 61, 63, 67, 69, 71, 75, 77, 79$
81	$67, 70, 76, 79, 85, 88, 94, 97$
	$106, 111, 116, 121, 131, 136, 141, 146$
128	$109, 111, 115, 117, 119, 121, 123, 125, 127,$
	$131, 133, 135, 137, 139, 141, 143, 147, 149$

Corollary 7.5 *(i) For q square, there exists a maximal plane cubic curve*

(ii) For q non-square, there exists an optimal plane cubic curve

Here are other examples of maximal and optimal curves.

Example 7.6 (a) The Hermitian curve, Example 4.1, is both $\mathbf{F}_q$-optimal and $\mathbf{F}_q$-maximal for $q = p^{2e}$.

(b) The DLS curve, Let $q_0 = 2^e$ and $q = 2q_0^2$. The irreducible plane curve

$$\mathcal{S} = \mathbf{v}(X^{2q_0}(X^q + X) + Y^q + Y)$$

is $\mathbf{F}_q$-optimal and $\mathbf{F}_{q^4}$-maximal for $q = 2^{2e+1}$ with $e \geq 1$. This curve is associated with the Suzuki–Tits ovoid in $\mathrm{PG}(3, q)$.

(c) the DLR curve, $p = 3$, $q = 3q_0^2$, with $q_0 = 3^s$, $s \geq 1$, and

$$\mathcal{R} = \mathbf{v}(Y^{q^2} - [1 + (X^q - X)^{q-1}]Y^q + (X^q - X)^{q-1}Y - X^q(X^q - X)^{q+3q_0})$$

is $\mathbf{F}_q$-optimal and $\mathbf{F}_{q^6}$-maximal for $q = 3^{2e+1}$ with $e \geq 1$. This curve is associated with the Ree–Lüneburg unital.

8   The Frobenius linear series of a maximal curve

It is more convenient from now on to use $\mathbf{F}_{q^2}$ as the underlying field. Thanks to the following linear equivalence of divisors, further theoretical results on maximal curves can be obtained; see, for example, [6], [19]:

$$qP + \Phi(P) \equiv (q+1)P_0; \qquad (8.1)$$

here $P_0 \in \mathcal{X}(\mathbf{F}_{q^2})$. The proof of this remark uses facts concerning Tate modules and can be seen in [21]. The linear series $\mathcal{D} = |(q+1)P_0|$ is the *Frobenius linear series* of $\mathcal{X}$ and it may be assumed that $\mathcal{X}$ is embedded in $\mathbf{P}^r$ [18], where r is the dimension of $\mathcal{D}$. From (8.1), $\dim |qP| = r - 1$ for every $P \in \mathcal{X}$. By the Weierstrass Gap Theorem, see [15, Section 6.6], an immediate consequence of (8.1) for the non-gap sequence of $\mathcal{X}$ at a point $P \in \mathcal{X}$ is the following:

$$0 < m_1(P) < \ldots < m_{r-1}(P) \leq q < m_r(P). \qquad (8.2)$$

For an $\mathbf{F}_{q^2}$-maximal curve, a number of basic facts are collected in the next proposition. For $P \in \mathcal{X}$, let $j_0(P) < j_1(P) < \ldots < j_r(P)$ denote the sequence of possible intersection multiplicities of $\mathcal{X}$ with hyperplanes of $\mathbf{P}^n$. For all but a finite number of points, the sequence above is constant. This generic sequence is denoted by replacing $j_i(P)$ by ϵ_i

In the case of the Hermitian curve $\mathcal{U}_{2,q^2}$, now written

$$\mathcal{H}_q = \mathbf{v}(Y^q + Y - X^{q+1}),$$

the linear series $\mathcal{D}$ is cut out by lines of the plane.

For an $\mathbf{F}_{q^2}$-maximal curve, a number of basic facts are collected in the next proposition.

Proposition 8.1 *With the notation above, the following hold.*

(I) *If P and Q are $\mathbf{F}_{q^2}$-rational points, then $(q+1)P \equiv (q+1)Q$, and $q+1$ is a non-gap at each $P \in \mathcal{X}(\mathbf{F}_{q^2})$.*

(II) *There exists $P_1 \in \mathcal{X}(\mathbf{F}_{q^2})$ such that both $q+1$ and q are non-gaps at P_1.*

(III) *The linear series $\mathcal{D}$ is complete, base-point-free, simple and defined over $\mathbf{F}_{q^2}$. it gives rise to an $\mathbf{F}_{q^2}$-rational curve Γ of $\mathrm{PG}(r,q)$ that is $\mathbf{F}_{q^2}$-birationally equivalent to $\mathcal{X}$.*

(IV) *The $(\mathcal{D}, P)$-orders at an $\mathbf{F}_{q^2}$-rational point P are precisely*

$$0 < q+1 - m_{r-1}(P) < \ldots < q+1 - m_1(P) < q+1;$$

that is, $j_{r-i}(P) + m_i(P) = q+1$ for $i = 0, \ldots, r-1$.

(V) *If $P \notin \mathcal{X}(\mathbf{F}_{q^2})$, then $j_1(P) = 1$ and so $\epsilon_1 = 1$.*

(VI) *The integer q is a $\mathcal{D}$-order, and so $r \geq 2$.*

(VII) *If $P \in \mathcal{X}(\mathbf{F}_{q^4}) \setminus \mathcal{X}(\mathbf{F}_{q^2})$ then $q - 1$ is a non-gap at P; if $P \notin \mathcal{X}(\mathbf{F}_{q^4})$ then q is a non-gap at P.*

(VIII) *If P is an $\mathbf{F}_{q^2}$-rational point of $\mathcal{X}$, then $j_{r-1}(P) < q$.*

(IX) *$\epsilon_r = \nu_{r-1} = q$, so Γ is Frobenius non-classical, and every $\mathbf{F}_{q^2}$-rational point of $\mathcal{X}$ is in the support of the ramification divisor R of Γ.*

(X) *If $N_1 \geq q^3 + 1$, then $m_1(P) = q$ for every $\mathbf{F}_{q^2}$-rational point P of $\mathcal{X}$.*

(XI) *If $P \in \mathcal{X}$ is not an $\mathbf{F}_{q^2}$-rational point, then*

$$0 \leq q - m_{r-1}(P) < \ldots < q - m_1(P) < q$$

are $(\mathcal{D}, P)$-orders at P. In particular, $j_r(P) = q$.

(XII) *If P is an $\mathbf{F}_{q^2}$-rational point then both q and $q + 1$ are non-gaps at P. In particular, $j_1(P) = 1$ for every $\mathbf{F}_{q^2}$-rational point P.*

(XIII) *Either $r = q - (g - 1)$ or $r \leq \frac{1}{2}(q + 1)$.*

9   Maximal curves of large genus

The number of rational points on a maximal curve over $\mathbf{F}_{q^2}$ is

$$q^2 + 1 + 2gq.$$

The Hermitian curve $\mathcal{U}_{2,q^2}$ is now written

$$\mathcal{H}_q = \mathbf{v}(Y^q + Y - X^{q+1}).$$

To avoid trivial cases, it is assumed that $g > 0$, unless otherwise stated.

Theorem 9.1 *If $\mathcal{X}'$ is an $\mathbf{F}_{q^2}$-rational curve covered by an $\mathbf{F}_{q^2}$-maximal curve $\mathcal{X}$, with an $\mathbf{F}_{q^2}$-rational covering, then $\mathcal{X}'$ is also $\mathbf{F}_{q^2}$-maximal.*

Example 9.2 (i) For q odd, the irreducible plane curve,

$$\mathcal{E}^1_{(q+1)/2} = \mathbf{v}(Y^q + Y - X^{(q+1)/2}), \tag{9.1}$$

is covered by the Hermitian curve $\mathcal{H}_q$, since $K(\mathcal{E}^1_{(q+1)/2})$ is the subfield $K(x^2, y)$ of $K(\mathcal{H}_q) = K(x, y)$. Note that $K(x^2, y)$ is a proper subfield of $K(\mathcal{H}_q)$ with

$$y^q + y - x^{q+1} = 0,$$

as the genus g' of $\mathcal{E}^1_{(q+1)/2}$ is smaller than $\frac{1}{2}(q^2 - q)$, the genus of $\mathcal{H}_q$. Therefore, $[K(\mathcal{H}_q) : K(\mathcal{E}^1_{(q+1)/2})] = 2$, and the rational transformation $\omega : (x, y) \mapsto (x^2, y)$ provides a two-fold covering of $\mathcal{E}^1_{(q+1)/2}$ by $\mathcal{H}_q$. Then, $g' = \frac{1}{4}(q - 1)^2$.

Since $x^2, y \in \mathbf{F}_{q^2}(\mathcal{H}_q)$, then $\mathcal{E}^1_{(q+1)/2}$, ω and the associated two-fold covering are $\mathbf{F}_{q^2}$-rational, as well. From Theorem 9.1, $\mathcal{E}^1_{(q+1)/2}$ is an $\mathbf{F}_{q^2}$-maximal curve.

Table 3: Families of curves $\mathcal{F} = \mathbf{v}(F)$ containing $\mathbf{F}_{q^2}$-maximal curves

$\mathcal{F}$	F
$\mathcal{D}_r$	$X^r + Y^r + 1$
$\mathcal{E}_m^1$	$Y^q + Y - X^m$
$\mathcal{T}_p$	$X^{q+1} - (Y + Y^p + Y^{p^2} + \ldots + Y^{q/p})$
$\mathcal{T}_3'$	$(Y + Y^3 + \ldots + Y^{q/3})^2 - X^q - X$
$\mathcal{T}_3''$	$Y + Y^3 + \ldots + Y^{q/3} + cX^{q+1}, \ c^{q-1} = -1$
$\mathcal{K}$	$Y^{q+1} - f(X)$
$\mathcal{A}$	$Y^q - Y - f(X)$
$\mathcal{F}_0$	$X^{(q+1)/3} + X^{2(q+1)/3} + Y^{q+1}$
$\mathcal{F}_0'$	$YX^{(q-2)/3} + Y^q + X^{(2q-1)/3}$
$\mathcal{G}$	$Y^q - YX^{2(q-1)/3} + \omega X^{(q-1)/3}, \ \omega^{q+1} = -1$
$\mathcal{C}_n$	$X^nY + Y^n + X$
$\mathcal{C}_{n,k}$	$X^nY^k + Y^n + X^k$
$\mathcal{C}_i^m$	$X^{mi+m} + X^{mi} + Y^{q+1}$
$\mathcal{X}_r$	$Y^{2^r} + a_1Y^{2^{r-1}} + \ldots + a_{r-1}Y^2 + Y + X^{q+1}$

(ii) The corresponding example for q even is the irreducible plane curve,

$$\mathcal{T}_2 = \mathbf{v}(X^{q+1} + Y + Y^2 + Y^4 + \ldots + Y^{q/2}), \tag{9.2}$$

whose genus is $\frac{1}{4}q(q-2)$. Since $K(\mathcal{T}_2)$ is the subfield $K(x, y^2 + y)$ of $K(\mathcal{H}_q)$, the same argument shows that $\mathcal{T}_2$ is two-fold covered by the Hermitian curve $\mathcal{H}_q$, and hence it is an $\mathbf{F}_{q^2}$-maximal curve.

(iii) The curve $\mathcal{F}_0 = \mathbf{v}(F_0)$ with

$$F_0 = X^{(q+1)/3} + X^{2(q+1)/3} + Y^{q+1};$$

its genus is $\frac{1}{6}(q^2 - q + 4)$; here, $q \equiv 2 \pmod 3$.

(iv) The curve $\mathcal{F}_0' = \mathbf{v}(F_0')$ with

$$F_0' = YX^{(q-2)/3} + Y^q + X^{(2q-1)/3};$$

its genus is $\frac{1}{6}(q^2 - q - 2)$; here, $q \equiv 2 \pmod 3$.

(v) For curves of genus $\frac{1}{6}(q^2 - q)$, examples are $\mathcal{T}_3' = \mathbf{v}(T_3')$ with

$$T_3' = \mathrm{T}(Y)^2 - X^q - X, \quad \mathrm{T}(Y) = Y + Y^3 + \ldots + Y^{q/3},$$

when $q \equiv 0 \pmod 3$, and $\mathcal{G} = \mathbf{v}(G)$ when $q \equiv 1 \pmod 3$, with

$$G = Y^q - YX^{2(q-1)/3} + \omega X^{(q-1)/3}, \quad \omega^{q+1} = -1,$$

Every non-trivial automorphism group G of $\mathcal{X}$ gives rise to a covering of $\mathcal{X}$. In Example 9.2 (i), $\mathcal{F} = \mathcal{H}_q/\langle\alpha\rangle$ with $\alpha : (X,Y) \mapsto (-X,Y)$. Such a covering and the corresponding quotient curve $\mathcal{X}' = \mathcal{X}/G$ are $\mathbf{F}_{q^2}$-rational if G is an $\mathbf{F}_{q^2}$-automorphism group; that is, G is the restriction to $\mathcal{X}$ of a subgroup of $\mathrm{PGL}(r, q^2)$. Each of the cases, (a), (b), (c), in Example 7.6 of $\mathbf{F}_{q^2}$-maximal curves has a large $\mathbf{F}_{q^2}$-automorphism group with many non-conjugate subgroups. From this, the existence of numerous $\mathbf{F}_{q^2}$-rational maximal curves is deduced. The genera of such quotient curves can often be computed using such theorems as Hurwitz's, but the problem of finding an explicit equation has been solved so far only in a few cases.

The existence of an $\mathbf{F}_{q^2}$-maximal curve which is not $\mathbf{F}_{q^2}$-covered by the Hermitian curve $\mathcal{H}_q$ is still unknown. Possible candidates are the DLS and DLR curves, or some of their quotient curves. In this vein, it would help to know if any quotient curves of the DLS and the DLR curves are quotient curves of $\mathcal{H}_q$.

A partial answer in the negative is given by the following example. Over $\mathbf{F}_{27^2}$, the Hermitian curve is

$$\mathcal{H}_{27} = \mathbf{v}(Y^{27} + Y - X^{28}).$$

Theorem 9.3 (Garcia and Stichtenoth) [8] *The curve $\mathcal{C}$ is $\mathbf{F}_{27^2}$-maximal but is not a Galois subcover of $\mathcal{H}_{27}$, where*

$$\mathcal{C} = \mathbf{v}(Y^9 - Y - X^7).$$

There are, however, $\mathbf{F}_{q^2}$-maximal curves with simple equations such as those of Kummer type,

$$\mathcal{K} = \mathbf{v}(Y^{q+1} - f(X)), \tag{9.3}$$

and those of Artin–Schreier type,

$$\mathcal{A} = \mathbf{v}(Y^q - Y - f(X)). \tag{9.4}$$

The classification of maximal curves is currently out of reach. However, for larger values of g for which there exists an $\mathbf{F}_{q^2}$-maximal curve, it seems that there are few curves: see Table 4. This has been shown so far for $g \geq \lfloor \frac{1}{6}(q^2 - q + 4) \rfloor$.

10 Non-isomorphic maximal curves

In this section, a 2-parameter family of curves $\mathcal{X}_i^m$ is presented; for each fixed m, there is a large number of non-isomorphic curves all with some identical properties.

With $K = \overline{\mathbf{F}}_q$, let $\mathcal{X}_i^m$ be a non-singular model over K of the plane curve,

$$\mathcal{C}_i^m = \mathbf{v}(X^{mi+m} + X^{mi} + Y^{q+1}), \tag{10.1}$$

where m is a positive divisor of $q+1$ for which $d = (q+1)/m > 3$ is prime. The curve $\mathcal{X}_i^m$ is the quotient curve of the Hermitian curve $\mathcal{H}_q$ arising from an automorphism group of $\mathcal{H}$ of the same order d. Let $\mathcal{D} = |(q+1)P|$ denote the associated complete linear series at a point P of $\mathcal{X}_i^m$.

Theorem 10.1 *Assume $1 \leq i \leq d - 2$.*

Table 4: Known $\mathbf{F}_{q^2}$-maximal curves of large genera

	Genus g	Condition on q	Curves
1.	$\frac{1}{2}q(q-1)$		$\mathcal{H}_q = \mathcal{D}_{q+1}$
2.	$\frac{1}{4}(q-1)^2$	$q \equiv 1 \pmod 2$	$\mathcal{E}^1_{(q+1)/2}$
3.	$\frac{1}{4}q(q-2)$	$q \equiv 0 \pmod 2$	$\mathcal{T}_2$
4.	$\frac{1}{6}(q^2 - q + 4)$	$q \equiv 2 \pmod 3$	$\mathcal{F}_0$
5.	$\frac{1}{6}(q^2 - q)$	$q \equiv 1 \pmod 3$	$\mathcal{G}$
6.	$\frac{1}{6}(q^2 - q)$	$q \equiv 0 \pmod 3$	$\mathcal{T}'_3$
7.	$\frac{1}{6}(q^2 - q - 2)$	$q \equiv 2 \pmod 3$	$\mathcal{F}'_0$
8.	$\frac{1}{6}(q-1)(q-2)$	$q \equiv 2 \pmod 3$	$\mathcal{E}^1_{(q+1)/3}$
9.	$\frac{1}{6}q(q-3)$	$q \equiv 0 \pmod 3$	$\mathcal{T}''_3$
10.	$\frac{1}{8}(q^2 - 2q + 5)$	$q \equiv 3 \pmod 4$	
11.	$\frac{1}{8}(q-1)^2$	$q \equiv 1 \pmod 4$	
12.	$\frac{1}{8}q(q-2)$	$q \equiv 0 \pmod 4$	
13.	$\frac{1}{8}(q-1)(q-3)$	$q \equiv 1 \pmod 4$	$\mathcal{E}^1_{(q+1)/4}$
14.	$\frac{1}{8}(q-1)(q-3)$	$q \equiv 3 \pmod 4$	$\mathcal{E}^1_{(q+1)/4}, \mathcal{D}_{(q+1)/2}$
15.	$\frac{1}{8}q(q-4)$	$q \equiv 0 \pmod 2$	$\mathcal{X}_2$

(i) (a) The curves $\mathcal{X}^m_i$ and $\mathcal{X}^m_j$ are K-equivalent if and only if one of the following equation holds modulo d :

$$i \equiv j, \qquad ij \equiv 1, \qquad ij + i + j \equiv 0,$$
$$i + j + 1 \equiv 0, \quad ij + i + 1 \equiv 0, \quad ij + j + 1 \equiv 0.$$

(b) The number of K-isomorphism classes of curves $\mathcal{X}^m_i$ is given by

$$n(d) = \begin{cases} \frac{1}{6}(d+1) & \text{if } d \equiv 2 \pmod 3, \\ \frac{1}{6}(d-1) + 1 & \text{if } d \equiv 1 \pmod 3. \end{cases} \qquad (10.2)$$

(c) Each of these classes consists of six curves, apart from two exceptions of sizes 2 and 3. The corresponding indices i are as follows:

(1) i_1, i_2, where i_1 and i_2 are the solutions of $t^2 + t + 1 = 0 \pmod d$, with $d \equiv 1 \pmod 3$;

(2) $1, \frac{1}{2}(d-1), d-2$.

(ii) The genus of $\mathcal{X}^m_i$ is $g = \frac{1}{2}m(q-2) + 1$.

(iii) *The K-automorphism group of $\mathcal{X}_i^m$ is*

$$\mathrm{Aut}(\mathcal{X}_i^m) = \begin{cases} \mathbf{Z}_3 \rtimes (\mathbf{Z}_{q+1} \times \mathbf{Z}_m) & \text{in case (c)(1)}, \\ \mathbf{Z}_2 \rtimes (\mathbf{Z}_{q+1} \times \mathbf{Z}_m) & \text{in case (c)(2)}, \\ \mathbf{Z}_{q+1} \times \mathbf{Z}_m & \text{otherwise}. \end{cases}$$

(iv) *When $m = 2$, the series $\mathcal{D}$ has projective dimension $\frac{1}{2}(d + 3)$. There are at least six K-rational points P such that, if*

$$(j_0 = 0, j_1 = 1, \ldots, j_{(d+1)/2}, j_{(d+3)/2})$$

is the $\mathcal{D}$-order sequence at P, then $j_{(d+1)/2} = d$, $j_{(d+3)/2} = q + 1$.

(v) *When $m = 2$ and q is prime, then the $\mathcal{D}$-order sequence at a generic point is $(0, 1, \ldots, \frac{1}{2}(d + 1), q)$.*

Theorem 10.2   (i) *The curves $\mathcal{X}_0^m$ and $\mathcal{X}_{d-1}^m$ are K-isomorphic.*

(ii) *$\mathcal{X}_0^m$ has genus $g = \frac{1}{2}(m - 1)(q - 1)$, and is hyperelliptic when $m = 2$.*

(iii) *The centre Z of the K-automorphism group $\mathrm{Aut}_K(\mathcal{X}_0^m)$ is a cyclic group of order m, and the factor group $\mathrm{Aut}_K(\mathcal{X}_0^m)/Z$ is isomorphic to $\mathrm{PGL}(2, q)$.*

(iv) *The complete linear series on $\mathcal{X}_0^2$ has projective dimension $d + 1$, and the $\mathcal{D}$-order sequence at a Weierstrass point is one of*

$$(0, 1, 2, \ldots, d, q), \quad (0, 1, 2, 4, 6, \ldots, q - 1, q + 1).$$

(v) *The $\mathcal{D}$-order sequence of $\mathcal{X}_0^2$ is $(0, 1, \ldots, d, q)$.*

11 Singular plane curves

Let $f \in \mathbf{F}_q[X, Y]$ be an irreducible polynomial of degree d, and let $\mathcal{F} = \mathbf{v}(f)$ be the corresponding irreducible, possibly singular, plane curve. The problem of counting the number R_q of points in $\mathrm{PG}(2, q)$ which lie on $\mathcal{F}$ is of interest not only in the present but also in other contexts.

Note that R_q counts the solutions of the equation $f(X, Y) = 0$ in $\mathbf{F}_q \times \mathbf{F}_q$ together with the homogeneous non-zero solutions of $\Phi(X, Y) = 0$, where $\Phi(X, Y)$ is the homogeneous polynomial of all terms of degree d in $f(X, Y)$.

It is natural to compare R_q with $S_q = N_1$, the number of $\mathbf{F}_q$-rational branch points. If B_q is the number of branches of $\mathcal{F}$ centred at points of $\mathrm{PG}(2, q)$, and $\widetilde{R}_q$ is similar to R_q but counts each r-fold point counted in R_q with multiplicity r, then

$$S_q \leq B_q \leq \widetilde{R}_q.$$

This shows that the problems of determining $S_q, R_q, B_q, \widetilde{R}_q$ are equivalent only when $\mathcal{F}$ is non-singular. Nevertheless, some results on $R_q, B_q, \widetilde{R}_q$ are similar to those on S_q. In fact, the proof of Theorem 11.1 remains valid when $\widetilde{R}_q$ replaces R_q. This also finds confirmation in Theorem 11.2, which extends Theorem 11.1, in itself a special case of Theorem 6.3, and Theorem 6.4 to singular plane curves.

Theorem 11.1 *Assume that K has odd characteristic. Let $\mathcal{F}$ be an irreducible plane curve of degree d defined over $\mathbf{F}_q$. If $\mathcal{F}$ has only finitely many points of inflexion, then the number S_q of $\mathbf{F}_q$-rational points of $\mathcal{F}$ satisfies the inequality,*

$$2S_q + N' \leq d(q + d - 1), \tag{11.1}$$

where N' counts the non-$\mathbf{F}_q$-rational points $Q \in \mathcal{F}$ such that the tangent line at Q contains the image Q^q of Q under the Frobenius collineation.

To see, in a simple case, the differences between $R_q, S_q, B_q, \widetilde{R}_q$, consider the three singular plane cubics, $\mathcal{N}_2, \mathcal{N}_1$, and $\mathcal{N}_0$, with two, one, and zero tangents over $\mathbf{F}_q$ at the singular point P; these are cubics with a node, a cusp, and an isolated double point at P. Then Table 5 is straightforward to verify.

Table 5: Numbers of points on singular cubics

	R_q	S_q	B_q	$\widetilde{R}_q$
$\mathcal{N}_2$	q	$q+1$	$q+1$	$q+1$
$\mathcal{N}_1$	$q+1$	$q+1$	$q+1$	$q+2$
$\mathcal{N}_0$	$q+2$	$q+1$	$q+3$	$q+3$

It may be noted that $B_q = \widetilde{R}_q$ when P is a node. This holds true for curves with only ordinary singularities centred at points in $\mathrm{PG}(2, q)$.

Theorem 11.2 *Let $\mathcal{F}$ be an irreducible plane curve of degree d and genus g defined over $\mathbf{F}_q$.*

(i) *If $\mathcal{F}$ is either classical, or non-classical but Frobenius classical, then*

$$B_q \leq \tfrac{1}{2}\{(2g - 2) + (q + 2)d\}. \tag{11.2}$$

(ii) *If $\mathcal{F}$ is Frobenius non-classical with order sequence $(0, q')$, then*

$$B_q \leq \tfrac{1}{2}\{q'(2g - 2) + (q + 2)d\}. \tag{11.3}$$

Theorem 11.3 *Let $\mathcal{F}$ be a non-classical irreducible plane curve of degree d and genus g defined over $\mathbf{F}_q$. If $\mathcal{F}$ is Frobenius non-classical, and has only tame branches, then*

$$B_q \geq (q - 1)d - (2g - 2). \tag{11.4}$$

Also, equality holds if and only if every non-linear branch of $\mathcal{F}$ is centred at a point in $\mathrm{PG}(2, q)$.

Example 11.4 (1) The Hermitian curve $\mathcal{H}_{\sqrt{q}} = \mathbf{v}(X^{\sqrt{q}+1} + Y^{\sqrt{q}+1} + 1)$ attains the upper bound (11.3) for $q' = \sqrt{q}$, and the lower bound (11.4).

(2) Another example which illustrates Theorems 11.2 and 11.3 for $q = p^3$ and p odd, is the dual curve $\mathcal{C}$ of the plane curve

$$\mathcal{F} = \mathbf{v}(Y^{p^2+p+1} - (X^{p+1}+1)^p X + (X^p + X)^p).$$

The main properties of $\mathcal{C}$ are as follows:

(i) $\mathcal{C}$ is a projective singular plane curve defined over $\mathbf{F}_q$ birationally equivalent to $\mathcal{C}$;

(ii) $\mathcal{C}$ has degree $(p^2 + p + 1)(p + 1)$, and genus $g = (p^2 + p)(p^2 + p - 1)/2$;

(iii) $\mathcal{C}$ is a Frobenius non-classical plane curve with $\epsilon_2 = \nu_1 = p^2$;

(iv) $\mathcal{C}$ has only one non-linear branch; it is centred at a point in $\mathrm{PG}(2, q)$ and has order $p + 1$.

Apply Theorem 11.2:

$$\begin{aligned}
B_{p^3} &= (p^2 + p + 1)(p + 1)(p^3 - 1) - (p^2 + p + 1)(p^2 + p - 2) \\
&= (p^2 + p + 1)(p - 1)(p^3 + 2p^2 + p - 1).
\end{aligned}$$

When $p = 3$, this gives $B_{27} = 1222$, $N_{27} = 208$.

(3) Now, let $q = p = 2$, and $\mathcal{F} = \mathbf{v}((X^2 + X)(Y^2 + Y) + 1)$. Then $\mathcal{F}$ is classical but Frobenius non-classical with $\epsilon_2 = \nu_1 = 2$. Here, $\mathcal{F}$ has two points in $\mathrm{PG}(2, q)$, namely X_∞ and Y_∞, both singular. Also, X_∞ is a double point and both branches of $\mathcal{F}$ centred at X_∞ are $\mathbf{F}_q$-rational; the similar property holds for Y_∞. Therefore, $\mathcal{F}$ has only linear and hence tame branches. Since $\mathcal{F}$ has genus $g = 1$, so $B_2 = 4$, as in (11.2).

(4) To illustrate Theorem 11.3, put $q = 2^{2e+1}$, $q_0 = 2^e$, with $e \geq 1$, and consider the DLS irreducible plane curve,

$$\mathcal{F} = \mathbf{v}(Y^q + Y + X^{q_0}(X^q + X)),$$

of genus $q_0(q - 1)$.

It has several interesting properties. First, $\mathcal{F}$ is Frobenius non-classical with $\epsilon = \nu = q_0$; in fact,

$$z_0(x, y) = x^{q/q_0 + 1} + y^{q/q_0}, \ z_1(x, y) = x, \ z_2(x, y) = 1.$$

Also, $\mathcal{F}$ has only one singular point, namely Y_∞. More precisely, Y_∞ is the centre of exactly one branch of $\mathcal{F}$, and hence $B_q = q^2 + 1$. This branch P has order $r = q_0$ and class $s = q$; in particular, P is a non-tame branch. Theorem 11.3 fails in the sense that here equality does not hold in (11.4); the unique singular branch of $\mathcal{F}$ is centred at a point in $\mathrm{PG}(2, q)$, but

$$q^2 + 1 > q^2 - qq_0 - q - q_0 + q_0 + 2 = (q - 1)(q + q_0) - (2g - 2).$$

12   Counting points on a plane curve

If $\mathcal{F}$ is a non-singular plane curve of degree n defined over $\mathbf{F}_q$, then $\mathcal{X}$ can be identified with $\mathcal{F}$, and the Hasse–Weil bound (3.3) reads as follows:

$$q + 1 - (d-1)(d-2)\sqrt{q} \leq S_q \leq q + 1 + (d-1)(d-2)\sqrt{q}, \qquad (12.1)$$

where d is the degree of $\mathcal{F}$ and $S_q = N_1$ is the number of all $\mathbf{F}_q$-rational points of $\mathcal{F}$. Now, (12.1) is extended to irreducible singular plane curves defined over $\mathbf{F}_q$.

If a curve is not identified with its non-singular model, then there is some ambiguity in the definition of an $\mathbf{F}_q$-rational point. For a non-singular plane curve $\mathcal{F}$ defined over $\mathbf{F}_q$, there is a one-to-one correspondence between the $\mathbf{F}_q$-rational points of $\mathcal{F}$ and the $\mathbf{F}_q$-rational places of the associated function field $K(\mathcal{F})$. If $\mathcal{F}$ is singular and $\mathcal{X}$ is a non-singular model of $\mathcal{F}$, defined over $\mathbf{F}_q$ as well, then $\mathcal{F}$ and $\mathcal{X}$ have the same function field and hence they have the same number N_1 of $\mathbf{F}_q$-rational points, but N_1 is, in general, different from the number R_q of points of $\mathcal{F}$ which lie in $\mathrm{PG}(2,q)$.

This already appeared in Section 11. For instance, if $\mathcal{C}_3$ is an irreducible cubic curve defined over $\mathbf{F}_q$ with an isolated double point, then, as for any plane singular cubic, a non-singular model is a twisted cubic in $\mathrm{PG}(3,q)$; so $S_q = q + 1$ but, from Table 5, $R_q = q + 2$.

From now on, $\mathcal{F}$ is any irreducible plane curve of degree d and genus g defined over $\mathbf{F}_q$. To prove the desired result

$$q + 1 - (d-1)(d-2) \leq R_q \leq q + 1 + (d-1)(d-2), \qquad (12.2)$$

it must be shown that (12.1) holds true when N_1 is replaced by R_q.

To compare R_q with $S_q = N_1$, certain other parameters for $\mathcal{F}$ must be defined, some of which have appeared previously:

R_q = number of points $P \in \mathcal{F}$ that lie in $\mathrm{PG}(2,q)$;

S_q = N_1 = number of $\mathbf{F}_q$-rational points of $\mathcal{F}$;

R_q^* = number of points $P \in \mathrm{PG}(2,q)$ which are centres of $\mathbf{F}_q$-rational branches;

$\widetilde{R}_q$ = number of points $P \in \mathcal{F}$ in $\mathrm{PG}(2,q)$

    with each r-fold point counted with multiplicity r;

B_q = number of branches of $\mathcal{F}$ centred at points of $\mathrm{PG}(2,q)$;

E_q = number of singular points of $\mathcal{F}$;

b_P = number of $\mathbf{F}_q$-rational branches of $\mathcal{F}$ with centre at $P \in \mathrm{PG}(2,q)$;

c_P = number of all branches of $\mathcal{F}$ with centre at $P \in \mathrm{PG}(2,q)$;

m_P = multiplicity of a point $P \in \mathcal{F}$.

In Section 11, bounds for B_q were obtained.

Here, $\mathcal{F}(\mathbf{F}_q)^*$ stands for the set of all points $P \in \mathrm{PG}(2,q)$ lying on $\mathcal{F}$. With this notation,

$$S_q = \sum_{P \in \mathcal{F}(\mathbf{F}_q)^*} b_P, \qquad R_q = |\mathcal{F}(\mathbf{F}_q)^*|.$$

Also, $R_q^* \leq S_q$ and equality holds if and only if no two distinct $\mathbf{F}_q$-rational branches of $\mathcal{F}$ have the same centre.

The starting point is an upper bound and a lower bound for $S_q - R_q$.

Lemma 12.1 *An upper bound:*

$$S_q - R_q \leq \tfrac{1}{2}(d-1)(d-2) - g. \tag{12.3}$$

Lemma 12.2 *A lower bound:*

$$R_q - S_q \leq \tfrac{1}{2}(d-1)(d-2) - g. \tag{12.4}$$

Theorem 12.3 *Let $\mathcal{F}$ be an irreducible plane curve defined over $\mathbf{F}_q$ of degree d and genus g. If R_q is the number of points of $\mathrm{PG}(2,q)$ lying on $\mathcal{F}$, then*

(i) $|R_q - (q+1)| \leq g\lfloor 2\sqrt{q} \rfloor + \tfrac{1}{2}(d-1)(d-2) - g$;

(ii) $|R_q - (q+1)| \leq \tfrac{1}{2}(d-1)(d-2)\lfloor 2\sqrt{q} \rfloor$; *(Serre bound)*

(iii) $|R_q - (q+1)| \leq (d-1)(d-2)\sqrt{q}$. *(Hasse–Weil bound)*

Under some additional hypotheses, equality can be attained in the above bounds.

Theorem 12.4 (i) *Equality occurs in Lemma 12.2 if and only if every singular point of $\mathcal{F}$ is an isolated double point lying in $\mathrm{PG}(2,q)$.*

(ii) *Equality occurs in Lemma 12.1 if and only if every singular point of $\mathcal{F}$ is a node.*

The following corollaries show that when equality occurs either in Lemma 12.1 or in Lemma 12.2, q cannot be too small compared to n.

Corollary 12.5 *If equality holds in Lemma 12.1, then*

(i) $(d-1)(d-2) - 2g \leq S_q$;

(ii) $(d-1)(d-2) - 2g \leq q + 1 + g\lfloor 2\sqrt{q} \rfloor$.

Corollary 12.6 *Let $\mathcal{F}$ be an irreducible singular plane curve of degree $d \geq 3$ defined over $\mathbf{F}_q$ for which equality holds in Lemma 12.2.*

(i) $q \geq d - 2 + (S_q - 2g - 2)/(d-2)$.

(ii) *If $g = 0$, then $q \geq d - 1$.*

(iii) *If d is odd and $d \geq 5$, then $q \geq d - 1 - 2g/(d-3)$.*

Remark The above results are sharp for curves of low degree and genus. For $g = 2$ this is illustrated by two examples. Let

$$\begin{aligned}
f(X_0, X_1, X_2) \\
&= X_0^2(X_1^2 + 2X_2^2) + X_0(X_1^3 + 2X_1X_2^2 + X_2^3) + 3X_1^3X_2 + 3X_1X_2^3, \\
g(X_0, X_1, X_2) \\
&= X_0^2(X_1^2 - X_2^2) + X_0(X_1^2X_2 - X_2^3) + (6X_3^4 + 6X_1^2X_2 - 4X_2^4).
\end{aligned}$$

Then the parameters are given in Table 6.

Table 6: Two curves

Curve	q	d	g	R_q	S_q	Bound on S_q	Bound on R_q
$\mathcal{F} = \mathbf{v}(f)$	5	4	2	13	12	$0 \le S_q \le 12$	$0 \le R_q \le 13$
$\mathcal{F} = \mathbf{v}(g)$	13	4	2	1	2	$2 \le S_q \le 26$	$1 \le R_q \le 27$

Proposition 12.7 *If $\mathcal{C}$ is an irreducible $\mathbf{F}_q$-rational curve of degree $d > 1$, then $R_q \le (d-1)q + \lfloor d/2 \rfloor$.*

Some applications require information on the number of points of $\mathrm{PG}(2,q)$ lying on a plane curve not defined over $\mathbf{F}_q$. An upper bound on this number is as follows.

Lemma 12.8 *If an irreducible plane curve of degree d is defined over $\mathbf{F}_{q^k}$ but not over $\mathbf{F}_q$, then the number N of its points lying in $\mathrm{PG}(2,q)$ does not exceed d^2.*

In conclusion, two more bounds are shown, which have applications to arcs in $\mathrm{PG}(2,q)$.

Theorem 12.9 *Let $\mathcal{C}$ be an algebraic plane curve of degree d defined over $\mathbf{F}_q$ with no $\mathbf{F}_q$-linear components, and let T_q be the number of rational simple points of $\mathcal{C}$. If*

$$\sqrt{q} > d - 1, \tag{12.5}$$

then

$$T_q < d(q + 2 - d). \tag{12.6}$$

Remark It is not true that $d^2 \le q + 1 + (d-1)(d-2)\sqrt{q}$ for all values of d and q.

Theorem 12.10 *Let $\mathcal{D}_{2t}$ be a plane algebraic curve of degree $2t$ defined over $\mathbf{F}_q$ and let $\mathcal{C}_d$ be a component of $\mathcal{D}_{2t}$, with degree $d \ge 3$ if $\mathcal{C}_d$ is $\mathbf{F}_q$-rational. Let T_q be the number of rational simple points of $\mathcal{C}_n$ and let S be the number of its rational singular points of which D are double points. If*

$$\sqrt{q} > 2t + d - 1, \tag{12.7}$$

then

$$T_q + D < \tfrac{1}{2}d(q + 2 - t). \tag{12.8}$$

References

[1] M. Abdón and F. Torres, On maximal curves in characteristic two, *Manuscripta Math.* **99** (1999), 39–53.

[2] M.L. Carlin and J.F. Voloch, Plane curves with many points over finite fields, *Rocky Mountain J. Math.* **34** (2004), 1255–1259.

[3] A. Cossidente, J.W.P. Hirschfeld, G. Korchmáros and F. Torres, On plane maximal curves, *Compositio Math.* **121** (2000), 163–181.

[4] A. Cossidente, G. Korchmáros and F. Torres, On curves covered by the Hermitian curve, *J. Algebra* **216** (1999), 56–76.

[5] A. Cossidente, G. Korchmáros and F. Torres, Curves of large genus covered by the Hermitian curve, *Comm. Algebra* **28** (2000), 4707–4728.

[6] R. Fuhrmann, A. Garcia and F. Torres, On maximal curves, *J. Number Theory* **67** (1997), 29–51.

[7] R. Fuhrmann and F. Torres, The genus of curves over finite fields with many rational points, *Manuscripta Math.* **89** (1996), 103–106.

[8] A. Garcia and H. Stichtenoth, A maximal curve which is not a Galois subcover of the Hermitian curve, *Bull. Braz. Math. Soc. (N.S.)* **37** (2006), 139–152.

[9] A. Garcia and P. Viana, Weierstrass points on certain non-classical curves, *Arch. Math.* **46** (1986), 315–322.

[10] M. Giulietti, J.W.P. Hirschfeld, G. Korchmáros and F. Torres, Curves covered by the Hermitian curve, *Finite Fields Appl.* **12** (2006), 539–564.

[11] M. Giulietti, J.W.P. Hirschfeld, G. Korchmáros and F. Torres, Families of curves covered by the Hermitian curve, *Sémin. Cong.*, in press.

[12] A. Hefez and J.F. Voloch, Frobenius nonclassical curves, *Arch. Math.* **54** (1990), 263–273.

[13] A. Hefez and J.F. Voloch, Correction to "Frobenius nonclassical curves", *Arch. Math.* **57** (1991), 416.

[14] J.W.P. Hirschfeld, *Projective Geometries over Finite Fields, second edition*, Oxford University Press, Oxford (1998).

[15] J.W.P. Hirschfeld, G. Korchmáros and F. Torres, *Algebraic Curves over a Finite Field*, Princeton University Press, Princeton (2007).

[16] J.W.P. Hirschfeld, L. Storme, J.A. Thas, and J.F. Voloch, A characterization of Hermitian curves, *J. Geom.* **41** (1991), 72–78.

[17] M.Q. Kawakita, A quotient curve of the Fermat curve of degree twelve attaining the Serre bound, *J. Algebra Appl.* **4** (2005), 173–178.

[18] G. Korchmáros and F. Torres, Embedding of a maximal curve in a Hermitian variety, *Composition Math.* **128** (2001), 95–113.

[19] G. Korchmáros and F. Torres, The genus of a maximal curve, *Math. Ann.* **323** (2002), 589–608.

[20] G. Lachaud, Sommes d'Eisenstein et nombre de points de certaines courbes algébriques sur les corps finis, *C.R. Acad. Sci. Paris Sér. I* **305** (1987), 729–732.

[21] H.G. Rück and H. Stichtenoth, A characterization of Hermitian function fields over finite fields, *J. Reine Angew. Math.* **457** (1994), 185–188.

[22] K.O. Stöhr and J.F. Voloch, Weierstrass points and curves over finite fields, *Proc. London Math. Soc.* **52** (1986), 1–19.

[23] J. Top, Curves of genus 3 over small finite fields, *Indag. Math. (N.S.)* **14** (2003), 275–283.

J.W.P. Hirschfeld
Department of Mathematics
University of Sussex
Brighton BN1 9RF
United Kingdom
jwph@sussex.ac.uk

G. Korchmáros
Dipartimento di Matematica
Università della Basilicata
85100 Potenza
Italy
korchmaros@unibas.it

F. Torres
IMECC-UNICAMP
Campinas-13083-970-SP
Brazil
ftorres@ime.unicamp.br

On the efficient approximability of constraint satisfaction problems

Johan Håstad

Abstract

We discuss some results about the approximability of constraint satisfaction problems. In particular we focus on the question of when an efficient algorithm can perform significantly better than the algorithm that picks a solution uniformly at random.

1 Introduction

The most famous problem in theoretical computer science is the question of whether the two complexity classes P and NP are equal.

Here P is the set of problems[1] that can be solved in time which is polynomial in the size of the input. This is the mathematical definition aimed to correspond to problems which can be solved efficiently in practice on fairly large instances. One might object that there are very large polynomials but this has rarely been a problem and most problems known to be in P are efficiently solvable in the everyday meaning of the concept.

The class NP is the set of decision problems such that for instances with a positive answer there is a short proof of this state of affairs that can be verified efficiently. One of the most famous problems in NP is the traveling salesman problem, TSP, in which we are given n cites and distances $d(c_i, c_j)$ between the cities. The task is, given an upper bound K, to find a tour that visits all the cities and returns to its origin and is of total length at most K. If there is such a tour then, given the tour, it is easy to verify that indeed it is of the desired quality. Formally "easy" should here be interpreted as computable in time which is polynomial in the input length. Formulated differently there is a proof, sometimes called a certificate, that the instance has a positive answer and this short proof can be verified efficiently.

The existence of an easily verifiable certificate says little about the difficulty of finding the certificate. Of course we can always try all certificates but apart from this general procedure, the existence of the certificate seems to be of little help and indeed there is no general method to find the optimal tour for TSP in time that is subexponential in the number of cities n. Note also that the problem is very non-symmetrical in that it is very hard, in general, to convince someone that indeed there is no tour shorter than K. In other words there are probably no short certificates proving tight lower bounds on tour lengths.

The question whether P equals NP is the question of whether, for any decision problem where a positive answer has a short proof that can be verified efficiently, this proof can be found efficiently. From an everyday perspective this would seem absurd in that it rules out the brilliant idea, the idea that is hard to come by but which can immediately be verified as being excellent. In the terms of mathematical

---

[1] To be precise P only contains decision problems but for simplicity let us ignore this formal definition in this informal discussion.

research NP=P would, informally, be the same as saying that whenever a theorem has a short proof then this proof can be found quickly. We all "disprove" this statement frequently but on the other hand one should not compare mathematicians to computers.

In any case, most people working in complexity theory have a strong conviction that NP does indeed contain problems not in P. It seems, however, that a proof of this requires new insights. To prove that a problem does not belong to P one is required to prove that any algorithm that solves the given problem must take a long time. Another way to phrase this is to say that any algorithm that runs quickly must make a mistake on some input. This quantification over algorithms is very difficult to handle. There are many ways to proceed in a computation and for some problems there are very counter-intuitive algorithms that turn out to be both efficient and correct. Before continuing with our main topic let us take a small detour.

Consider ordinary arithmetic with large integers. It is easy to add two n-bit numbers with $O(n)$ operations. This is clearly, up to constants, the best we can do as each input bit needs to participate in some operation.

Multiplication done the standard, grade school, way requires $O(n^2)$ operations to multiply two n-bit numbers. There are more efficient methods and the most efficient algorithm [24] is based on the discrete Fourier transform and runs in time $O(n \log n \log \log n)$. It is unknown if this is best possible or even more embarrassingly it is unknown whether multiplication can be done in time $O(n)$. In other words we do not know whether multiplication is a more difficult operation to perform than addition or whether it is simply that we have not found the best way to do it. It is not obvious which is the most difficult problem, to prove that NP$\neq$ P, or to prove that multiplication needs super-linear time, but with our inability to prove lower bounds for computation a major new idea seems to be needed to succeed with either task. Let us return to our main line of reasoning.

Even if we have not been able to prove that NP contains difficult problem we have identified the best candidates for such problems; the NP-complete problems. Loosely speaking a problem in NP is NP-complete if it belonging to P is equivalent to NP=P. Thus, as we strongly believe that NP$\neq$ P, being NP-complete is strong evidence that a problem is computationally difficult. A slight variant is to prove that a problem is NP-hard. Such a problem has the property that it belonging to P implies that NP=P, but as opposed to the NP-complete problem such a problem need not itself belong to NP. For instance it might not be a decision problem or it might be a decision problem of even higher complexity.

Many problems are known to be NP-complete and in particular TSP discussed above is NP-complete. Ever since Cook [5], in 1971, first defined the class of NP-complete problems, one problem, Satisfiability, has turned out to play a central role. In Satisfiability we are given a Boolean formula and the question is to find an assignment that satisfies the formula. One example is given by the formula

$$\varphi = (x_1 \vee \bar{x}_2 \vee x_3) \wedge (x_2 \vee \bar{x}_3 \vee \bar{x}_4) \wedge (\bar{x}_1 \vee \bar{x}_2 \vee x_4).$$

This formula has the further property that it is a conjunction of disjunctions of literals[2] and this type of formula is usually called 3-CNF, where 3 is a bound on the number of literals in each disjunction (or clause) and CNF is short for Conjunctive Normal Form. It was proved by Cook that satisfiability of 3-CNF formulas is NP-complete. One might note that satisfiability of 2-CNF formulas is decidable in polynomial time and we invite the reader to find an efficient algorithm in this case when each disjunction only contains at most 2 literals.

The problem of deciding satisfiability of 3-CNF formulas is known as 3-Sat. It is a prime example of what it is called a Constraint Satisfaction Problem (CSP) as it is a collection of constraints each on a small set of variables. As 3-Sat would seem like a rather simple CSP one might expect that most CSPs are NP-complete and this is indeed correct. Already in 1978 Schaefer [24] constructed the short list of classes[3] of Boolean CSPs for which the problem is in P while in all other cases it is NP-complete. Over larger domains the situation is more complicated and a complete characterization over the domain of three values was found only recently by Bulatov [4]. The general case is still not resolved.

In this paper, however, we focus on the Boolean case where each variable only takes two values. As mentioned above almost all problems in their basic form are NP-complete but we turn to a more refined question. Above we described a black or white world where we want to satisfy all constraints. We now turn to a world of shades of gray where we want to satisfy as many constraints as possible. Let us try to formulate the central question.

Consider an instance φ of 3-Sat where each clause is of length exactly 3. We know that it is NP-complete to decide whether we can satisfy all the constraints, but maybe we are happy if we can satisfy almost all the constraints. It is easy to see that a random assignment satisfies a fraction $\frac{7}{8}$ of the constraints and hence we want to find an assignment that satisfies significantly more than this fraction. There are instances where no assignment satisfies more than a fraction $\frac{7}{8}$ of the constraints and hence the question is most interesting when we, for one reason or another, are guaranteed that there is some unknown assignment that satisfies all, or almost all of the constraints.

It turns out that for 3-Sat we cannot efficiently find a good assignment even in this case while for some other NP-complete problems we can find an assignment that does asymptotically better than a random assignment. The goal of the current paper is to survey these results.

2 Efficient computation

As our arguments are quite informal almost any definition of efficient computation would be sufficient. A reader desiring a precise definition can think of the Turing machine or consult any standard text such as [21].

Many of our efficient algorithms are randomized. Again the precise details are not important but as we in some cases count the number of possible random choices let us fix a model.

---

[2] A literal is a variable or a negated variable.

[3] 0-valid, 1-valid, weakly positive, weakly negative, affine or 2-CNF: consult [24] for definitions.

We assume that the algorithms have the ability to choose independent random bits to use in their computations. These random bits are recorded on the work tape of the Turing machine. As is commonly done we sometimes refer to these bits as the "random coin flips" made by the algorithm.

3   Maximal constraint satisfaction problems

On the input side it turns out that it is more convenient to use $\{-1, 1\}$ rather than $\{0, 1\}$ for our two possible values. A predicate P of arity k is a mapping $\{-1, 1\}^k \to \{0, 1\}$. An instance of Max-CSP-P is given by a collection $(C_i)_{i=1}^m$ of k-tuples of literals [4]. When we want to emphasize the arity of P we call Max-CSP-P a k-CSP.

For an assignment to the variables, a particular k-tuple is satisfied if P, when applied to values of the literals, returns 1. For an instance I and an assignment x we let $N(I, x, P)$ be the number of constraints of I satisfied by x under the predicate P.

We could allow positive weights giving different constraints different importance. As we do allow repetition of the same constraints we can, to a large extent, simulate weights and the existence of weights turn out not to be of any significant importance for our discussion.

Definition 3.1 Max-CSP-P is the problem of, given an instance I, to find the assignment x that maximizes $N(I, x, P)$.

A key parameter for a Max-CSP is the number of assignments that satisfy the predicate P.

Definition 3.2 The *density*, $d(P)$, of a predicate P on k Boolean variables is defined as $p2^{-k}$ where p is the number of assignments in $\{-1, 1\}^k$ that satisfy P.

Definition 3.3 A k-CSP where each constraint is a disjunction of at most k literals is an instance of *Max-k-Sat*. The subproblem where each constraint is the disjunction of exactly k literals is denoted by *Max-Ek-Sat*.

Next we look at linear equations modulo 2. As we use $\{-1, 1\}$ as our two values with -1 corresponding to true, addition modulo 2 is in fact multiplication.

Definition 3.4 A k-CSP where each constraint is that a product of at most k literals equals a constant is an instance of *Max-k-Lin*. If each product contains exactly k variables we denote it by *Max-Ek-Lin*.

Note that, by Gaussian elimination, if all equations can be satisfied then such an assignment can be found in polynomial time. Thus the interesting case of the problem is when we cannot satisfy all the constraints.

In this paper we mostly consider linear equations modulo 2 but many results apply to linear equations modulo m for larger values of m which need not even be prime. This problem is most natural in the case when the variables are also

---

[4] Note that we allow both variables and negated variables.

allowed to take values modulo m and in such a case we would speak of the problems
Max-k-Lin-m and Max-Ek-Lin-m.

Normally we allow negation of variables for free but to make the next problem
both a Max-CSP and a graph problem we make an exception.

Definition 3.5 A 2-CSP where each constraint is an inequality $x_i \neq x_j$ is an instance of *Max-Cut*.

To see that Max-Cut is a graph problem think of each pair (i, j) that appears
as a constraint as an edge between vertices i and j in a graph. The task is now to
divide the vertices into two parts in such a way as the number of edges between the
two classes is maximized. Note also that Max-Cut is a special case of Max-E2-Lin
with each equation of the form

$$x_i x_j = -1.$$

4   Approximation algorithms

In the best of all worlds we would, given a Max-CSP, efficiently find the optimal
solution. As stated in the introduction, however, for almost all Max-CSPs this is an
NP-hard task and hence we have to ask for less. We focus on algorithms that are
guaranteed to return a reasonably good solution.

Definition 4.1 Let O be a maximization problem and let $C \leq 1$ be a real number. For an instance x of O let $OPT(x)$ be the optimal value. A *C-approximation
algorithm* is an algorithm that on each input x outputs a number V such that
$C \cdot OPT(x) \leq V \leq OPT(x)$.

One might require the algorithm to return a proof, in the case of a Max-CSP an
assignment of the given quality, that indeed its output satisfies the given condition.
In fact all algorithms we discuss will have this property and thus they do more than
required. On the other hand it turns out that when we prove that some computational problem is hard we usually prove that already finding the approximate answer
in the above sense is hard.

We have of course a similar definition for minimization problems but as we here
only deal with maximization problems we do not state it formally.

Definition 4.2 An *efficient* C-approximation algorithm is a C-approximation algorithm that runs in worst case polynomial time.

We also allow randomized approximation algorithms in which case we require
the upper bound $V \leq OPT(x)$ to always hold while $V \geq C \cdot OPT(x)$ is relaxed to
hold only in expectation. Note that the expectation is taken only over the random
coin flips of the algorithm and is assumed to hold for each individual input. In
particular, there is no randomness associated with the input.

Most of the algorithms described can, at an arbitrarily small loss, be derandomized and in any case running the algorithm repeatedly can make sure that, with very
high probability, we get an output within (almost) a factor C of optimal.

The formulation "having approximation ratio C" is sometimes used as an alternative to saying "being a C-approximation algorithm".

Any Max-CSP-P has an approximation algorithm with constant approximation ratio.

Theorem 4.3 *Max-CSP-P admits a polynomial time approximation algorithm with approximation ratio $d(P)$.*

Proof Assume that we are given an instance with m constraints. A random assignment satisfies any given constraint with probability $d(P)$ and thus it satisfied $d(P)m$ constraints on the average. As the optimal assignment can satisfy at most all m constraints we get a randomized $d(P)$-approximation algorithm.   □

Let us note that it is not difficult to deterministically find an assignment that satisfies a fraction $d(P)$ of the constraints by the method of conditional expectation. We leave the details to the reader.

The random assignment algorithm finds an assignment that satisfies $d(P)m$ constraints independent of $OPT(x)$. If we want to have a better approximation ratio then it is sufficient to do better in the case when $OPT(x)$ is close to the maximal value m.

The main classification we study in this paper is inspired by this fact and is slightly different from approximation ratio. We concentrate on what is needed from the optimal solution in order for an efficient algorithm to find an assignment that does significantly better than the naive randomized algorithm used above which simply picks a random assignment.

Definition 4.4 A Max-CSP given by predicate P on k variables is *approximation resistant on satisfiable instances* if for any $\epsilon > 0$ it is NP-hard to distinguish instances where all constraints can be simultaneously satisfied from those where only a fraction $d(P) + \epsilon$ of the constraints can be simultaneously satisfied.

The existence of the arbitrarily small constant ϵ is somewhat annoying. It is not difficult to see that it cannot be the case that it is NP-hard to distinguish satisfiable instances from instances where exactly a fraction $d(P)$ of the constraints can be simultaneously satisfied, and thus some weakening is needed. The chosen weakening turns out to be convenient but there are other alternatives and in particular one could ask for ϵ to be a function of the size of the input and tend to 0 as this size increases.

While the previous definition does not exactly correspond to a statement about approximation ratio, the next definition is equivalent to saying that, again up to an arbitrary $\epsilon > 0$, the approximation ratio given by Theorem 4.3 is the best possible.

Definition 4.5 A Max-CSP given by predicate P on k variables is *approximation resistant* if for any $\epsilon > 0$ it is NP-hard to distinguish instances where a fraction $(1 - \epsilon)$ of the constraints can be simultaneously satisfied from those where only a fraction $d(P) + \epsilon$ of the constraints can be simultaneously satisfied.

Next we have a class of problems that are of intermediate complexity. It is possible to find an assignment of non-trivial quality for almost satisfiable instances but when the quality of the optimal solution is below a certain threshold this is no longer possible.

Definition 4.6 A Max-CSP given by predicate P on k variables is *somewhat approximation resistant* if it is not approximation resistant and there is some $\delta > 0$ such that for any $\epsilon > 0$ it is NP-hard to distinguish instances where a fraction $d(P) + \delta$ of the constraints can be simultaneously satisfied from those where only a fraction $d(P) + \epsilon$ of the constraints can be simultaneously satisfied.

Finally we have the class of problems where we can, as soon as the optimum is significantly better than the random assignment, find an assignment that also does significantly better than the random assignment.

Definition 4.7 A Max-CSP given by predicate P on k variables is *always approximable* if for each $\delta > 0$ there is an $\epsilon_\delta > 0$ and an efficient algorithm that given an instance where a fraction $d(P) + \delta$ of the constraints can be simultaneously satisfied finds an assignment that satisfies at least a fraction $d(P) + \epsilon_\delta$ of the constraints.

We proceed to give some positive results in the next section.

5   Constraints on two variables

The main technique used for deriving efficient approximation algorithms for Max-CSPs is semi-definite programming which was introduced in this context by Goemans and Williamson [9] as a tool to attack several problems. As the basic algorithm is very beautiful and quite easy to state, especially in the case of Max-Cut, we describe their approach for this problem here.

Let us formalize Max-Cut as a quadratic program

$$\max_{x \in \{-1,1\}^n} \sum_{(i,j) \in E} \frac{1 - x_i x_j}{2}. \tag{5.1}$$

This sum measures exactly the size of the maximal cut as each term is one if the edge is cut and zero otherwise. Let us relax (5.1) by introducing variables y_{ij} for the products $x_i x_j$ giving the program

$$\max_y \sum_{(i,j) \in E} \frac{1 - y_{ij}}{2}.$$

Allowing the variables y_{ij} to be completely independent would, of course, make the problem uninteresting and the key is to require that the numbers y_{ij} form a positive symmetric semidefinite matrix with ones on the diagonal. Let us write this as follows

$$\max_{y \succeq 0, y_{ii} = 1} \sum_{(i,j) \in E} \frac{1 - y_{ij}}{2}. \tag{5.2}$$

Note that this is a relaxation of the original problems as if $y_{ij} = x_i x_j$ then indeed this matrix fulfills the conditions.

The magic now comes from the fact that the optimization problem (5.2) can, by a result by Alizadeh [1], be solved to any desired accuracy in polynomial time. For simplicity of discussion we ignore that we can only find an almost optimal solution and assume that we find the true optimum. This slight inaccuracy does not affect our results as we state them.

Let us phrase the problem (5.2) slightly differently. Remember that an $n \times n$ matrix Y is symmetric positive semidefinite iff there is another $n \times n$ matrix V such that

$$Y = V^T V.$$

This implies that there are vectors (in fact the columns of V) such that

$$y_{ij} = (v_i, v_j)$$

and the requirement that $y_{ii} = 1$ is equivalent to v_i being a unit vector. Thus (5.2) is equivalent to

$$\max_{(\|v_i\|=1)_{i=1}^n} \sum_{(i,j) \in E} \frac{1 - (v_i, v_j)}{2}. \tag{5.3}$$

In this formulation it is obvious that (5.3) is a strict generalization of (5.1) as we can interpret the x_i as vectors forced to lie in a single dimension.

While it is easy to interpret a one-dimensional solution as a high dimensional solution the challenging and interesting part is to take a high dimensional solution and produce a good one-dimensional solution. The inspired solution by Goemans and Williamson is to pick a random vector $r \in \mathbf{R}^n$ and set

$$x_i = \text{sign}((r, v_i)), \tag{5.4}$$

where, in the probability 0 event that $(r, v_i) = 0$, we set $x_i = 1$.

Let us analyze this rounding from the approximation ratio perspective. Assume that the angle between v_i and v_j is θ_{ij}. The contribution to the objective function is then

$$\frac{1 - \cos \theta_{ij}}{2}$$

while the probability that the edge is cut, i.e. that $\text{sign}((r, v_i)) \neq \text{sign}((r, v_j))$ is exactly $\frac{\theta_{ij}}{\pi}$. Define the real number α_{GW} by

$$\alpha_{GW} = \min_{0 \leq \theta \leq \pi} \frac{\frac{\theta}{\pi}}{\frac{1 - \cos \theta}{2}}.$$

The numeric value of α_{GW} is approximately .878. We get the following chain of inequalities

$$E\left[\sum_{(i,j) \in E} \frac{1 - x_i x_j}{2} \right] = \sum_{(i,j) \in E} \frac{\theta_{ij}}{\pi} \geq \alpha_{GW} \sum_{(i,j) \in E} \frac{1 - \cos \theta_{ij}}{2} \geq \alpha_{GW} \cdot OPT.$$

The last inequality follows as the maximum of the relaxed problem is at least the true maximum. We conclude that the given algorithm is an α_{GW}-approximation algorithm. It is randomized but it can be derandomized [20] with an arbitrarily small loss in the quality of the obtained solution.

Let us turn to Max-2-Sat. Remember that we are using -1 to denote true and hence a clause $x_i \lor x_j$ is equivalent to the quadratic expression

$$\frac{3 - x_i - x_j - x_i x_j}{4}. \tag{5.5}$$

Negations are handled in the natural way and a clause $\bar{x}_i \lor x_j$ corresponds to

$$\frac{3 + x_i - x_j + x_i x_j}{4}, \tag{5.6}$$

with similar formulas for the other types of clauses.

The expressions (5.5) and (5.6) look essentially different from the corresponding terms for Max-Cut in that they contain linear terms and to remedy this we introduce an extra variable x_0 which will always be true. With this trick (5.5) turns into

$$\frac{3 + x_0 x_i + x_0 x_j - x_i x_j}{4}, \tag{5.7}$$

and we can relax this to

$$\frac{3 + (v_0, v_i) + (v_0, v_j) - (v_i, v_j)}{4}, \tag{5.8}$$

with unit length vectors v_i. This suggests the following approximation algorithm for Max-2-Sat.

Make an objective function by summing all quadratic expressions corresponding to clauses and solve the resulting semi-definite program. To retrieve Boolean values pick a random vector r and set x_i to true if $\text{sign}((v_i, r)) = \text{sign}((v_0, r))$.

It turns out that this algorithm gives an approximation ratio of α_{GW} for Max-2-Sat, i.e. the same constant as obtained for Max-Cut. To see this note that (5.8) can be written as

$$\frac{1 + (v_0, v_i)}{4} + \frac{1 + (v_0, v_j)}{4} + \frac{1 - (v_i, v_j)}{4}, \tag{5.9}$$

and the old argument can be applied to each term separately. That we have denominator 4 instead of 2 and that we might have plus signs instead of minus signs does not affect that analysis.

The fact that signs do not matter implies that the algorithm approximating Max-Cut can, essentially without change, be applied to Max-E2-Lin giving the same constant α_{GW} also for this problem.

Let us make a couple of observations about the algorithm for Max-2-Sat. It is clearly needed to have a non-symmetrical relation between "true" and "false" for Max-2-Sat while there is complete symmetry between the two sides in Max-Cut. Thus there is some need for v_0 or some other mechanism for breaking the symmetry between "true" and "false".

Secondly it turns out that the direction of v_0 is very special and this can be used to obtain better approximation ratios [8, 19]. Using a search over a large set of rounding procedures Lewin et al [19] obtain an algorithm whose approximation ratio probably[5] is approximately .940.

---

[5] This value has only been determined numerically and has not been proved analytically.

Although it does not follow from the given analysis one can modify the rounding to show that any Boolean 2-CSP is always approximable according to our definition. Furthermore, it turns out that this result generalizes to any constant size domain [13] and thus in our terminology any Max-CSP where each constraint depends on two variables is always approximable.

6   Some approximation resistant predicates

Let us start by stating one result. We then discuss some consequences and only later, very briefly, discuss some of the ideas behind the proof.

Theorem 6.1 *[12] For any $\epsilon > 0$ it is NP-hard to approximate Max-E3-Lin within a factor $\frac{1}{2} + \epsilon$. In other words, Max-E3-Lin is approximation resistant.*

Stated in other terms, for any $\epsilon > 0$ it is NP-hard, given a system of linear equations modulo 2, to determine whether there is a solution that satisfies a fraction $1 - \epsilon$ of the equations or if no assignment satisfies more than a fraction $\frac{1}{2} + \epsilon$ of the equations.

Note that we cannot strengthen Theorem 6.1 to prove Max-Lin approximation resistant on satisfiable instances as Gaussian elimination gives an efficient procedure to determine whether all equations are simultaneously satisfiable. It is possible to prove a variant of Theorem 6.1 with a sub-constant value for ϵ. Results along those lines have been obtained by Khot and Ponnuswami [18].

One can also note that, if variables are allowed to take values in $[m]$, Theorem 6.1 can be extended [12] to give inapproximability $\frac{1}{m} + \epsilon$ of Max-E3-Lin-m and the result even extends to equations over non-Abelian groups [7].

Let us postpone the discussion of the proof of Theorem 6.1 and first use it to obtain results for some other Max-CSPs of interest.

Theorem 6.2 *[12] For any $\epsilon > 0$ it is NP-hard to approximate Max-E3-Sat within a factor $\frac{7}{8} + \epsilon$. In other words, Max-E3-Sat is approximation resistant.*

Proof We give a reduction from Max-E3-Lin. We are given a system of equations modulo 2 with three variables in each equation and we want to produce an instance of Max-E3-Sat. Since we are using $\{-1, 1\}$, addition modulo 2 is conveniently represented as multiplication and a linear equation containing three variables can be written as

$$x_i x_j x_k = b \tag{6.1}$$

for some indices i, j, and k and $b \in \{-1, 1\}$. Assume for the time being that $b = 1$ and consider the following clauses

$$(x_i \vee x_j \vee \bar{x}_k)$$
$$(x_i \vee \bar{x}_j \vee x_k)$$
$$(\bar{x}_i \vee x_j \vee x_k)$$
$$(\bar{x}_i \vee \bar{x}_j \vee \bar{x}_k).$$

Then if (6.1) is satisfied so are all four clauses while if (6.1) is not satisfied then exactly three clauses are true. If $b = -1$ then we use the four clauses

$$(x_i \vee x_j \vee x_k)$$
$$(x_i \vee \bar{x}_j \vee \bar{x}_k)$$
$$(\bar{x}_i \vee x_j \vee \bar{x}_k)$$
$$(\bar{x}_i \vee \bar{x}_j \vee x_k)$$

with the same property. Thus if we start with a system of m equations we get $4m$ clauses and an assignment that satisfies v of the equations satisfies exactly $3m + v$ clauses. It is now easy to check that the existence of an algorithm which gives a $\frac{7}{8} + \epsilon$ approximation for Max-E3-Sat with $\epsilon > 0$ implies a $\frac{1}{2} + \epsilon'$ approximation for Max-E3-Lin with $\epsilon' > 0$. Theorem 6.2 follows from Theorem 6.1. $\square$

For Max-E3-Sat we could "hope" for approximation resistance on satisfiable instances and this is true.

Theorem 6.3 *[12] Max-E3-Sat is approximation resistant on satisfiable instances.*

There does not seem to be any fast way of deriving Theorem 6.3 from Theorem 6.1 and major parts of the proof have to be modified in a substantial way.

We will only briefly touch upon the ideas of these theorems but before we do even this we need to discuss proof systems in general.

7   Proof systems

The complexity class NP can be seen as a proof system. We have a prover P and a verifier V. P finds the witness of membership and sends it to V who then can efficiently check the proof. As an example, to prove that a formula φ is satisfiable P would supply the satisfying assignment which V then would verify by evaluating the formula.

In this standard situation V reads the entire proof, but we want to model a situation where V does spot checks and only reads a small fraction of the proof. We count the number of bits V reads from the proof and to make this easy to formalize we envision the proof in the shape of an oracle. Let Σ^* be the set of all finite binary strings.

Definition 7.1 An *oracle* is a function $\Sigma^* \rightarrow \{0, 1\}$.

Written proofs are identical with proofs using oracles where reading the i'th bit of the written proof corresponds to evaluating the oracle function on the binary representation of i. The entire concept of oracle Turing machines is just to be able to formally count the numbers of bits accessed by V when checking the proof. The reader unfamiliar with oracle Turing machine might be more comfortable by disregarding the notion and simply count the number of bits read by V in a less formal way.

A typical verifier $V^\pi(x, r)$ is a probabilistic Turing machine where π is the oracle, x the input and r the (internal) random coins of V. We say that the verifier *accepts* if it outputs 1 (written as $V^\pi(x,r) = 1$) and otherwise it *rejects*. As is standard in complexity theory we identify a decision problems with a language which is simply the set of inputs with the answer "yes". We can thus speak of the "language of satisfiable formulas" and use membership as the primitive notion.

Definition 7.2 Let c and s be real numbers such that $1 \geq c > s \geq 0$. A probabilistic polynomial time Turing machine V is a verifier in a *Probabilistically Checkable Proof (PCP)* with soundness s and completeness c for a language L iff

- For $x \in L$ there exists an oracle π such that $Pr_r[V^\pi(x,r) = 1] \geq c$.

- For $x \notin L$, for all π $Pr_r[V^\pi(x,r) = 1] \leq s$.

An important property turns out to be whether the identity of later bits read by V depends on the values obtained for earlier bits read. If they do, V is called "adaptive", with the opposite being called "non-adaptive".

In the current paper we always have perfect completeness in that a correct proof of a correct statement is always accepted and hence c is always equal to one in the above definition. Using smaller values of c might seem counterintuitive but this is used in the proof of Theorem 6.1.

We are interested in a number of properties of the verifier and one property that is crucial to us is that V does not use too much randomness.

Definition 7.3 The verifier V uses *logarithmic randomness* if there is an absolute constant c such that on each input x and proof π, the length of the random string r used by V^π is bounded by $c \log |x|$.

Using logarithmic randomness makes the total number of possible sets of coin flips for V polynomial in $|x|$ and hence all such sets can be enumerated in polynomial time.

We also care about the number of bits V reads from the proof.

Definition 7.4 The verifier V reads c bits in a PCP if, for each outcome of its random coins and each proof π, V^π asks at most c questions to the oracle.

To get a feeling for why we discuss PCPs let us envision a proof of Theorem 6.3. An NP-hardness proof is essentially always a reduction. To prove that it is NP-hard to distinguish objects of class X from objects of class Y one describes a polynomial time algorithm that takes a Boolean formula φ as input and produces an output which is of class X if φ is satisfiable while it is of class Y if φ is not satisfiable.

In particular to prove Theorem 6.3 we would expect to have a polynomial time reduction which on input φ and a number $\epsilon > 0$, produces another formula ψ in 3-CNF with the following properties.

- If φ is satisfiable so is ψ.

- If φ is not satisfiable then any assignment satisfies only a fraction $\frac{7}{8} + \epsilon$ of the clauses of ψ.

Clearly this would prove Theorem 6.3 as an algorithm A distinguishing satisfiable formulas from formulas where only a fraction ($\frac{7}{8} + \epsilon$) of the clauses can be simultaneously satisfied could be used to determine satisfiability. Given φ one could simply compute ψ and then run the algorithm A on input ψ.

Let us see that what we have in our hands is in fact a PCP. The written proof would now not be a satisfying assignment for φ but a satisfying assignment of ψ. This would be checked by a verifier V that given φ first constructs ψ, picks a random clause of ψ, reads the values of the three variables in the proof and accepts if the clause is satisfied. Let us check the properties of this verifier.

If φ is satisfiable then so is ψ and the prover can specify the oracle accordingly and hence V would accept with probability 1 giving perfect completeness $c = 1$.

If φ is not satisfiable then no assignment satisfies more than a fraction $\frac{7}{8} + \epsilon$ of the clauses of ψ and hence independently of the proof, the verifier would accept with probability at most $\frac{7}{8} + \epsilon$.

Note that the only randomness used by V is to pick a random clause of ψ and since ψ is of polynomial size this can be done with a logarithmic number of random coins. Finally also note that V only reads three bits of the proof.

Our type of reduction is more or less equivalent to the existence of a PCP which is limited to reading three bits. Let us sketch the reverse reduction in the case when the verifier is non-adaptive. The correspondence also exists in the adaptive case but is less tight.

Suppose there is a PCP to determine if φ is satisfiable, which reads 3 bits of the proof, has completeness one, soundness s, and where the verifier uses a logarithmic number of random bits and is non-adaptive.

Consider the "proof optimization problem" where we put ourselves in the shoes of a prover that wants to find the proof that maximizes the probability that the verifier accepts. Having not determined the proof yet we use a Boolean variable x_i to be determined as the value of the i'th bit of the proof. For each set of coinflips, r, of V it reads three bits i_1^r, i_2^r and i_3^r and accepts given a condition $C_r(x_{i_1^r}, x_{i_2^r}, x_{i_3^r})$ on these bits. Any condition on three bits can be written as a formula that is a 3-CNF and hence let us assume that C_r is of this form. Then the probability that V accepts is essentially given by the number of clauses satisfied in the formula

$$\psi = \wedge_r C_r(x_{i_1^r}, x_{i_2^r}, x_{i_3^r}).$$

We encourage the reader to work out the details of this reduction. Note that it is important that the resulting formula is of polynomial size and for this it is crucial that we only have a polynomial number of different r's. This is equivalent to saying that we only have a logarithmic number of random coins.

The existence of a PCP with even the gross behaviour of what we need, i.e. reading a constant number of bits, completeness one and soundness s, a constant strictly smaller than 1 and using logarithmic randomness is already mind-boggling. It is remarkable that it is possible to efficiently verify an arbitrary NP-statement of arbitrary size reading only a constant number of bits of the proof. The existence of such a PCP was established in a sequence of papers and the final construction was given by Arora et el [2].

Theorem 7.5 *[2] There is a universal integer c such that any language in NP has a*

PCP with soundness $1/2$ *and completeness* 1 *where* V *uses logarithmic randomness and reads at most* c *bits of the proof.*

Remark This theorem is often called the "PCP-theorem" or referred to as "ALMSS" after the initials of the authors.

Remark Although the number of bits read is independent of which language in NP we are considering, this is not true for the amount of randomness. The number of random bits is $d \log n$ for any language L, but the constant d depends on L.

To establish Theorem 6.1 and Theorem 6.3 one needs to find PCPs with very good constants. As the construction builds on a number of results it is not possible to present them here and we refer the interested reader to [12]. Let us give some results for other Max-CSPs.

8 Constraints on three Boolean variables

Zwick has determined good bounds for the approximation constants for all predicates on three variables [25]. We focus here on the classification into our four groups which gives us fewer details to consider.

To discuss the results it is convenient to write the predicate as a multilinear polynomial. We already used this representation when we discussed 2-CSPs but let us now do it more formally.

Theorem 8.1 *Any predicate* P *on* k *variables* $x_1, x_2 \ldots x_k$, *can in a unique way be written as a real sum*

$$P(x) = \sum_{S \subseteq [k]} c_S^P \prod_{i \in S} x_i.$$

There are a number ways to prove this theorem and readers familiar with the discrete Fourier transform might realize that the coefficients c_S^P are exactly the Fourier coefficients. Let us give an example. If P accepts the strings $(-1, -1, -1)$, $(-1, -1, 1)$, $(-1, 1, -1)$ $(1, -1, -1)$ and $(1, -1, 1)$ then

$$P(x) = \frac{1}{8}(5 - x_1 - 3x_2 - x_3 - x_1x_2 + x_1x_3 - x_2x_3 + x_1x_2x_3) \qquad (8.1)$$

For predicates on three variables the question whether $c_{\{1,2,3\}}$ equals 0 is of crucial importance.

Theorem 8.2 *A predicate* P *on three variables is always approximable iff* $c_{\{1,2,3\}}^P = 0$.

Sketch of proof. If the coefficient is 0 then P can be written as a real weighted sum of predicates each depending on two variables. Such sums can be efficiently approximated as discussed in Section 5.

The proof of the reverse direction uses essentially the same reduction as used in our sketch of proof of Theorem 6.2 from Theorem 6.1. The fact that $c_{\{1,2,3\}}^P \neq 0$ is equivalent to P not accepting the same number of strings of even parity and odd parity. Assume that P accepts a strings of even parity and b strings of odd parity where $a > b$. In our example (8.1) above $a = 3$ and $b = 2$.

Now given an equation

$$x_i x_j x_k = 1 \qquad (8.2)$$

we write down the four constraints

$$P(x_i, x_j, x_k)$$
$$P(x_i, \bar{x}_j, \bar{x}_k)$$
$$P(\bar{x}_i, x_j, \bar{x}_k)$$
$$P(\bar{x}_i, \bar{x}_j, x_k).$$

An assignment that satisfies (8.2) satisfies a of these constraints while an assignment that does not satisfy (8.2) satisfies b constraints. Equations of the form $x_i x_j x_k = -1$ are handled by adding one negation. We produce an instance with $4m$ constraints where an assignment that satisfies v of the linear equations satisfies $bm + v(a - b)m$ of the P-constraints. As $d(P) = (a + b)/8$ a small calculation is sufficient to prove that P is at least somewhat approximation resistant. We leave the details to the reader.

Next we look at approximation resistance.

Theorem 8.3 *A predicate P on three Boolean inputs is approximation resistant iff it is implied by parity or the negation of parity.*

Sketch of proof. Looking more closely at the previous proof, and working out the numbers, if $a = 4$ then the given reduction establishes approximation resistance.

The proof that no other predicates are approximation resistant goes by giving an efficient approximation algorithm for each of the other predicates. We refer to [25] for the details.

As stated in the introductory sections we allow negation for free. Another degree of freedom is the order of the inputs to P and hence any two predicates that can be transformed into each other by negations of inputs together with a permutation of the inputs are, in our eyes, equivalent. Viewed this way, for $d = 4, 5, 6, 7$ there is only one predicate accepting d inputs proved to be approximation resistant by Theorem 8.3. For each of these predicates we can ask what happens on satisfiable instances.

As stated several times, parity itself is not approximation resistant on satisfiable instances while Theorem 6.3 states that the predicate we get that accepts 7 inputs does indeed have the property. This result can be extended to the predicate that accepts 6 inputs.

Theorem 8.4 *Let P be a predicate on three Boolean inputs implied by parity which accepts 6 or 7 inputs. Then P is approximation resistant on satisfiable instances.*

The only remaining question for predicates on three Boolean inputs with regards to our classification is whether a predicate P that is implied by parity and which accepts 5 inputs is approximation resistant on satisfiable instances. Such a predicate

P has several equivalent formulations but one convenient way is to define it to accept the input unless exactly one of the three input bits is true. To determine whether this predicate is approximation resistant on satisfiable instances is still an open question. We known that it is approximation resistant and there is no obvious way to make use of the fact that an instance is satisfiable and not only almost satisfiable. In particular it is NP-complete to determine whether an instance is satisfiable or not.

9   Max-CSP on Boolean variables of higher width

The approximation resistance of predicates on more than three variables has been studied but results are far less complete.

Some of the results extend without problems and in particular we have the following.

Theorem 9.1 *[12] For any $k \geq 3$, parity on k variables as well as any predicate implied by parity on k variables is approximation resistant.*

From this theorem it possible to conclude that always approximable predicates are very few.

Theorem 9.2 *A predicate P is always approximable iff $c_S^P = 0$ for any S of size at least 3.*

The proof of this is a very slight generalization of the proof of Theorem 8.2. If indeed all the coefficients are 0 then we again can write P as a weighted sum of 2-CSPs.

If we have some S of size $k \geq 3$ with $c_S^P \neq 0$ we can use this as a basis of a reduction in a similar way as was sketched in the proof of Theorem 8.2. We leave the details to the interested reader.

Theorem 9.2 leads to a full characterization of the always approximable predicates. The only predicate that depends on four variables that has this property is

$$P(x) = \frac{2 + x_1 x_3 + x_1 x_4 + x_2 x_3 - x_2 x_4}{4}$$

while there is no predicate that depends on at least 5 variables and that is always approximable.

Hast [11] classified many predicates on four variables as to whether they were approximation resistant, ignoring whether approximation resistance held for satisfiable instances.

When we identify predicates that can be made equal by permuting and negating inputs and ignore the constant predicates there are are 400 different predicates on four Boolean variables. Of these 79 were determined to be approximation resistant, 275 were found not to be approximation resistant while Hast was not able to determine the status of the remaining 46 predicates.

It is interesting to look at the division into groups depending on the number of accepting 4-tuples.

Accepted inputs	1	2	3	4	5	6	7	8	9	10	11	12	13	14	15
Non-resistant	1	4	6	19	27	50	50	52	27	26	9	3	1	0	0
Resistant	0	0	0	0	0	0	0	16	6	22	11	15	4	4	1
Unknown status	0	0	0	0	0	0	6	6	23	2	7	1	1	0	0

From this table it seems evident that the more inputs a predicate accepts the more likely it is to be resistant. This is partly true but not fully true as the following theorem indicates.

Theorem 9.3 *[11] There are predicates on four variables P and Q such that P implies Q, P is approximation resistant while Q is not approximation resistant.*

Setting
$$P = ((x_1 \vee (x_2 = x_3)) \wedge (\bar{x}_1 \vee (x_2 = x_4)))$$
and
$$Q = ((x_2 = x_3) \vee (x_2 = x_4))$$
gives an example. Approximation resistance of P was proved in [10] while an approximation for Q was given in [25]. The fact that Q only depends on three variables might be taken as a drawback of this example but this can be remedied. In fact, if we make Q accept also the string $(1, 1, -1 - 1)$ then it remains only somewhat approximation resistant.

The intuition that predicates that accept few inputs are easy to approximate while predicates that accept most inputs are hard to approximate, can, however be given some formal support.

Theorem 9.4 *[11] Any predicate on k variables accepting at most $2\lfloor k/2 \rfloor + 1$ inputs is not approximation resistant.*

There are, however, some predicates that accept rather few inputs but are still approximation resistant. A prime example was given by Samarodnitsky and Trevisan [22].

Theorem 9.5 *[22] Assume $l_1 + l_2 + l_1 l_2 = k$ then there is a predicate P_{ST} on k variables which accepts $2^{l_1 + l_2}$ inputs and is approximation resistant.*

The predicate is the conjunction of $l_1 l_2$ linear constraints. It is possible to prove [11] that any predicate implied by P_{ST} is also approximation resistant. This can be used to prove the following:

Theorem 9.6 *[11]Assume $l_1 + l_2 + l_1 l_2 = k$ then any predicate on k variables that accepts at least $2^k + 1 - 2^{l_1 l_2}$ inputs is approximation resistant.*

Håstad and Khot [15] extended, at the cost of slightly worse bounds, the work of Samarodnitsky and Trevisan to achieve approximation resistance on satisfiable instances.

Theorem 9.7 *[15] Assume $2l_1 + 2l_2 + l_1 l_2 = k$ then there is a predicate P_{HK} on k variables which accepts $2^{2l_1 + 2l_2}$ inputs and is approximation resistant on satisfiable instances.*

10   Exact constants and the unique games conjecture

Many basic questions with regards to approximability of NP-hard questions remain. Due to lack of space let us not introduce more problems but instead mention some recent developments.

The approximation constant α_{GW} obtained for Max-Cut has not been improved since the original Goemans-Williamson paper over a decade ago. There is now evidence that it might be the correct constant for Max-Cut. To be more precise Khot [16] formalized in 2002 a conjecture about a game characterization of NP known as the "Unique Games Conjecture", (UGC) . This has turned out to be a strong conjecture with many important consequences, one [17] being that the Max-Cut constant α_{GW} is indeed correct.

Of the problems we discussed the approximability of Max-2-Sat can also be more or less resolved using UGC. Austrin [3] proved that, again assuming UGC, the numerically found approximation ratio for the algorithm by Lewin et al [19] (around .940) is an upper bound for the approximation ratio of any efficient approximation algorithm. Thus in the likely case that [19] did find the correct approximation ratio for their algorithm the rather natural question of the best approximation constant for Max-2-Sat might be determined and the true answer is the optimal value of an ugly but well defined optimization problem and happens to be around .940.

Samorodnitsky and Trevisan [23] proved that if d is the smallest number such that $k \leq 2^d - 1$ then, based on the UGC, there is an approximation resistant predicate on k inputs that accepts 2^d inputs and thus Theorem 9.4 might be quite close to the truth. Håstad [14] used this result to show that, again based on UGC, for large k, a random predicate P of width k is with high probability approximation resistant.

The truth of the UGC is uncertain and there seems to be no compelling evidence either to believe it or doubt it. As the number of interesting consequences of UGC increases, the urgency of proving or disproving it is mounting and UGC is now one of the main open problems of the area of PCPs and approximability of NP-hard optimization problems.

11   Conclusions and open problem

We have discussed the classification of Max-CSPs mainly in the Boolean case. Our knowledge of Max-CSPs over larger domains is far less complete[6] and a lot of work remains to be done. There are results also for larger size domain, in particular Theorem 6.1, Theorem 9.5 and Theorem 9.7 do extend [12, 6, 15] but as we do not have space to discuss these problems here we refer to those papers for a discussion.

Approximation resistance on satisfiable instances is possibly the ultimate hardness condition for a CSP. Even though there is some assignment that satisfies all the constraints, efficient computation cannot do essentially better than picking an assignment at random. We do believe that already approximation resistance is a central property of a CSP and much better evidence of hardness than the standard NP-completeness that is abundant and hence not very informative.

A few open questions have been mentioned in the text. In the best of all worlds

---

[6]Indeed, even classical NP-hardness is just resolved for size three domains [4].

one would have a complete characterization of which predicates fall into which category. Currently there is such a characterization only for the miniature class of always approximable predicates. The optimist could hope for a full characterization of all our classes. Of course once we have such a characterization there are many more detailed questions to study.

Acknowledgment. I am grateful to Per Austrin and an anonymous referee for comments on a preliminary version of this manuscript.

References

[1] F. Alizadeh. Interior point methods in semidefinite programming with applications to combinatorial optimization. *SIAM Journal on Optimization*, 5:13–51, (1995).

[2] S. Arora, C. Lund, R. Motwani, M. Sudan, and M. Szegedy. Proof verification and the hardness of approximation problems. *JACM: Journal of the ACM*, 45:501–555, (1998).

[3] P. Austrin. Balanced Max-2-Sat might not be the hardest. ECCC Technical report 06/088, (2006).

[4] A. Bulatov. A dichotomy theorem for constraint satisfaction problems on a 3-element set. *JACM: Journal of the ACM*, 53:66–120, (2006).

[5] S. Cook. The complexity of theorem proving procedures. In *3rd Annual ACM Symposium on Theory of Computing*, pages 151–158, (1971).

[6] L. Engebretsen. The nonapproximability of non-boolean predicates. *SIJDM: SIAM Journal on Discrete Mathematics*, 18:114–129, (2004).

[7] L. Engebretsen, J. Holmerin, and A. Russell. Inapproximability results for equations over finite groups. *Theoretical Computer Science*, 312:17–45, (2004).

[8] U. Feige and M. Goemans. Approximating the value of two prover proof systems, with applications to Max-2-Sat and Max Dicut. In *3rd Israeli Symposium on the theory of Computing and Systems*, pages 182–189, (1995).

[9] M. Goemans and D. Williamson. Improved approximation algorithms for maximum cut and satisfiability problems, using semi-definite programming. *Journal of the ACM*, 42:1115–1145, (1995).

[10] V. Guruswami, D. Lewin, M. Sudan, and L. Trevisan. A new characterization of NP with 3 query PCPs. In *Proceedings of 39th Annual IEEE Symposium of Foundations of Computer Science*, pages 8–17, (1998).

[11] G. Hast. Beating a random assignment. Ph.D. Thesis, Royal Institute of Technology, (2005).

[12] J. Håstad. Some optimal inapproximability results. *Journal of the ACM*, 48:798–859, (2001).

[13] J. Håstad. Every 2-CSP allows nontrivial approximation. In *Proceedings of the 37th Annual ACM Symposium on Theory of Computation*, pages 740–746, (2005).

[14] J. Håstad. On the approximation resistance of a random predicate. unpublished manuscript, (2006).

[15] J. Håstad and S. Khot. Query efficient PCPs with perfect completeness. *Theory of Computing*, 1:119–149, (2005).

[16] S. Khot. On the power of unique 2-prover 1 round games. In *Proceedings of the 34th Annual ACM Symposium on Theory of Computing*, pages 767–775, (2002).

[17] S. Khot, G. Kindler, E. Mossel, and R. O'Donnell. Optimal inapproximability results for Max-Cut and other 2-variable CSPs. In *Proceedings of the 45th Annual IEEE Symposium on Foundations of Computer Science*, pages 146–154, (2004).

[18] S. Khot and A. K. Ponnuswami. Better inapproximability results for Max-Clique, Chromatic number and Min-3-Lin-Deletion. In *ICALP (1)*, pages 226–237, (2006).

[19] M. Lewin, D. Livnat, and U. Zwick. Improved rounding techniques for the Max 2-Sat and Max Di-Cut problems. In *Proceedings of the 9th IPCP, Lecture Notes in Computer Science 2337*, pages 67–82, (2002).

[20] S. Mahajan and H. Ramesh. Derandomizing approximation algorithms based on semidefinite programming. *SIAM Journal on Computing*, 28:1641–1663, (1999).

[21] C. Papadimitriou. *Computational Complexity*. Addison-Wesley, (1994).

[22] A. Samorodnitsky and L. Trevisan. A PCP characterization of NP with optimal amortized query complexity. In *Proceedings of the 32nd Annual ACM Symposium on Theory of Computing*, pages 191–199, (2000).

[23] A. Samorodnitsky and L. Trevisan. Gowers uniformity, influence of variables and PCPs. In *Proceedings of the 38th Annual ACM Symposium on Theory of Computing*, pages 11–20, (2006).

[24] T. Schaefer. The complexity of satisfiability problems. In *Conference record of the Tenth annual ACM Symposium on Theory of Computing*, pages 216–226, (1978).

[25] U. Zwick. Approximation algorithms for constraint satisfaction problems involving at most three variables per constraint. In *SODA*, pages 551–560, (1998).

Johan Håstad

School of Computer Science and Communication

KTH-Royal Institute of Technology

S-100 44 Stockholm

SWEDEN

johanh@kth.se

The Combinatorics of Cryptographic Key Establishment

Keith M. Martin

Abstract

One of the most important processes involved in securing a cryptographic system is establishing the keys on which the system will rely. In this article we review the significant contribution of combinatorial mathematics to the development of the theory of cryptographic key establishment. We will describe relevant applications, review current research and, where appropriate, identify areas where further research is required.

1 Introduction

Cryptography provides the core information security services that are necessary to safeguard electronic communications. The sound management of cryptographic keys is the fundamental supporting activity that underpins the secure implementation of cryptography. The purpose of this paper is to demonstrate the significant contribution of combinatorial mathematics to the development of the theory of cryptographic key establishment.

- **Scope**: This paper surveys areas of key establishment where combinatorial models or construction techniques have proven of value. Our aim is not to provide a comprehensive survey of the literature, but rather to provide sufficient coverage that most relevant work will be (to use the terminology of Section 7.3.3) at most a "two-hop path" from this review. This paper is not an attempt to survey the vast research on key establishment in general.

- **Detail**: The primary aim is to bring these applications of combinatorics to the attention of the mathematical community within a sensible unifying framework. We thus focus on introducing concepts and providing pointers for further study. This paper contains no proofs. Combinatorial modelling typically involves the establishment of bounds and constructions. For illustrative purposes we will tend to focus on constructions in this review.

- **Novelty**: This article largely describes existing research and will contain few surprises for those already familiar with the field. That said, as far as we are aware, the full range of applications covered in this review have not previously all been presented within a common framework and so it is hoped that this may be of interest in its own right.

- **Applicability**: While the schemes in this paper are all of potential interest to a designer of a real cryptographic system, most are proposed under more rigorous mathematical security requirements than are demanded by the "real world", where security is often (validly) traded off against efficiency and practicality. Most of the key establishment schemes discussed in this paper are unlikely to be currently employed in commercial applications. This does not,

223

however, preclude them from influencing real designs or prevent them from being used in the future. They are all of interest in their own right as theoretical models of what is possible.

The remainder of the paper is structured as follows. In Section 2 we provide some background to cryptography and key management. We present a framework for key establishment in Section 3, which sets the context for comparison of schemes presented elsewhere in the paper. Section 4 contains some brief mathematical preliminaries. Our main review is spread over the subsequent three sections, where in Section 5 we look at key predistribution, in Section 6 we look at key distribution, and in Section 7 we look at key agreement. In Section 8 we provide some concluding remarks.

2   Cryptographic key management

We live in a society where electronic communication has become indispensable and ubiquitous. Electronic networks pervade all aspects of our professional and private lives. Many people, however, fail to appreciate that well-established and understood security safeguards that apply to traditional communication media are often absent in their electronic counterparts. In fact many of the features of electronic communication that we most value potentially expose information to previously unimaginable vulnerabilities.

The simple act of writing a letter suffices to illustrate this well. A traditional hand-written letter is normally posted in a sealed envelope and delivered to the specified address by a postal service. Interception of the contents requires physical access to the letter during the delivery service and breaking of the protective seal. The recipient can inspect the envelope for damage and may well gain assurance of the integrity of the contents through physical means, such as inspection of the postmark and recognition of handwriting. In contrast, an email is normally unprotected. In order to reach the specified address it is sent over a series of computer networks, passing through numerous computer servers and network routers on its journey. At any point its contents could be inspected, copied, forwarded, changed, and even the name of its sender could be forged. The recipient gains only cosmetic levels of assurance that the content is genuine and unaltered. Security in this electronic environment relies more on luck and lack of motivation for attack. If someone really wants to learn the content of an email then with very little technological expertise they probably can. With just a few clicks of their mouse button they can also share it with a significant percentage of the world's population.

There are of course solutions to most of these electronic security problems, as it is inconceivable that some of the earliest adopters of commercial electronic networks, such as the banking industry, could have developed electronic business without suitable security mechanisms in place. The science of *cryptography* underpins the bulk of these solutions. Cryptography is essentially a toolkit of mathematical techniques, algorithms and protocols that provide the core security services that are required in electronic communications. These services include *confidentiality* (restricting access to the contents of communicated data), *data integrity* (protecting data from manipulation), *data origin authentication* (correctly attributing the originator of some data) and *non-repudiation* (providing evidence of the occurrence of a data exchange that cannot later be denied). We have all used cryptography, even if we are not always aware that we are doing so, as cryptographic mechanisms are used to protect banking transactions (for example ATM transactions, Internet banking, SWIFT transfers), mobile telephone communications, secure web transactions (by means of the SSL protocol), password storage on computer operating systems, etc. Most modern computers have the facility to encrypt email (even if we tend not to use it) and almost everyone carries around at least one plastic card with a chip on it, whose purpose is primarily to allow cryptographic computations to be performed when that card is placed in contact with a reader.

Regardless of their purpose or application, most cryptographic mechanisms critically rely on the use of *keys*, which are essentially numbers selected at random from a large space. As the majority of cryptographic mechanisms are published processes that can be analysed by anyone, the entire security of a cryptographic mechanism typically relies on the protection of the relevant keys. The nature of these keys provides a natural broad classification of cryptographic mechanisms into *symmetric* mechanisms, where the secret keys employed by the sender and the receiver of data must be identical, and *public-key* mechanisms, where only one of the keys needs to be secret, and the other key can be made public. While for many applications both symmetric and public-key mechanisms are used in tandem, the fact that symmetric mechanisms tend to be faster and require shorter keys means that for a range of applications, symmetric key mechanisms are favoured. We will encounter several such applications during this paper.

Assuming that strong cryptographic mechanisms are employed and implemented correctly, it is fair to say that the security of cryptographic mechanisms relies almost entirely on the secure management of the relevant keys. The phrase *key management* tends to be associated with the entire lifecycle of a cryptographic key, including its creation (*key generation*), the methods by which it is sent to the relevant users of the system (*key establishment*), the techniques that are used to change or refresh it (*key update*) and ultimately the means by which it is deleted at the end of its usage period (*key destruction*).

The purpose of this paper is to review a number of interesting areas where combinatorics has found application in aspects of key management, and in particular key establishment. We will generally not need to concern ourselves with the purpose, or indeed even the algorithms, for which these keys are needed. The key establishment problems that we will look at in detail are mostly intended to support applications of symmetric cryptography. The reason for this is quite simple. The fundamental key management challenge in symmetric cryptography is one of key establishment. We need to arrange for every group of users who wish to engage in a secure communication exchange to have a common key. It should already be self-evident that this lends itself to a combinatorial setting. This fundamental problem does not always exist for public-key cryptography since one of the keys is public. Key management of public-key cryptography involves quite different challenges, which are mainly beyond our scope.

There are many introductory texts that provide a basic primer in cryptography. For a short mathematics-free background read, we recommend [63]. For a more comprehensive coverage of techniques and methodology we highly recommend [73].

A good survey of some of the topics considered in this paper is [72]. More generally, [10] provides an excellent survey on key establishment that goes beyond areas of combinatorial interest and [18] is probably the definitive work on cryptographic protocols relating to key establishment. Finally we note that both [7] and [26] include good reviews of other combinatorial applications to problems arising in information security.

3   Key establishment framework

In this section we will propose a framework within which the various schemes that we study can be meaningfully compared. In the remainder of the paper we will review schemes that have been proposed for a range of applications within this framework.

We use the term *key establishment* to indicate that this framework primarily covers the key management processes directly related to ensuring that the right keys are established in the right places within the network. We normally assume the existence of a *trusted authority* (or *TA*), which is an entity that is regarded as trustworthy and secure by all users in the network and that is relied on for various security critical operations, in particular during initialisation. We will not be particularly concerned with operations such as key generation, which in most case we leave to the TA, and key destruction, which in most cases we need to leave to individual users.

We represent the set of *users* of our network by $\mathcal{U} = \{U_1, \ldots, U_n\}$ and the TA by $\mathcal{T}$. It is probably most intuitive if we assume that we are establishing keys in this network for confidentiality purposes (in other words our keys are encryption keys), however this need not be the case.

Let $\mathcal{C}$ be a collection of subsets of $\mathcal{U}$, which we refer to as a *communication structure*, that consists of the collection of subsets of users for whom we wish to establish common keys. Note that many treatments of key management assume that cryptographic keys only need to be established between pairs of users. We make no such restriction here and will often refer to *group keys* in order to emphasise that we are establishing keys for general subsets. A group key k_A for a set $A \in \mathcal{C}$ is a value that all members of A can compute and use to secure joint communication within the group.

Definition 3.1 Informally, a *key establishment scheme* for communication structure $\mathcal{C}$ is a set of protocols that allow any set $A \in \mathcal{C}$ to establish a group key k_A. It consists of the following operational phases:

1. **Initialisation**. In this phase $\mathcal{T}$ generates all the data required to initialise the scheme. More precisely, this comprises:

   - Secret data specific to each user. We denote the secret data specific to user U_i by u_i. This value is only known to $\mathcal{T}$ and U_i and we assume that there exists some secure channel by which u_i can be transported from $\mathcal{T}$ to U_i (this channel is regarded as something outside of the key establishment scheme and could include, for example, physical delivery).

On receiving u_i, user U_i is responsible for ensuring that u_i is suitably protected.

- Public system-wide data, which we denote by Pub. This is made available by $\mathcal{T}$ to all users in $\mathcal{U}$ by means of an authenticated channel, the details of which do not concern us here.

2. **Key establishment**. In this phase a group of users $A \in \mathcal{C}$ establish their common key k_A. Whether this process involves the TA, communication between users, or no communication between scheme entities, is a major distinguisher between schemes in this paper. We return to this issue shortly (Section 3.1).

3. **Update**. In this optional phase, the secret and public data are modified. This may be because the communication structure has changed (for example users have left the scheme or new users have joined) or because the original keys have *expired* (all cryptographic keys have a finite lifetime and eventually need to be renewed). The simplest update operation is key *refreshment*, where existing group keys are simply replaced by new keys.

In the following subsections we specify our framework by identifying issues that can be used to define specific types of key establishment scheme.

3.1 Broad classification of key establishment schemes

A major distinguisher between different key establishment schemes is the extent to which communication between entities occurs during the key establishment phase. Note that the costly secure channels between the TA and users that were employed during the initialisation phase are not normally regarded as being readily available throughout the scheme lifetime (if they were available then one easy way to establish a common key would simply be for the TA to generate one at the time of request and distribute it over these same secure channels). We identify three potential operational environments during the key establishment phase:

1. Users have no communication channels available to support key establishment and thus must be able to do so on their own. We refer to such schemes as *group key predistribution schemes*.

2. The TA has some ability to communicate with users during the key establishment phase. We refer to such schemes as *group key distribution schemes*.

3. Users have some ability to communicate with one another during the key establishment phase. We refer to such schemes as *group key agreement schemes*.

Note that these environments apply strictly to the key establishment phase. Most group key predistribution schemes, for example, require involvement of an online TA during any update phase.

3.2 Secondary distinguishers

The next set of issues are *secondary distinguishers* in the sense that they subdivide schemes within the broad categories of Section 3.1.

3.2.1 Security The security model within which a key establishment scheme operates is a secondary distinguisher. The main threat to the security of a key establishment scheme that we consider is the ability of users (or outside parties) to obtain a key that they are not entitled to. There are two different aspects to this security issue that need to be identified for any given solution:

1. **Type of security**: The most common two types of security that we will encounter are:

   - *Unconditional security*: where the security of the scheme is independent of the resources available to an attacker.

   - *Computational security*; where the scheme can only be broken by an attacker with sufficient computational resources.

2. **Resilience**: This specifies the degree of resilience of the scheme to collusion between users. We will refer to the collection $\mathcal{X}$ of subsets of $\mathcal{U}$ who, even if they collude and share all their secret data, are unable to obtain any group keys to which they are not entitled, as the *exclusion structure*. This is always a monotone decreasing set (if $B_1 \in \mathcal{X}$ and $B_2 \subseteq B_1$ then $B_2 \in \mathcal{X}$). While general exclusion structures will be considered, the two most common degrees of resilience we will encounter are:

   - *Full collusion security*: $\mathcal{X}$ consists of all subsets of $\mathcal{U}$, meaning that no collusion of users should be able to determine a key that they are not entitled to.

   - *w-security*: $\mathcal{X}$ consists of all subsets of $\mathcal{U}$ of at most size w, meaning that no collusion of up to w users should be able to determine a key that they are not entitled to.

3.2.2 Deterministic v probabilistic An important secondary distinguisher between key establishment schemes is whether they are:

- *Deterministic*: we can guarantee that a group $A \in \mathcal{C}$ is able to establish a common key.

- *Probabilistic*: we can only guarantee that a group $A \in \mathcal{C}$ is able to establish a common key with a certain probability.

3.2.3 Communication channels Schemes also differ in the types of communication channel that exists between entities involved in the scheme. Two particular types of channel that we will regularly encounter are:

- *Secure*: we assume that any information exchanged on such a channel is totally protected, both in terms of being kept confidential and authentic (unchanged and from an identified originator).

- *Broadcast*: we assume that any information exchanged on such a channel is authentic, but not confidential.

Broadcast channels are much less costly and easier to maintain than secure channels. For example, publishing some data on an authenticated public noticeboard would realise a broadcast channel.

3.2.4   Properties of keys   A number of subtle secondary distinguishers concern the nature and structure of the the group keys. The following definitions will be useful in this regard. A group key k_A established by a group of users $A \in \mathcal{C}$ is:

- *Predistributed*: if k_A is a function only of the values $\{u_i \,|\, U_i \in A\}$ and *Pub*. In other words, k_A is computed only from data made available to the group members during the initialisation phase (this is necessarily the case for group key predistribution schemes).

- *Independent*: if knowledge of other group keys provides no information about the value of k_A.

- *Combinatorial*: if k_A can be represented as a subset of the collective secret user data of users belonging to A.

3.2.5   Extended capabilities   Further secondary distinguishers arise from additional properties that may be required by specific applications. Examples include:

- **Flexibility**: the extent to which a key establishment scheme is able to efficiently accommodate an update phase.

- **Computational capability**: the extent to which entities (particularly users) have the ability to perform computations.

- **Decentralisation**: whether roles normally conducted by the TA are required to be distributed amongst a number of separate entities. This can be for reasons of scalability, security or reliability.

- **Collaboration**: the degree of collaboration that is required (or permitted) to take place between users in order to establish a group key.

- **Robustness**: a stronger security model might be required for applications where either the TA or users are not trusted to perform their operations honestly.

- **Temporal restrictions**: whether key establishment for certain groups is restricted to specific time intervals or limited to a finite number of key establishment events.

- **Traceability**: whether it is possible to identify fraudulent users who abuse the key establishment scheme.

3.3   Evaluation criteria

The previous criteria that we have discussed are largely distinguishers based on scheme functionality. The following are the most common evaluation criteria that allow comparisons to be made between functionally similar key establishment schemes.

- **Secret storage**: the amount of information that a user needs to keep secure. As secure storage is expensive, this is an important quantity to minimise.

- **Public storage**: the amount of public information that needs to be maintained in order to operate the scheme. While it not so important to reduce this as it is to reduce secret storage, maintaining authenticated public data induces a cost and keeping this as small as possible is desirable.

- **Communication costs**: the quantity of data that needs to be exchanged (whether by expensive secure channels or less expensive broadcast channels) between entities in the key establishment scheme is something we would like to minimise.

- **Computational costs**: we would like to minimise the computational requirements for users in the scheme. Efficient computation is particularly important for applications where users are represented by low-memory devices with limited computational capabilities.

4   Preliminaries

In this section we briefly review some definitions and notation that we will employ later. We refer the reader to the combinatorial literature for further details.

4.1   Designs

A *set system* $(\mathcal{I}, \mathcal{B})$ consists of a set $\mathcal{I}$ of v elements (*points*) and a collection $\mathcal{B}$ of subsets (*blocks*) of $\mathcal{I}$. The *degree* of $x \in \mathcal{I}$ is the number of blocks of $\mathcal{B}$ containing x and $(\mathcal{I}, \mathcal{B})$ is *regular* if all points have the same degree r. The *rank* k of $(\mathcal{I}, \mathcal{B})$ is the size of the largest block in $\mathcal{B}$ and we say that $(\mathcal{I}, \mathcal{B})$ is *uniform* if all blocks have size k.

A regular, uniform set system with $|\mathcal{I}| = v$, $|\mathcal{B}| = b$, and with every t points occurring on precisely λ blocks is known as a t-(v, b, r, k, λ)-*design* (we often just refer to a t-(v, k, λ)-*design* since b and r can then be uniquely derived). The following special cases are of particular interest:

- A 2-$(s^2 + s + 1, s^2 + s + 1, s + 1, s + 1, 1)$-design is known as a *projective plane*.

- A 1-(v, b, r, k, λ)-design (which by definition has $\lambda = r$) with the further property that any pair of points occur in at most one block is called a (v, b, r, k)-*configuration*.

- A t-(v, k, λ)-design whose blocks can be partitioned into parallel classes is said to be *resolvable*.

A set system is a *group-divisible design* $\mathrm{GD}(n^u, k)$ if $v = nu$ and there exists a partition $\mathcal{H}$ of $\mathcal{I}$ into u *groups* of size n such that:

1. Every $H \in \mathcal{H}$ intersects a block $B \in \mathcal{B}$ in at most one point;

2. Every pair of points from different groups occur together in precisely one block.

A *transversal design* $\mathrm{TD}(k, n)$ is a $\mathrm{GD}(n^k, k)$.

4.2 Arrays

An *orthogonal array* $\mathrm{OA}_\lambda(t,k,v)$ is a $\lambda v^t \times k$ array with entries from a set of size v such that for any tuple $(x_1, \ldots, x_t)$ and any columns $C_1, \ldots, C_t$ there are precisely λ rows of the array in which the entry x_i occurs in column C_i (for all $1 \le i \le t$).

4.3 Graphs

A *graph* $\mathcal{G} = (\mathcal{I}, \mathcal{E})$ consists of a set of of vertices (or *nodes*) $\mathcal{I}$ joined by *edges* in $\mathcal{E}$, where $\mathcal{E} \subseteq \mathcal{I} \times \mathcal{I}$. We say that a pair of vertices U and V are *adjacent* if $\{U, V\} \in \mathcal{E}$ (we will also say that V is a *neighbour* of U). The *degree* of a vertex U is the number of vertices adjacent to U. A graph is *regular of degree r* if all vertices have degree r. If the order of adjacent vertices $\{U, V\}$ matters then we write (U, V) (if an edge connects U to V) and we say that $\mathcal{G}$ is a *directed graph*.

A *path* of *length* L from U_0 to U_L is a sequence of edges and vertices of the form $U_0, e_1, U_1, e_2, \ldots, U_{L-1}, e_L, U_L$, where the vertices U_i and the edges e_j are all distinct and U_{i-1} and U_i are adjacent and connected by e_i. A *cycle* is a path from a vertex to itself of length more than one (a cycle of length one is called a *loop*). A graph is *connected* is every pair of vertices are joined by at least one path.

A *complete t-partite* graph is a graph whose vertices can be partitioned into t disjoint subsets such that two vertices are adjacent if and only if they belong to distinct subsets.

An (n, r, λ, μ)-*strongly regular graph* is a regular graph on n vertices with degree r and any two distinct vertices have λ common neighbours if they are adjacent and μ common neighbours if they are not adjacent.

A *tree* is a connected graph with no cycles, loops or multiple edges. There thus exists a unique path between any two vertices. Any vertex of a tree can be chosen to be the *root* of the tree, with all edges and vertices descending from this root. We call this a *rooted* tree and can interpret it is a directed graph with a natural ordering induced from the root. Every vertex U in a rooted tree (except the root) has a unique *parent* and any other vertex adjacent to U is said to be a *child* of U. Any vertex of degree one with no children is called a *leaf*. A *binary* tree is a tree where every vertex has at most two child nodes (in general an *a-ary* tree is one where every vertex has at most a child nodes). A *chain* is a tree consisting of a single path, where each intermediate vertex has precisely one parent and one child. A *starlike subgraph* is a tree in which every path has length at most two.

4.4 Posets

A *partially ordered set* (*poset*) is a pair $(\mathcal{L}, \le)$, where $\le$ is a reflexive, anti-symmetric, transitive binary relation on $\mathcal{L}$. We say that x *covers* y, denoted $y \lessdot x$, if $y < x$ and there does not exist $z \in \mathcal{L}$ such that $y < z < x$ (in this case we also refer to y as a *child* of x and x as a *parent* of y). The *Hasse diagram* $(\mathcal{L}, \lessdot)$ of a poset is the directed graph $(\mathcal{L}, \mathcal{E})$ where $(x, y) \in \mathcal{E}$ if and only if $x \lessdot y$. Note that every rooted tree is a Hasse diagram for the poset defined by $U \lessdot V$ if and only if U is a parent of V in the rooted tree.

4.5 Cryptographic primitives

We will use a number of cryptographic primitives as building blocks in some of the schemes in this paper. We briefly mention three that will see repeated use.

A *symmetric encryption algorithm* E is a function that converts binary strings of *plaintext* into binary strings of *ciphertext*. More precisely, the ciphertext is a function of the plaintext and a symmetric key K, which is shared between sender and receiver. The receiver of the ciphertext is able to use a related decryption algorithm to recover the plaintext from the ciphertext using the same key K. Symmetric encryption algorithms, applied directly as described, provide data confidentiality. They can be applied in other ways to establish other security services.

A *hash function* is a function that converts an arbitrary long input into a fixed length compressed output. A hash function should have the properties that it is *one-way* (it is hard to recover an input from a given output) and *collision-free* (it is hard to find two inputs with the same output, even though there will be many such pairs). Hash functions are extremely versatile cryptographic primitives and are employed widely in cryptographic protocols.

A *secret sharing scheme*, which is a method of sharing a secret value amongst a group of participants by distributing related information (*shares*) in such a way that only certain specified subsets of the participants (defined by the *access structure* Γ) can reconstruct the secret from their shares. If subsets of participants not in the access structure learn nothing about the secret from their shares then the scheme is referred to as being *perfect*. We make the natural restriction that Γ is *monotone* (in other words, if $X \in \Gamma$ and $X \subseteq Y$ then $Y \in \Gamma$). If Γ consist of all subsets of at least t out of n participants then we refer to a secret sharing scheme for Γ as being a (t,n)-*threshold scheme*.

Secret sharing schemes were first proposed in [8, 69] and are of significant combinatorial interest in their own right (see [71] for a review). It can be shown that in perfect secret sharing schemes each participant's share must be at least as large as the secret it is protecting. Secret sharing schemes in which each share has this minimal size are called *ideal*. Ideal secret sharing schemes are closely related to matroids [19] and ideal threshold schemes correspond to orthogonal arrays [39].

5 Key predistribution schemes

The first class of key establishment schemes that we will look at are group key predistribution schemes. Applications suitable for group key predistribution are those where during key establishment the TA cannot be accessed in any capacity (it may have ceased to exist, or be impractical or too costly to communicate with it) and users cannot employ secure communication channels amongst themselves (they may not be able to afford the computational costs of establishing such channels).

Definition 5.1 A $(\mathcal{C}, \mathcal{X})$-*key predistribution scheme* (KPS) is a key establishment scheme with communication structure $\mathcal{C}$ and exclusion structure $\mathcal{X}$ such that:

1. Given $A \in \mathcal{C}$, any $U_i \in A$ can compute the group key k_A from knowledge of u_i and Pub.

2. Given disjoint sets $B \in \mathcal{X}$ and $A \in \mathcal{C}$, it is not possible to compute the group key k_A from knowledge of u_B and Pub (where $u_B = \{u_i \,|\, U_i \in B\}$).

Note that the precise meaning of property (2) in Definition 5.1 depends on the security model within which we are operating. If a KPS is unconditionally secure then these conditions can be stated information theoretically (see, for example [10]).

The literature contains a wide variety of key predistribution schemes. We will begin this section by identifying a number of (generic) fundamental KPSs, most of which have manifested themselves on numerous occasions as published schemes. We then discuss several different types of KPS that are of combinatorial interest.

5.1 Fundamental schemes

In this section we identify seven fundamental key predistribution schemes, divided into two different classes. These fundamental schemes are a combination of generic schemes that help to illustrate some of our definitions as well as extremal schemes that provide useful performance benchmarks for comparison.

5.1.1 Fundamental edge-based KPSs

We identify four fundamental schemes in this class. All four schemes are deterministic, have independent keys, offer full collusion security and can be established for arbitrary communication structures.

Scheme 5.2 *A trivial key predistribution scheme (TKPS) has the following properties:*

- $u_i = \{k_A \,|\, U_i \in A, A \in \mathcal{C}\}$;

- $Pub = \emptyset$;

- $k_A \in u_i$ *if and only if* $U_i \in A$.

A TKPS offers unconditional security. The most obvious problem with a TKPS is that the secret information u_i that each user has to store is potentially very large. A further problem with this type of scheme is that if group keys have to be refreshed during a key update phase then this requires the initialisation phase to be rerun.

This motivates our next fundamental scheme, where E is a secure symmetric encryption algorithm with key size l and $E_k(m)$ denotes the encryption of plaintext m using key k.

Scheme 5.3 *A trivial key encrypting key predistribution scheme (TKEKPS) has the following properties:*

- $u_i = \{K_A \,|\, U_i \in A, A \in \mathcal{C}\}$, *where each K_A is randomly chosen from $\{0,1\}^l$*;

- $Pub = \{E_{K_A}(k_A) \,|\, A \in \mathcal{C}\}$;

- $K_A \in u_i$ *if and only if* $U_i \in A$, *with k_A obtained by decrypting $E_{K_A}(k_A)$.*

A TKEKPS offers computational security, since any attacker with the computational resources to break the encryption algorithm can obtain group keys. Users have to store as much secret information as in a TKPS, but refreshing group keys can now easily be done by the TA updating Pub (more precisely, by replacing $E_{K_A}(k_A)$ by $E_{K_A}(k'_A)$, where k'_A is the refreshed version of k_A).

Our next fundamental scheme offers the minimum possible secret storage and is, in some sense, the opposite "extreme" to a TKPS.

Scheme 5.4 *A* direct key encrypting key predistribution scheme *(DKEKPS) has the following properties:*

- $u_i = k_i$, *where each k_i is randomly chosen from $\{0, 1\}^l$;*

- $Pub = \{E_{k_i}(k_A) \mid U_i \in A, A \in \mathcal{C}\};$

- $E_{k_i}(k_A) \in Pub$ *if and only if $U_i \in A$, with k_A obtained by decrypting $E_{k_i}(k_A)$.*

A TKEKPS also offers computational security. The secret storage is as small as possible, but this comes at the expense of potentially large public storage requirements (as large as the secret storage in a TKPS and TKEKPS).

Our fourth fundamental scheme is essentially a refinement of a TKEKPS that reduces the public storage at the expense of an iterated key derivation process. Let $(\mathcal{C}, \leqslant)$ be the poset induced by set containment, where for $A, B \in \mathcal{C}$, $A \leqslant B$ if and only if $B \subseteq A$. For any U_i let $roots_i = \{C \in \mathcal{C} \mid U_i \in C \text{ and } U_i \notin B \text{ for any } B > C\}$.

Scheme 5.5 *An* iterative key encrypting key predistribution scheme *(IKEKPS) has the following properties:*

- $u_i = k_i$, *where each k_i is randomly chosen from $\{0, 1\}^l$;*

- $Pub = Pub_1 \cup Pub_2$, *where $Pub_1 = \{E_{k_i}(k_C) \mid C \in roots_i\}$ and $Pub_2 = \{E_{k_B}(k_C) \mid B, C \in \mathcal{C}, B > C\};$*

- $U_i \in A$ *if and only if there exists a path (in the Hasse diagram of $(\mathcal{C}, \leqslant)$) $(Z_0, Z_1), \ldots, (Z_{m-1}, Z_m)$, where $Z_0 \in roots_i$ and $Z_m = A$. In this case U_i obtains k_{Z_0} from Pub_1 by decrypting $E_{k_i}(k_{Z_0})$ and then iteratively obtains k_{Z_i} from Pub_2 by decrypting $E_{k_{Z_{i-1}}}(k_{Z_i})$.*

Thus an IKEKPS offers computational security, has minimal secret storage and reduced public storage compared to a TKEKPS. This reduction comes at the expense of greater computational effort to iteratively derive a group key.

We refer to these four schemes as *edge-based* key predistribution schemes because they all make use, either in the public or secret data, of the set of edges in the Hasse diagram of the poset $(\mathcal{C}, \leqslant)$.

5.1.2 Fundamental node-based KPSs Our next fundamental schemes encode the structure of the poset $(\mathcal{C}, \leqslant)$ into the public information by assigning an item of public data Pub_i to each user. For this reason we refer to them as *node-based*. Unlike for the edge-based schemes, which were all distinct, the three fundamental node-based schemes are "nested", with the first being the most generic and each subsequent scheme being a special case of the previous scheme.

Scheme 5.6 *A node-based key predistribution scheme (NBKPS) has the following properties:*

- $Pub = \cup_{1 \leq i \leq n} Pub_i$, *where Pub_i is associated with user U_i;*

- $u_i = f(Pub_i)$ *for some secret function f known only to the TA, which is chosen in such a way that there exists a public function g such that for any $A \in \mathcal{C}$ and any pair $U_i, U_j \in A$ we have that $g(u_i, Pub_A) = g(u_j, Pub_A) = k_A$ (where $Pub_A = \cup_{U_i \in A} Pub_i$).*

- *By choice of f and g it follows that any $U_i \in A$ can compute k_A.*

Scheme 5.6 is clearly only the blueprint of a concept and precise properties of actual NBKPSs will depend on specific instances. It should be clear however that NBKPSs demand careful choice of functions and internal structure and are clearly ripe for combinatorial application.

Our next fundamental scheme represents one particular type of NBKPS. Let $\mathcal{I} = \{x_i \,|\, 1 \leq i \leq v\}$ be a set of v identifiers, each of which is associated by means of a secret function f with a randomly chosen key $k_i = f(x_i)$ from a set $\mathcal{K}$. Let $\mathcal{B}$ be a collection of subsets of $\mathcal{I}$. We will let $\mathcal{R} = (\mathcal{I}, \mathcal{B})$ collectively be referred to as a *key ring*.

Scheme 5.7 *A key ring predistribution scheme (KRPS) based on key ring $\mathcal{R} = (\mathcal{I}, \mathcal{B})$ is a node-based key predistribution scheme with the following properties:*

1. *$Pub_i = B_i$ is randomly chosen from $\mathcal{B}$ (such that $u_i \neq u_j$ if $i \neq j$);*

2. *$u_i = \{k_j \,|\, x_j \in B_i\}$;*

3. *$\mathcal{C} \subseteq \{A \subseteq \mathcal{U} \,|\, \cap_{U_i \in A} u_i \neq \emptyset\}$;*

4. *For $A \in \mathcal{C}$, group key $k_A = g(\cap_{U_i \in A} u_i)$ for some public combining function g. In other words, a group $A = \{U_1, \ldots, U_t\} \in \mathcal{C}$ of users check their public identifier sets $Pub_1, \ldots, Pub_t$ to see which common identifiers they share. They then establish a group key k_A by applying g to the keys k_i that correspond the identifiers in $\cap_{j=1}^{t} Pub_j$.*

KRPSs are examples of group key establishment schemes with combinatorial keys (see Section 3.2.4). Whether they offer unconditional or computational security depends on the combining function g. For example, an unconditionally secure scheme can be obtained if $k_A = \oplus_{k_i \in X} k_i$, where $X = \cap_{U_i \in A} u_i$.

Our final fundamental node-based scheme is a particular type of KRPS.

Scheme 5.8 *A random key predistribution scheme (RKPS) is a key ring predistribution scheme based on key ring $\mathcal{R} = (\mathcal{I}, \mathcal{B})$, where $\mathcal{B} = 2^{\mathcal{I}}$ (the collection of all subsets of $\mathcal{I}$).*

In other words, an RKPS involves issuing each user with a set of random keys from $\mathcal{K}$. An RKPS is thus an example of a probabilistic key establishment scheme. This may seem like a very strange way of constructing a KPS for a specific communication structure $\mathcal{C}$, since it involves "getting lucky" with regard to the intersection properties of the resulting blocks. However the idea behind a KRPS can be useful in situations where certain properties of a KPS (such as user storage) have higher priority than establishing a desired communication structure precisely (we will see examples of such applications in Section 7.3).

5.2   The BDVHKY scheme

In this section we present an important benchmark NBKPS. If $\mathcal{C} = \{A \subseteq \mathcal{U} \mid |A| = t\}$ and $\mathcal{X} = \{A \subseteq \mathcal{U} \mid |A| \leq w\}$ then we will also refer to a $(\mathcal{C}, \mathcal{X})$-KPS as a (t, w)-KPS. We will also refer to communication structures $\mathcal{C}$ of this type as *threshold* communication structures. The following (t, w)-KPS was proposed by Blundo, De Santis, Vaccaro *et al* in [17] and is a generalisation of a much earlier $(2, w)$-KPS proposed in [9].

Scheme 5.9 *The* BDVHKY *key predistribution scheme (BDVHKY-KPS) is defined as follows, where $q \geq n$:*

- *$Pub_i = s_i$, where $s_i \in GF(q)$ and $Pub_i \neq Pub_j$ if $i \neq j$;*

- *The TA (randomly) constructs a secret t-variate polynomial f with coefficients from $GF(q)$,*

$$f(x_1, \ldots, x_t) = \sum_{i_1=0}^{w} \cdots \sum_{i_t=0}^{w} a_{i_1 \ldots i_t} x_1^{i_1} \ldots x_t^{i_t},$$

where $a_{i_1 \ldots i_t} = a_{j_1 \ldots j_t}$ for any permutation $(j_1 \ldots j_t)$ of the indices $\{i_1, \ldots, i_t\}$.

- *$u_i = f(Pub_i, x_2, \ldots, x_t) = f(s_i, x_2, \ldots, x_t)$, a $(t-1)$-variate polynomial with coefficients from $GF(q)$;*

- *For any $A = \{U_{z_1}, \ldots, U_{z_t}\} \in \mathcal{C}$, the user U_{z_i} computes*

$$k_A = u_{z_i}(s_{z_1}, \ldots, s_{z_{i-1}}, s_{z_{i+1}}, \ldots, s_{z_t}) = f(s_{z_1}, \ldots, s_{z_t}).$$

The BDVHKY-KPS is an example of a deterministic NBKPS that is not a KRPS. It offers unconditional w-security. We note that in the BDVHKY-KPS, each user needs to store a secret $t-1$ variate polynomial of degree w of a special form. It can be shown that this involves the equivalent of storing $\binom{t+w-1}{t-1}$ elements of $GF(q)$. The BDVHKY-KPS is of particular interest because it is shown in [17] that this is the optimally small user storage for any unconditionally secure (t, w)-KPS.

The following variant of Scheme 5.9 is a generalisation of a scheme proposed in [52], which uses the random key predistribution scheme (Scheme 5.8) to obtain some interesting tradeoffs.

Scheme 5.10 *The* randomised BDVHKY-KPS *is similar to Scheme 5.9 except that:*

- *The TA (randomly) constructs r secret t-variate polynomials $f_1, \ldots, f_r$ with coefficients from $GF(q)$, each with the property required for Scheme 5.9;*

- *For each U_i, the TA generates a random subset $U[i] = \{i_1, \ldots, i_{r'}\}$ of the set $\{1, \ldots, r\}$, which is made public;*

- $u_i = \{f_{i_1}(s_i, x_2, \ldots, x_t), \ldots, f_{i_{r'}}(s_i, x_2, \ldots, x_t)\};$

- *For any* $A = \{U_{z_1}, \ldots, U_{z_t}\} \in \mathcal{C}$, *if* $\cap_{j=1}^t U[z_j] \neq \emptyset$ *then for some* $l \in \cap_{j=1}^t U[z_j]$ *user* U_{z_i} *computes* $k_A = f_l(s_{z_1}, \ldots, s_{z_t})$. *(Note that since the sets* $U[i]$ *are public, the choice of* l *can be publicly predetermined.)*

Scheme 5.10 is an example of a probabilistic NBKPS, since there is no guarantee that $\cap_{j=1}^t U[z_j] \neq \emptyset$. Compared to the BDVHKY-KPS, the scheme also involves an increased user storage by a magnitude of r'. However the significant gain is in resilience. The BDVHKY-KPS is only w-secure, whereas in [52] it is shown that careful selection of the parameters r and r' in Scheme 5.10 can result in very good resilience.

We will discuss a further variant of the BDVHKY-KPS in Section 7.3.4. We note that in [62] it was shown that a number of key predistribution schemes, including Scheme 5.9 (under certain constraints on the combining function used to determine the final key), are examples of a wider family of *linear key predistribution schemes*, which can be described in linear algebraic terms and permit an inherent duality.

5.3 Key distribution patterns

In this section we look at an interesting family of key ring predistribution schemes that have arisen in the literature in a number of different guises.

Definition 5.11 Let $(\mathcal{C}, \mathcal{X})$ be a communication and exclusion structure defined on n users. A $(\mathcal{C}, \mathcal{X})$-*key distribution pattern* (KDP) is a set system $(\mathcal{I}, \mathcal{B})$ with $|\mathcal{B}| = n$, where each user U_i is associated with a block B_i, such that for any disjoint pair $A \in \mathcal{C}$ and $B \in \mathcal{X}$ we have:

$$\bigcap_{U_i \in A} B_i \not\subseteq \bigcup_{U_j \in B} B_j.$$

When $\mathcal{C}$ consists of all t-subsets of users and $\mathcal{X}$ consists of all subsets of at most w users, we will refer to a (t, w)-KDP. In [76], (t, w)-KDPs were noted to correspond to the following more granularly defined family of set systems:

Definition 5.12 A (t, w, d)-*cover-free family* (CFF) is a set system $(\mathcal{I}, \mathcal{B})$ such that for any disjoint sets of t blocks A and w blocks B we have:

$$\left| \bigcap_{B_i \in A} B_i \setminus \bigcup_{B_j \in B} B_j \right| \geq d.$$

The motivation for Definition 5.11 is that a KDP can be used as a key ring to form a KRPS.

Scheme 5.13 A $(\mathcal{C}, \mathcal{X})$-*key distribution pattern predistribution scheme (KDPPS) is a* $(\mathcal{C}, \mathcal{X})$-*KRPS that arises by applying Scheme 5.7 with a* $(\mathcal{C}, \mathcal{X})$-*KDP as the key ring.*

KDPs were first introduced in [56, 57], where (t, w)-KDPs were proposed and analysed. These structures have subsequently been investigated by a number of authors who have investigated bounds and constructions for efficient KDPs, particularly of uniform (t, w)-KDPs of rank k. We now briefly mention some of the work that has been undertaken on KDP constructions and KDP efficiency.

5.3.1 KDP constructions   We first define two fundamental KDPs.

Scheme 5.14 *Given a communication structure $\mathcal{C}$ defined on a user set $\mathcal{U}$, a $(\mathcal{C}, 2^{\mathcal{U}})$-trivial inclusion KDP (TIKDP) is defined as follows.*

- *For each $A \in \mathcal{C}$, associate a point $x_A \in \mathcal{I}$;*

- *For each user $U_i \in \mathcal{U}$, define a block $B_i = \{x_A \mid U_i \in A\}$.*

Given any $A \in \mathcal{C}$, the blocks B_j such that $U_j \in A$ have the unique point x_A in common. As no B disjoint from A contains x_A, we see that $(\mathcal{I}, \mathcal{B})$ is a $(\mathcal{C}, 2^{\mathcal{U}})$-KDP.

Note that the KDPPS arising from applying a TIKDP in Scheme 5.13 is essentially Scheme 5.2, the trivial KPS.

Scheme 5.15 *Given an exclusion structure $\mathcal{X}$ defined on a user set $\mathcal{U}$, a $(2^{\mathcal{U}}, \mathcal{X})$-trivial exclusion KDP (TEKDP) is defined as follows.*

- *For each $B \in \mathcal{X}$, associate a point $x_B \in \mathcal{I}$;*

- *For each user $U_i \in \mathcal{U}$, define a block $B_i = \{x_B \mid U_i \notin B\}$.*

Given any subset $B \in \mathcal{X}$, none of the blocks B_j such that $U_j \in B$ contain point $x_{\mathcal{U} \setminus B}$, and thus $(\mathcal{I}, \mathcal{B})$ is a $(2^{\mathcal{U}}, \mathcal{X})$-KDP.

The TEKDP for the case where $\mathcal{X}$ consists of all subsets of users of size at most w was first defined in [35].

Both Scheme 5.14 and Scheme 5.15 result in users potentially having to store a large amount of secret data. A number of combinatorial objects have been used to construct (t, w)-KDPs that perform much better than these fundamental KDPs.

- In [72] it is shown that a $(t+1)$-(n, k, λ) design with $w < (n-t)/(k-t)$ is the dual of a (t, w)-KDP.

- In [60], [61] and [67] special finite geometrical structures have been used to construct KDPs.

- In [64] KDPs were constructed from conics arising from finite projective planes and affine planes.

- In [74], KDPs are defined from orthogonal and perpendicular arrays.

We note that in [32] a non-constructive existence result for very efficient (t, w)-KDPs was proven which, when applied to Scheme 5.13, generates a KDPPS that is essentially a manifestation of the RKPS.

5.3.2 Efficiency of KDPs Given a fixed number of users n we are particularly interested in trying to find KDPs of low rank (small block size), since the resulting KDPPS produced using Scheme 5.13 will have relatively low user storage. A different (but related) optimisation problem is to minimise v, which corresponds to the number of different keys in the system. In [65] several lower bounds on the information storage of KDPs were determined. Subsequently several bounds on (t, w, d)-cover free families have been proven in [76]. These all indicate that, in general, KDPPSs are not particularly efficient. However there are several generic techniques in which KDPPSs can be made more efficient. One such technique was proposed in [72], based on the following concept:

Definition 5.16 An (n, m, t, q)-*resilient function* is a function $f : [GF(q)]^n \to GF(q)$ such that if t input bits are fixed and the remaining $n - t$ chosen independently at random, then every possible element of $GF(q)$ occurs as output with equal probability.

Let $(\mathcal{I}, \mathcal{B})$ be a $(\mathcal{C}, \mathcal{X})$-KDP. For any $A \in \mathcal{C}$ let $I_A = \cap_{U_i \in A} B_i$. Denote $c_A = |I_A|$ and $d_A = \max\{|I_A \cap B| \mid B \in \mathcal{X} \text{ and } A \cap B = \emptyset\}$. In other words, each set A in the communication structure is associated with at least $c_A - d_A$ identifiers (keys) that are unknown to any disjoint set in the exclusion structure. The following refinement to Scheme 5.13 was observed in [72].

Scheme 5.17 *Let $(\mathcal{I}, \mathcal{B})$ be a $(\mathcal{C}, \mathcal{X})$-KDP and $m = \min\{c_A - d_A \mid A \in \mathcal{C}\}$.*

1. *For each $A \in \mathcal{C}$ choose a public (c_A, m, d_A, q)-resilient function f_A. (Such a function always exists for suitable large q [72].)*

2. *Now construct a $(\mathcal{C}, \mathcal{X})$-KDPPS by applying Scheme 5.13 with the $(\mathcal{C}, \mathcal{X})$-KDP as the key ring and f_A as the public combining function for group key k_A. (In other words, using the notation of Scheme 5.7, $k_A = f_A(\cap_{U_i \in A} u_i)$.)*

A KPS arising from a KDP is likely to benefit from the refinement proposed in Scheme 5.17 if the KDP has a relatively high value of m (or, in the case of (t, w)-KDPs, if the KDP is a (t, w, d)-CFF for a high value of d). In [74] some (t, w)-KDPs were constructed from orthogonal and perpendicular arrays that lend themselves to this improvement and result in KPSs with good user storage. In [65] an alternative technique for improving the efficiency of a KDP was proposed, based on the idea of using an *information map* to reduce the information content of the keys k_i held by each user.

5.4 Hash-tree key predistribution schemes

We now describe a family of key predistribution schemes whose security is based on repeated iterations of a cryptographic hash function (see Section 4.5). In a similar manner to the construction of KDPPSs from KDPs, we first define a combinatorial object (in this case an array) from which KPSs can be generated. Recall from Section 4.4 that a rooted tree $\mathcal{T}$ has a natural partial ordering $\leqslant$ defined by parenthood.

Definition 5.18 Let $(\mathcal{C}, \mathcal{X})$ be a communication and exclusion structure defined on n users. Let $\mathcal{T}$ be a rooted tree with vertices labelled by $\{0, 1, \ldots, d-1\}$ and let b be a positive integer. A $(\mathcal{C}, \mathcal{X}, \mathcal{T})$-*hash-tree key predistribution pattern* (HTKDP) is a $b \times n$ matrix $M = (\alpha_{ij})$, where each column is associated with a unique user, with entries from $\{0, 1, \ldots, d-1\}$, such that for any disjoint $A \in \mathcal{C}$ and $B \in \mathcal{X}$ there exists indices (i^{AB}, j^{AB}) such that $j^{AB} \in A$ and:

1. $\alpha_{(i^{AB})j} \leqslant \alpha_{(i^{AB})(j^{AB})}$ for all $j \in A$;

2. $\alpha_{(i^{AB})j} \not\leqslant \alpha_{(i^{AB})(j^{AB})}$ for all $j \in B$.

In [49] it was shown how a KPS can be constructed from a $(\mathcal{C}, \mathcal{X}, \mathcal{T})$-HTKDP.

Scheme 5.19 *Let the $b \times n$ matrix $M = (\alpha_{ij})$ with entries from $\{0, 1, \ldots, d-1\}$ be a $(\mathcal{C}, \mathcal{X}, \mathcal{T})$-HTKDP. A $(\mathcal{C}, \mathcal{X}, \mathcal{T})$-hash-tree key predistribution scheme (HTKPS) can be constructed from M and a suitable hash function h as follows:*

1. *The TA publishes M, $\mathcal{T}$ and h as public system parameters.*

2. *For each $1 \leq i \leq b$ the TA chooses a secret random seed value s_i^0. For each $1 \leq j \leq d-1$ a hash value can then be iteratively computed such that if j is a child of l in $\mathcal{T}$ then $s_i^j = h(s_i^l, j)$.*

3. *The TA securely delivers $u_j = \{s_1^{\alpha_{1j}}, \ldots, s_b^{\alpha_{bj}}\}$ to user U_j.*

4. *For $A \in \mathcal{C}$, define*

$$I_A = \{1 \leq i \leq b \mid \text{ there exists } m_j \in A \text{ such that } \alpha_{ij} \leqslant \alpha_{im_j} \text{ for all } j \in A\}.$$

   Then $k_A = \sum_{i \in I_A} s_i^{\alpha_{im_j}}$.

Thus we see that in an HTKPS, any user U_l belonging to A can compute k_A since for each $i \in I_A$ they can iteratively compute $s_i^{\alpha_{im_j}}$ from their component $s_i^{\alpha_{il}}$ of u_l. On the other hand, for any $B \in \mathcal{X}$, Definition 5.18 guarantees that $\alpha_{(i^{AB})m_j} \not\leqslant \alpha_{(i^{AB})(j^{AB})}$. Thus $s_i^{\alpha_{(i^{AB})m_j}}$, and hence k_A, cannot be computed by any user in B.

Scheme 5.19 is thus a deterministic KPS that offers computational security, since the security of group keys k_A depends on the security of the underlying hash function.

Following the convention of previous sections, we will refer to a $(t, w, \mathcal{T})$-HTKDP and $(t, w, \mathcal{T})$-HTKPS respectively when $\mathcal{C}$ consists of all t-subsets of users and $\mathcal{X}$ consists of all subsets of at most w users.

Example 5.20 Let $\mathcal{T}$ be a starlike tree with 7 leaves (where the centre is labelled 0 and the leaves labelled $1, \ldots, 7$). The following $(2, 2, \mathcal{T})$-HTKDP on 7 users was given in [49]:

$$M = \begin{matrix} 0 & 0 & 3 & 0 & 5 & 6 & 7 \\ 1 & 0 & 0 & 4 & 0 & 6 & 7 \\ 0 & 2 & 0 & 4 & 5 & 0 & 7 \\ 1 & 2 & 3 & 0 & 0 & 0 & 7 \\ 0 & 2 & 3 & 4 & 0 & 6 & 0 \\ 1 & 0 & 3 & 4 & 5 & 0 & 0 \\ 1 & 2 & 0 & 0 & 5 & 6 & 0 \end{matrix}.$$

To construct a $(2, 2, \mathcal{T})$-HTKPS, the TA first generates secret seeds $s_1^0, \ldots, s_7^0$. Seven copies of $\mathcal{T}$ are then labeled with iterations of h as indicated in Figure 1. User U_1

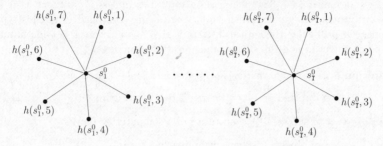

Figure 1: Hash iterations based on the starlike tree with 7 leaves

then receives:

$$
\begin{aligned}
u_1 &= \{s_1^{\alpha_{11}}, s_2^{\alpha_{21}}, s_3^{\alpha_{31}}, s_4^{\alpha_{41}}, s_5^{\alpha_{51}}, s_6^{\alpha_{61}}, s_7^{\alpha_{71}}\} \\
&= \{s_1^0, s_2^1, s_3^0, s_4^1, s_5^0, s_6^1, s_7^1\} \\
&= \{s_1^0, h(s_2^0, 1), s_3^0, h(s_4^0, 1), s_5^0, h(s_6^0, 1), h(s_7^0, 1)\}.
\end{aligned}
$$

Similarly, U_2 receives:

$$
u_2 = \{s_1^0, s_2^0, h(s_3^0, 2), h(s_4^0, 2), h(s_5^0, 2), s_6^0, h(s_7^0, 2)\}.
$$

The group key $k_{\{U_1, U_2\}}$ is constructed by first noting that $I_{\{U_1, U_2\}} = \{1, 2, 3, 5, 6\}$ and thus that $k_{\{U_1, U_2\}} = s_1^0 + h(s_2^0, 1) + h(s_3^0, 2) + h(s_5^0, 2) + h(s_6^0, 1)$.

There are three special cases worth mentioning.

1. If $\mathcal{T}$ degenerately consists of just one vertex then a $(\mathcal{C}, \mathcal{X}, \mathcal{T})$-HTKDP is a $(\mathcal{C}, \mathcal{X})$-KDP, as defined in Section 5.3.

2. Scheme 5.19 was motivated by an earlier scheme in [50] that constructed a $(2, w, \mathcal{T})$-HTKDP, where $\mathcal{T}$ was a chain and the underlying matrix M was generated randomly, resulting only in a probabilistic scheme.

3. In [66] a further variant (called HARPS) was proposed for use in wireless sensor networks (see Section 7.3). This combines the scheme of [50] and the idea behind the RKPS (Scheme 5.8) by only allocating to each user a value on a random subset of the b hash chains (instead of all the chains). This reduces user storage at the expense of a poorer probability that a group will be able to construct a group key.

5.5   Key assignment schemes

A very interesting class of key predistribution schemes arise from what are known as information flow policies. These have largely been investigated by researchers in computer security since they define a type of access control mechanism, but they can be considered as a class of group key predistribution schemes. The technique of key predistribution is particularly appropriate for this type of application because it allows an information flow policy to be applied "seamlessly", without users even necessarily being aware that their actions are being controlled using this type of scheme. Of particular theoretical interest is that these schemes provide naturally arising examples of non-threshold communication structures.

Definition 5.21 An *information flow policy* is a tuple $(\mathcal{L}, \mathcal{E}, \mathcal{S}, \mathcal{O}, \lambda)$, where:

- $(\mathcal{L}, \mathcal{E})$ is a directed graph of security labels (see Figure 2);

- $\mathcal{S}$ is a set of subjects (perhaps users of a computer system);

- $\mathcal{O}$ is a set of objects (perhaps computer files);

- $\lambda : \mathcal{S} \cup \mathcal{O} \to \mathcal{L}$ is a security function that associates subjects and objects with security labels.

Figure 2: Directed graph of security labels

An information flow policy is used to model the access of subjects in $\mathcal{S}$ to a set of objects in $\mathcal{O}$ in a hierarchical system, where the directed graph indicates when a subject can read an object. More precisely, subject S can read object O if and only if $(\lambda(S), \lambda(O)) \in \mathcal{E}$. One way of implementing this policy is to use what is known as a key assignment scheme.

Scheme 5.22 A key assignment scheme *for information flow policy* $(\mathcal{L}, \mathcal{E}, \mathcal{S}, \mathcal{O}, \lambda)$ *is a scheme initialised by a TA as follows:*

- *The TA identifies each label* $x \in \mathcal{L}$ *with a cryptographic key* k_x.

- *For each label* $x \in \mathcal{L}$ *the TA generates secret information* $\sigma(x)$ *and securely distributes it to all subjects with security label* x.

- *The TA generates some system-wide public data Pub that is made available to all subjects using an authenticated channel.*

- *There exists a function that takes as input labels* $x, y \in \mathcal{L}$, $\sigma(x)$ *and Pub and outputs* k_y *if and only if* $(x, y) \in \mathcal{E}$.

A key assignment scheme implements the information flow policy since k_y can be used to encrypt objects with security label y, and only subjects with label x, where $(x, y) \in \mathcal{E}$ can compute k_y and hence decrypt the encrypted object.

It is worth observing at this stage that the vast majority of information flow policies, and hence key assignment schemes, are defined for hierarchies where the security labels in $\mathcal{L}$ form a poset. Not only are such *poset-based* schemes easier to design, but they are also by far the most natural policies to implement in real applications. In this case we can represent the policy by $(\mathcal{L}, \leqslant, \mathcal{S}, \mathcal{O}, \lambda)$ (more commonly just denoted by $(\mathcal{L}, \leqslant)$ when the context is obvious). In this case subject S can read object O if and only if $\lambda(S) \geqslant \lambda(O)$.

For any $y \in \mathcal{L}$, let $\uparrow y = \{x \in \mathcal{L} \mid (x, y) \in \mathcal{E}\}$ and $\downarrow y = \{z \in \mathcal{L} \mid (y, z) \in \mathcal{E}\}$. The following result is immediate from the relevant definitions.

Theorem 5.23 *A key assignment scheme for information flow policy* $(\mathcal{L}, \mathcal{E}, \mathcal{S}, \mathcal{O}, \lambda)$ *is a* $(\mathcal{C}, \mathcal{X})$*-key predistribution scheme where:*

- $\mathcal{U} = \mathcal{L}$;

- $\mathcal{C} = \{\uparrow y \mid y \in \mathcal{L}\}$;

- $\mathcal{X}$ *is inherited from the degree of collusion security of the underlying key assignment scheme.*

- $u_x = \sigma(x)$;

- *Pub is the same as for the key assignment scheme;*

- *For* $A = \uparrow y \in \mathcal{C}$, $k_A = k_y$.

A key assignment scheme can thus be thought of as a special type of deterministic computationally secure KPS, where there are as many groups in the communication structure as there are users, and where the groups can be derived from the vertices of a directed graph defined on the set of users. Note that it is perhaps more appropriate to consider a key assignment scheme as a KPS defined on the set $\mathcal{S}$ of subjects. In this case each subject S with security label $\lambda(S) = x$ is given the same piece of secret information $\sigma(x)$ in the resulting KPS. A subject S is thus able to compute all the group keys $k_{\uparrow y}$ for each $y \in \downarrow x$.

A review of key assignment schemes can be found in [27]. In the remaining sections we provide examples of some of the techniques used to construct them.

5.5.1 Unconditionally secure key assignment

We first observe that unconditionally secure key assignment schemes are not very interesting from either a theoretical or a practical perspective. One obvious example is the *trivial key assignment scheme* (TKAS) based on letting $\sigma(x) = \{k_y \mid (x, y) \in \mathcal{E}\}$, which gives rise to Scheme 5.2 when interpreted as a KPS. We have already observed in Section 5.1 that such a scheme has unacceptably high secret storage. The unconditional secure setting was modelled formally in [34] and it was shown that the TKAS is essentially the best possible (more precisely it was shown that it can only be slightly improved by first compressing the representation of the information flow policy and then generating a TKAS for this slightly simpler policy). As a result, the only key assignment schemes of real interest are necessarily computationally secure.

5.5.2 Key assignment for poset policies An extraordinary variety of key assignment schemes for poset policies have been proposed in the literature. In [28] it was shown that they all fall into five broad classes. When interpreted as KPSs, these five classes coincide with five of the fundamental KPSs identified in Section 5.1. As the majority are either IKEKPSs (Scheme 5.5) or NBKPSs (Scheme 5.6) we will present one example from each of these classes here. While these schemes are not strictly combinatorial, the fact that they implement communication structures with interesting combinatorial structure merits their inclusion in this review.

The following key assignment scheme was first proposed in [1] (our version is based on an observation in [28]) and gives rise to a NBKPS.

Scheme 5.24 *The* Akl-Taylor *key assignment scheme (ATKAS) for a poset-based policy* $(\mathcal{L}, \leqslant)$ *is defined as follows:*

- *Let* $n = pq$ *be the product of two large primes and* $m \in Z_n^*$ *(all subsequent calculations are modulo n). The value n is public, but p, q and m are kept secret by the TA.*

- *For each* $x \in \mathcal{L}$, *let* $Pub_x = p_x$, *where* p_x *is a small prime and* $Pub_x \neq Pub_y$ *if* $x \neq y$ *(it suffices for* $\{p_x \mid x \in \mathcal{L}\}$ *to be chosen to be the first* $|\mathcal{L}|$ *primes). Let* $Pub = \cup_{x \in \mathcal{L}} Pub_x$.

- *For each* $x \in \mathcal{L}$, *let* $e(x) = \prod_{y \not\leqslant x} p_x$ *and* $\sigma(x) = k_x = m^{e(x)}$.

- *If* $y \leqslant x$ *then given* x, y, $\sigma(x) = k_x$ *and* Pub, *we can calculate* k_y *as follows:*

$$k_y = k_x^{p(x,y)}, \text{ where } p(x,y) = \prod_{z \in (\mathcal{L} \setminus \downarrow y) \setminus (\mathcal{L} \setminus \downarrow x)} p_z.$$

Thus the ATKAS uses a public labeling of the nodes of the poset $(\mathcal{L}, \leqslant)$ to generate a set of exponents $e(x)$ that have the property that $e(x)|e(y)$ if and only if $y \leqslant x$. This allows keys k_x associated with a higher level in the poset to compute keys k_y at lower levels. If $y \not\leqslant x$ then it is impossible to compute k_y from k_x without knowledge of m. Calculating m from any k_x is believed to be a hard computational problem known as the *RSA problem* (see, for example [73], for more information about the RSA cryptosystem on which this is based). In fact it is possible to show that any collusion of nodes cannot determine a key that they are not entitled to, assuming that the RSA problem is hard, and so the ATKAS (and thus its resulting KPS) is computationally secure with full collusion security.

There have been many variants of the ATKAS proposed (for example [37, 53]) and [27] contains a comprehensive list. Most of these either attempt to optimise the poset labelling in some way or change its performance with respect to an update phase. The principle behind all these schemes remain the same.

The next key assignment scheme was proposed by [4] and gives rise to an IKEKPS.

Scheme 5.25 *The* AFB *key assignment scheme (AFBKAS) for a poset-based policy* $(\mathcal{L}, \leqslant)$ *is defined as follows:*

- *Let h be a one-way hash function such that $h\colon \{0,1\}^* \to \{0,1\}^l$ for some integer l.*

- *For each $x \in \mathcal{L}$, let $\sigma(x) = k_x$ be randomly selected from $\{0,1\}^l$.*

- *For each $x \in \mathcal{L}$, let $Pub = \{k_z - h(k_x, z) \mid z \lessdot x\}$.*

- *If $y \leqslant x$ then given x, y, $\sigma(x) = k_x$ and Pub, we can calculate k_y since there exists a path $(z_0, z_1), \ldots, (z_{m-1}, z_m)$, where $z_0 = x$ and $z_m = y$. The key k_{z_i} can be iteratively obtained from $k_{z_{i-1}}$ and Pub by computing $h(k_{z_{i-1}}, z_i)$, from which $k_{z_i} = (k_{z_i} - h(k_{z_{i-1}}, z_i)) + h(k_{z_{i-1}}, z_i)$.*

To see that the KPS arising from Scheme 5.25 is an IKEKPS (Scheme 5.5), we observe that there exists an isomorphism between the poset $(\mathcal{L}, \leqslant)$ and the poset $(\mathcal{C}, \leqslant^*)$ associated with the resulting KPS, resulting in the following correspondences between Scheme 5.5 and Scheme 5.25:

- $\uparrow x \in \mathcal{C}$ corresponds to $x \in \mathcal{L}$;

- $roots_x$ for security label x correspond to $\{x\}$;

- u_x in Scheme 5.5 corresponds to k_x;

- $k_{\uparrow x}$ in Scheme 5.5 also corresponds to k_x;

- If $(\uparrow z) \lessdot^* (\uparrow x)$ then $E_{k_{\uparrow z}}(k_{\uparrow x})$ in Scheme 5.5 is defined by $k_z - h(k_x, z)$.

Note that Pub_1 in Scheme 5.5 corresponds to $\{k_x - h(k_x, x) \mid x \in \mathcal{L}\}$. This serves no purpose, as it is essentially an encryption of key k_x using key k_x and therefore has been omitted from the description of Scheme 5.25. (In fact Pub_1 is redundant in any KPS arising from a poset-based key assignment scheme.)

There have been many proposals for key assignment schemes that give rise to IKEKPs, for example [51, 80, 81].

5.5.3 Key assignment for directed graphs

We have already observed that most key assignment schemes are designed for information flow policies based on posets. It is at least of theoretical interest to investigate schemes for general information flow policies (general directed graphs).

One method of constructing a key assignment scheme for a general information flow policy is to embed the policy into a poset and then use a poset-based key assignment scheme. In [68] such an embedding was exhibited that enables the poset-based scheme of Akl-Taylor [1] to be extended to a general information flow policy. The majority of poset-based key assignment schemes are *simple*, which means that for any $x \in \mathcal{L}$ we have $\sigma(x) = k_x$. The embedding of [68] works by embedding $(\mathcal{L}, \mathcal{E})$ in a poset $(\mathcal{L}^*, \leqslant)$, creating a simple Akl-Taylor poset-based key assignment scheme for $(\mathcal{L}^*, \leqslant)$, and interpreting this as a non-simple scheme for $(\mathcal{L}, \mathcal{E})$. In [27] it is shown that this *De Santis decoupling*, presented as Scheme 5.26, can be applied to any simple poset-based key assignment scheme.

Scheme 5.26 *The De Santis decoupling generates a key assignment scheme for the information flow policy $(\mathcal{L}, \mathcal{E})$ as follows:*

- *Define a poset $(\mathcal{L}^*, \leqslant)$, where*

  - $\mathcal{L}^* = \{x_l \mid x \in \mathcal{L}\} \cup \{x_u \mid x \in \mathcal{L}\}$,
  - $x_l \leqslant x_u$,
  - $(y, x) \in \mathcal{E}$ *implies* $y_l \leqslant x_u$.

- *Establish any simple poset-based key assignment scheme for $(\mathcal{L}^*, \leqslant)$, with key k_x^* for each $x \in \mathcal{L}^*$.*

- *Interpret this as a key assignment scheme for $(\mathcal{L}, \mathcal{E})$ where $k_x = k_{x_l}^*$ and $\sigma(x) = k_{x_u}^*$.*

An illustration of how the De Santis decoupling works is shown in Figure 3. The upper row of nodes in $(\mathcal{L}^*, \leqslant)$ represents $a_u, \ldots, f_u$, while the lower row of nodes represents $a_l, \ldots, f_l$.

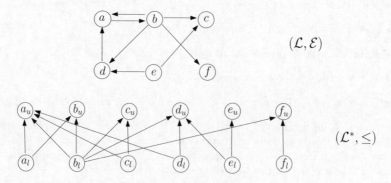

Figure 3: The construction of $(\mathcal{L}^*, \leqslant)$ from $(\mathcal{L}, \mathcal{E})$

6   Group key distribution schemes

Our next class of key establishment schemes, group key distribution schemes, are appropriate for applications where it is possible (and practical) to communicate in some way with a trusted entity throughout the lifetime of the scheme. This scenario is desirable for applications where group keys k_A are necessarily generated at the time of request (not during the initialisation phase as is the case for most key predistribution schemes).

Definition 6.1 A $(\mathcal{C}, \mathcal{X})$-*key distribution scheme* (KDS) is a key establishment scheme with communication structure $\mathcal{C}$ and exclusion structure $\mathcal{X}$ such that:

1. Given $A \in \mathcal{C}$, any $U_i \in A$ can compute the group key k_A from knowledge of u_i and $v_{i,A}$, where $v_{i,A}$ is some information obtained by U_i from the TA during the key establishment phase for key k_A.

2. Given disjoint sets $B \in \mathcal{X}$ and $A \in \mathcal{C}$, it is not possible to compute the group key k_A from knowledge of u_B and v_B (where $u_B = \{u_i \mid U_i \in B\}$ and $v_B = \{v_{i,A} \mid U_i \in B\}$).

Note that Definition 6.1 includes the case where the secure channels that were used to distribute the initial user secret data u_i are still available and $v_{i,A} = k_A$ if $U_i \in A$, otherwise $v_{i,A} = \emptyset$. This "trivial" solution is in fact one that is often adopted in real applications where such secure channels exist throughout the scheme lifetime, however it is of little mathematical interest and so we do not consider it further here.

6.1 Broadcast encryption

We now look at a well-studied family of group key distribution schemes where, although the TA is online during the key establishment phase, it no longer maintains secure channels to the users and must rely on broadcast channels to establish group keys.

Definition 6.2 A $(\mathcal{C}, \mathcal{X})$-*broadcast encryption scheme* (BES) is a key distribution scheme with communication structure $\mathcal{C}$ and exclusion structure $\mathcal{X}$ such that $v_{i,A} = B_A$ for every user $U_i \in \mathcal{U}$, where B_A is a public message broadcast to all users in $\mathcal{U}$ at the start of the key establishment phase for k_A.

Broadcast encryption schemes were first proposed with applications such as access to streamed multimedia services in mind. In this type of application some digital content, such as a film, is encrypted using k_A (where A is the group of users permitted to access the service) and then B_A is broadcast as a header that allows an authorised user U_i in A to determine k_A and hence decrypt the service. There are two slightly different applications of broadcast encryption, which we illustrate using the above multimedia service scenario:

1. **General broadcast encryption**: These schemes are usually designed for as large a communication structure as possible, since this maximises the possible number of different groups for whom group keys can be generated. These are suitable for *pay-per-view* services, where the groups of users receiving content are highly variable (for example only a small group from the set of all users may want to pay to watch a particular football match).

2. **Long term group management**: These schemes are characterised by a single large group of users that may change gradually over time. These are suitable for *subscription* services, where we only ever want to broadcast to the entire group of subscribed users, but the make-up of this group is dynamic.

Note that these two applications are far from being mutually exclusive. The main difference is that while schemes designed for the first scenario should be able to efficiently broadcast to user groups of all sizes, schemes designed for the second scenario initially associate a group key k_H with a single group of users H (from the universe $\mathcal{U}$ of possible users) and should be specifically designed to efficiently cope with relatively small changes to H over time. This scenario is often described in terms of maintaining a *multicast group*, where the term "multicast" arises from internet-related technology for sending a single message to a designated set of recipients [22].

One significant difference between proposed broadcast encryption schemes relates to the computational capabilities of users in the scheme. We say that a broadcast encryption scheme is suitable for:

- *stateless receivers* if the users cannot retain information from previous broadcasts (or have ability to write to memory). This might be the case for example if the user is a set-top decoder. The decoder is preloaded with decryption keys that cannot be changed over time. Each time a broadcast message is sent, the decoder can use these keys to decrypt the broadcast, but it will not retain any memory of the information it receives (if the same group key is used twice, the decoder will have to decrypt it on each occasion, as it cannot store any information supplied to it during a key establishment event).

- *stateful receivers* if the users can retain information from previous broadcasts (or have the ability to write to memory). The critical difference in this case is that if new keys are broadcast to them then users can use these to replace the keys that were distributed to them on initialisation (in other words users have the ability to update their secret data).

The motivation for considering a stateless receiver model is that this greatly simplifies the software or hardware needed by the users. Almost all the schemes that we discuss in this paper are suitable for stateless receivers.(Whether real human users are stateless or stateful will be left as an open problem!)

6.1.1 Benchmark broadcast encryption schemes

We now define two benchmark broadcast encryption schemes against which others need to be compared. Both are suitable for stateless receivers. These are analogues of Scheme 5.3 and Scheme 5.4 respectively. Throughout the remainder of this section we assume that E is a secure symmetric encryption algorithm with key size l and $E_k(m)$ denotes the encryption of plaintext m using key k.

Scheme 6.3 *A trivial broadcast encryption scheme (TBES) has the following properties:*

- $u_i = \{K_A \mid U_i \in A, A \in \mathcal{C}\}$, *where each K_A is randomly chosen from $\{0,1\}^l$;*

- $B_A = E_{K_A}(k_A)$;

- $K_A \in u_i$ *if and only if $U_i \in A$, with k_A obtained by decrypting $E_{K_A}(k_A)$.*

Scheme 6.4 *A direct broadcast encryption scheme (DBES) has the following properties:*

- $u_i = k_i$, *where each k_i is randomly chosen from $\{0,1\}^l$;*

- $B_A = \{E_{k_i}(k_A) \mid U_i \in A\}$;

- $E_{k_i}(k_A) \in B_A$ *if and only if $U_i \in A$, with k_A obtained by decrypting $E_{k_i}(k_A)$.*

We thus see that Schemes 6.3 and 6.4 offer extreme ends of the tradeoff between the size of user secret data u_i and the size of the broadcast header B_A. Both schemes are deterministic, computationally secure and have independent keys. In general the hunt for good broadcast encryption schemes is about finding schemes with reasonable parameter tradeoffs between these two benchmark schemes.

6.1.2 Broadcast encryption schemes from key predistribution schemes

One simple way of establishing a broadcast encryption scheme suitable for stateless receivers is to build it onto an existing key predistribution scheme.

Scheme 6.5 *If we have a $(\mathcal{C}, \mathcal{X})$-KPS then we can realise a $(\mathcal{C}, \mathcal{X})$-BES as follows:*

- *u_i is the same for both the KPS and the BES;*

- *$B_A = E_{k_A^*}(k_A)$, where k_A^* is the group key for $A \in \mathcal{C}$ in the KPS and k_A is a freshly generated group key for A in the BES;*

- *Only a user U_i in A can establish k_A^* from u_i and hence decrypt the new group key k_A.*

While Scheme 6.5 is attractively simplistic, the main problem with it is that for broadcast encryption we generally want $\mathcal{C}$ to be large, and a KPS for a large $\mathcal{C}$ typically has high user storage requirements.

In [15] a broadcast encryption scheme was suggested that employs a KPS that establishes keys for groups of l users in order to construct a BES that establishes keys for groups of $t = \lambda l$ users. This scheme uses a resolvable design defined on t points to partition the t users into blocks of size l, which are then used to define a broadcast message. The advantage of this idea is that the user storage required for the KPS on group size l is smaller than that for group size t. This scheme is not particularly efficient with respect to broadcast size if we are planning to use a BES to distribute a group key k_A (which in this paper we are), however it has some merits if the information to be broadcast to the group is longer.

In [72] another interesting family of broadcast encryption schemes were proposed, which combine a KPS with an ideal secret sharing scheme (see Section 4.5). The idea is the following:

Scheme 6.6 *Let $\mathcal{U}$ be a set of users and $\mathcal{X}$ be an exclusion structure defined on $\mathcal{U}$. Suppose that we can find:*

1. *A set system $(\mathcal{U}, \mathcal{B})$, where $|\mathcal{B}| = b$ and for each block $B_j \in \mathcal{B}$ we can construct a $(2^{B_j}, \mathcal{X}_j)$-KPS on user set B_j, where:*

    (a) *The user secret for each $U_i \in B_j$ is denoted by u_i^j.*

    (b) *The group key for each $A \subseteq B_j$ is denoted by k_A^j and is an element of $\mathcal{K}$.*

2. *An ideal secret sharing scheme (with shares and secrets from $\mathcal{K}$) on participant set $\mathcal{B}$ with access structure Γ such that:*

    (a) *For every $U_i \in \mathcal{U}$, we have $\{B_j \mid U_i \in B_j\} \in \Gamma$;*

    (b) *For every $X \in \mathcal{X}$, we have $\{B_j \mid X \cap B_j \notin \mathcal{X}_j\} \notin \Gamma$.*

Then a KIO $(2^{\mathcal{U}}, \mathcal{X})$*-broadcast encryption scheme is defined by:*

- *For each U_i let $u_i = \{u_i^j \mid U_i \in B_j\}$.*

- *$B_A = (E_{k_1}(y_1), \ldots, E_{k_b}(y_b))$, where:*

  1. *$(y_1, \ldots, y_b)$ are shares of the ideal secret sharing scheme corresponding to secret k_A;*

  2. *$k_j = k_{A \cap B_j}^j$ for each $1 \leq j \leq b$;*

  3. *E is a symmetric encryption algorithm with keys from set $\mathcal{K}$.*

Although at first glance complex, the intuition behind Scheme 6.6 is straightforward. If U_i is a member of set A then they can use their secret information u_i^j to determine the group keys $k_{A \cap B_j}^j$ for each KPS that U_i is a member of. These then allow U_i to decrypt a set of shares y_j that correspond to a set in the access structure of the secret sharing scheme, which means that the shares can be used to reconstruct k_A. On the other hand, a set of users X in the exclusion structure can, in the worst case, determine a set of group keys that decrypt a set of shares not in the access structure, hence they obtain no information about the group key k_A.

Thus we need to find combinations of set systems, KPSs and ideal secret sharing schemes that allow Scheme 6.6 to be enabled. The KIO broadcast encryption scheme construction was first proposed in [72] using the Trivial Exclusion KDPs (Scheme 5.15) from [35] with exclusion parameter w as the KPSs. In particular it has been shown that the following combinations result in KIO broadcast encryption schemes:

1. Let $(\mathcal{U}, \mathcal{B})$ be a $2 - (n, b, r, k, \lambda)$ design with $r > \lambda\binom{w}{2}$ and choose an ideal $(\lambda\binom{w}{2} + 1, b)$-threshold scheme [72].

2. An improved scheme is obtained by letting $(\mathcal{U}, \mathcal{B})$ be an (n, b, r, λ)-*broadcast key distribution pattern* (BKDP) [73], which is a set system of n points, b blocks, every point on r blocks, every pair of points in at most λ blocks and $r > \lambda\binom{w}{2}$. These structures were first defined in [74] using the name *threshold designs* and several constructions based on Steiner systems and orthogonal arrays were provided.

3. A further improvement was made in [75], where it was observed that the KIO construction technique can be further generalised to allow the ideal secret sharing scheme to be replaced by a *ramp scheme* (see [38]), which is a type of secret sharing scheme that permits smaller share sizes.

Example 6.7 In order to see the kinds of parameter tradeoff that are possible using KIO, we note that in [74] it was shown that an orthogonal array $OA_1(t, q, q)$ can be used to construct a $(q^t, q^2, q, t - 1)$-BKDP. This gives rise to a BES for q^t users, where each user stores at most $q + (t - 1)(q^t - 1)$ values and the broadcast B_A to enable any group key k_A is of length q^2. This compares with user storage of $2^{q^t - 1}$ and broadcast length 1 for the TBES, and user storage of 1 and broadcast length $|A|$ for the DBES. Given that this construction works for any $t < q$, it is clear that the KIO construction provides a balance between the extremes of TBES and DBES, especially when t is close to q and A is very large (as is likely to be the case in many of the applications envisaged for broadcast encryption).

6.1.3   Logical key hierarchies for stateful receivers   Using a tree of keys to manage a long term group key was first suggested in [77] and [79]. The broadcast encryption scheme that they independently proposed is suitable for stateful receivers. The basic idea is as follows.

Scheme 6.8 *Let H be a subset of m users (chosen from a universe $\mathcal{U}$) who wish to establish a group key k_H. For simplicity, assume that $m = 2^h$. To establish a logical key hierarchy:*

1. *Define a (complete) binary tree with m leaves, each associated with a user from H. Iteratively label this tree with independent keys as follows: root by $k_{0,0}$; the left child of $k_{i,j}$ by $k_{i+1,2j}$; and right child of $k_{i,j}$ by $k_{i+1,2j+1}$. Associate the users, which we label $U_0, \ldots, U_{m-1}$, with the nodes labeled by keys $k_{h,0}, \ldots, k_{h,m-1}$.*

2. *For each user U_j let $u_j = \{k_{x,y} \mid k_{x,y}$ is on the path from $k_{h,j}$ to $k_{0,0}\}$. Each user thus holds $h + 1$ keys.*

3. *The key $k_{0,0}$ is held by every user in H. In order to establish a group key k_H the TA could broadcast $B_H = E_{k_{0,0}}(k_H)$. Note however that since this scheme is intended for stateful receivers (and thus users have the ability to refresh their keys) it is also possible just to let $k_H = k_{0,0}$, in which case this scheme can actually be considered as a type of key predistribution scheme.*

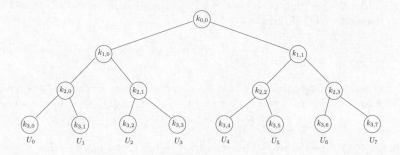

Figure 4: Logical key hierarchy tree for eight users

Example 6.9 Figure 4 shows the binary tree for a logical key hierarchy for eight users H. Each user stores four keys. Thus, for example, user U_3 stores $u_3 = \{k_{3,3}, k_{2,1}, k_{1,0}, k_{0,0}\}$. The group key can be decrypted using (or is) $k_{0,0}$, which is stored by every user, and the remaining keys that a user stores all facilitate group changes. We illustrate this process by an example. Suppose that U_5 leaves the group. It is necessary to replace all the keys held by U_5 that are also held by any other user. The most efficient process for doing this is as follows:

1. The TA generates new keys $k'_{2,2}, k'_{1,1}, k'_{0,0}$;

2. The TA encrypts $k'_{2,2}$ using key $k_{3,4}$;

3. The TA encrypts $k'_{1,1}$ using keys $k_{3,4}$ and $k_{2,3}$;

4. The TA encrypts $k'_{0,0}$ using keys $k_{3,4}$, $k_{2,3}$ and $k_{1,0}$;

5. The TA broadcasts all these encrypted keys;

6. User U_4 decrypts $k'_{2,2}$ and replaces $k_{2,2}$ with this new key (user U_4 can do this because it is a stateful receiver);

7. Similarly, users U_4, U_6 and U_7 replace $k_{1,1}$ by $k'_{1,1}$;

8. Similarly, all users except U_5 replace $k_{0,0}$ by $k'_{0,0}$.

The scheme has now been updated in such a way that the new group key is determined using $k'_{0,0}$, which is a key that the departing user U_5 does not know.

It is straightforward to generalise Scheme 6.8 to use a-ary trees rather than binary trees. Example 6.9 should be sufficient to illustrate how general protocols for leaving or joining groups of users can be derived.

6.1.4   Schemes based on covers for stateless receivers

We now look at a family of broadcast encryption schemes designed for stateless receivers.

Definition 6.10 Let $(\mathcal{I}, \mathcal{B})$ be a set system and for each $x \in \mathcal{I}$ let $\beta(x) = \{B \in \mathcal{B} \,|\, x \in B\}$. We say that $(\mathcal{I}, \mathcal{B})$ is a *cover-based revocation system* (CBRS) if for every non-empty $\mathcal{A} \subseteq \mathcal{B}$ there exists $\mathcal{I}_A \subseteq \mathcal{I}$ such that

$$\bigcup_{x \in \mathcal{I}_A} \beta(x) = \mathcal{A}.$$

In other words, a set system is a CBRS if for every non-empty collection $\mathcal{A}$ of blocks there exists a subset $\mathcal{H}$ of points such that the subsets $\{\beta(x) \,|\, x \in \mathcal{H}\}$ form a cover of $\mathcal{A}$.

Scheme 6.11 *Given a cover-based revocation system $(\mathcal{I}, \mathcal{B})$ we can define a broadcast encryption scheme for stateless receivers as follows:*

- *Associate each point $x \in \mathcal{I}$ with a key k_x, and associate each block $B_i \in \mathcal{B}$ with a user U_i;*

- $u_i = \{k_x \,|\, x \in B_i\}$;

- *For any subset A of users (corresponding to the set of blocks $\mathcal{A}$), $B_A = \{E_{k_x}(k_A) \,|\, x \in \mathcal{I}_A\}$;*

- *By definition of a CBRS, the only users holding at least one of the keys k_x are those in A.*

Broadcast encryption schemes arising from Scheme 6.11 allow group keys to be established for any subset of users (hence $\mathcal{C} = 2^{\mathcal{U}}$), while not permitting any unauthorised subset access to the group key (full collusion security). It is worth noting however that such schemes are only practical if there also exists an efficient algorithm for determining the appropriate cover of keys, given a particular subset of users.

We now describe a manifestation of Scheme 6.11 from [58], based on the binary tree exhibited in Scheme 6.8.

Scheme 6.12 *For simplicity, assume that* $|\mathcal{U}| = m = 2^h$. *To establish a* complete subtree revocation scheme:

1. *Define a (complete) binary tree with m leaves as in Scheme 6.8.*

2. *As in Scheme 6.8, let* $u_j = \{k_{x,y} \mid k_{x,y}$ *is on the path from* $k_{h,j}$ *to* $k_{0,0}\}$. *Each user thus holds* $h + 1$ *keys.*

3. *In order to establish a group key* k_A, *where* $A = \mathcal{U} \setminus R$:

   • *Form the subtree* $ST(R)$ *consisting of the paths from the nodes in* R *to the root (this is sometimes called the* Steiner *Tree of nodes* R).

   • *Identify the set* $\mathcal{K}(A)$ *of nodes* $k_{x,y}$ *of the main tree such that* $k_{x,y}$ *is not a node of* $ST(R)$ *but the parent of* $k_{x,y}$ *is a node of* $ST(R)$ *(such nodes are sometimes referred to as* hanging off $ST(R)$, *and form a cover of* A).

   • *Let* $B_A = \{E_{k_{x,y}}(k_A) \mid k_{x,y} \in \mathcal{K}(A)\}$.

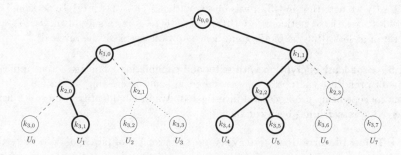

Figure 5: Complete subtree revocation of users U_1, U_4 and U_5

Example 6.13 Figure 5 shows the binary tree for a complete subtree revocation scheme for eight users in which users $R = \{U_1, U_4, U_5\}$ are being revoked (in other words, a group key is being established for $A = \{U_0, U_2, U_3, U_6, U_7\}$). The edges in bold form $ST(R)$ and the keys connected to $ST(R)$ by dashed edges form $\mathcal{K}(A)$. In this case $B_A = \{E_{k_{3,0}}(k_A), E_{k_{2,1}}(k_A), E_{k_{2,3}}(k_A)\}$.

Scheme 6.12 was generalised in [2] to a-ary trees and some interesting compression techniques were proposed for further reducing the user storage.

In [58] it was shown that the following alternative CBRS, which we describe informally, can be extracted from a different labeling of a complete binary tree.

Scheme 6.14 *For simplicity, assume that* $|\mathcal{U}| = m = 2^h$. *To establish a subset* difference revocation scheme:

1. *Define a (complete) binary tree with* m *leaves whose nodes are labeled from the root downwards by* $v_1, \ldots, v_{2^{h+1}-1}$. *Associate the users, which we label* $U_0, \ldots, U_{m-1}$, *with the leaf nodes (labeled by keys* $v_{2^h}, \ldots, v_{2^{h+1}-1}$).

2. *Associate a key* $k_{x,y}$ *with any pair of nodes* v_x *and* v_y *of the tree such that* v_y *is a descendent of* v_x.

3. *For each user* U_j *let* $u_j = \{k_{x,y} \mid U_j$ *is a descendent of* v_x *but not of* $v_y\}$.

4. *In order to establish a group key* k_A, *where* $A = \mathcal{U} \setminus R$, *run a simple algorithm (defined in [58]) to find a subset of keys that cover* A.

Without going into further details it should be evident that Scheme 6.14 results in a broadcast encryption scheme with much greater user storage than Scheme 6.12. However in general Scheme 6.14 has a much smaller broadcast message (resulting from a smaller cover) than Scheme 6.12 and so represents an alternative tradeoff.

Several variants of these schemes have been proposed in the literature. For example: in [36] a modification of the subset difference scheme based on defining layers of the underlying tree was proposed; in [3] a compression technique for reducing the user storage of the subset difference scheme was identified; in [55] it was shown that the complete subtree and subset difference schemes can be combined to obtain further examples of attractive tradeoffs.

Lastly we note that in [43] it was observed that a $(1, w, d)$-CFF (see Section 5.3) provides a restricted notion of a CBRS, where there exists a suitable cover for any subset of at least $|\mathcal{B}| - w$ blocks (and hence up to w users can be revoked).

6.1.5 Broadcast encryption with extended capabilities The attractive applications of broadcast encryption have resulted in some interesting extensions to the basic concept being proposed and investigated. We briefly identify two areas where interesting research has been conducted:

- **Traceable broadcast encryption**: The problem of piracy of decoder boxes for commercial information services has led to an interest in incorporating *traceability* into the keys allocated to a user in a broadcast encryption scheme. This means that any group of users who combine their keys to forge a new decoder can have at least one of their identities revealed if that decoder is later captured and analysed. This idea was first proposed in [25] and has subsequently been extensively investigated. Creating suitable distributions of keys presents a number of interesting combinatorial problems, which were comprehensively reviewed in [7].

- **Self-healing broadcast encryption**: If the broadcast channels being used are unreliable then it is possible that some users may not reliably receive the broadcast information that allows them to determine a given group key. The idea behind a *self-healing* broadcast encryption scheme is that additional information is broadcast on each occasion that allows valid group members to recover any missing group key from a combination of previous and subsequent broadcast messages. This idea is most appropriate for applications where regular group keys are established over a series of discrete time intervals. Self-healing broadcast encryption was first proposed in [70] and subsequently investigated in, for example, [13] and [14].

6.2 Decentralised schemes

One of the concerns of relying on a TA to play a major role during the key establishment phase is that it becomes a potential central point of failure. This even applies to the (trivial) key distribution scheme when the TA maintains secure channels with users throughout the scheme lifetime. One method of mediating against this, which was first studied in [59], is to have a set $TA_1, \ldots, TA_r$ of r different TAs, of which at least m must be involved in the establishment of any group key. This idea was first suggested for decentralising key predistribution schemes in [44]. We capture this concept informally in the following definition.

Definition 6.15 An $(m, r, \mathcal{C}, \mathcal{X})$-*distributed key distribution scheme* (DKDS) is a $(\mathcal{C}, \mathcal{X})$-KDS with the stronger properties that:

1. Given $A \in \mathcal{C}$, any $U_i \in A$ can compute the group key k_A from knowledge of u_i and $\{v_{i,A,j_1}, \ldots, v_{i,A,j_m}\}$, where v_{i,A,j_l} is some information obtained by U_i from TA_l during the key establishment phase for key k_A.

2. Given $A \in \mathcal{C}$, $B \in \mathcal{X}$ and a set of $m-1$ TAs, it is not possible to determine k_A from u_A, the private information held by the $m-1$ TAs and any information sent to any user in a previous key establishment event.

Defintion 6.15 was formalised in [11] in an information theoretically secure model and several bounds on scheme parameters, including TA storage, were established. These essentially show that a scheme suggested in [59] is optimal. We briefly outline this optimal DKDS.

Scheme 6.16 *Given* $(\mathcal{C}, \mathcal{X})$, *we let* $\lambda = \max_{B \in \mathcal{X}} |\{A \in \mathcal{C} \mid A \cap B \neq \emptyset\}|$ *(in other words, the maximum number of group keys that any set* $B \in \mathcal{X}$ *can compute) and associate each* $A \in \mathcal{C}$ *with an element* $h_A \in GF(q)$. *Initialise the* $(m, r, \mathcal{C}, \mathcal{X})$-*DKDS as follows:*

- *Each user* U_i *is issued with a set of keys that allow them to communicate securely with each of the* r *TAs.*

- TA_i *$(1 \leq i \leq m)$ constructs a random bivariate polynomial* $f_i(x, y)$ *of degree* $m - 1$ *in* x, *degree* $\lambda - 1$ *in* y, *and with coefficients from* $GF(q)$.

- TA_i *securely sends the univariate polynomial* $f_i(j, y)$ *to* TA_j *$(1 \leq j \leq r)$.*

- TA_j *computes and stores the univariate polynomial* $t_j(y) = \sum_{i=1}^{m} f_i(j, y)$ *as their private information.*

When user $U_i \in A$ *wants to establish* k_A:

- U_i sends a request to a set of m TAs, say $TA_{i_1}, \ldots, TA_{i_m}$.

- TA_{i_j} sends $v_{i,A,i_j} = t_{i_j}(h_A)$ to U_i.

- U_i uses Lagrange interpolation to recover $k_A = \sum_{l=1}^{m} f_l(0, h_A)$ from the values $\{t_{i_1}(h_A), \ldots, t_{i_m}(h_A)\}$.

Various generalisations and extensions to the basic idea of a DKDS have been studied in the literature. For example in [12] a tradeoff between storage and security was exhibited by employing ramp schemes and in [29] a DKDS was proposed which is more robust against users and TAs who do not follow the specified protocols correctly.

7   Group key agreement schemes

The third class of key establishment schemes that we look at are those where users can communicate with one another during the key establishment phase. We assume that, as for the reasons given at the start of Section 5, there is no trusted authority available to assist with key establishment after the initialisation phase. The majority of group key agreement schemes are particularly suited to environments where the nature of the communication structure is not known in advance. A group key agreement scheme then allows an ad hoc group to create a group key amongst themselves. For this reason (motivated by potential applications to secure teleconferencing) they are sometimes referred to as *conference key* schemes. In [15] they are referred to as *interactive key distribution schemes*.

Our classification of key agreement schemes as any scheme that involves user interaction unassisted by a trusted authority is highly generic and allows us to group together several very different types of group key establishment schemes. The vast majority of group key agreement schemes are based on public key cryptographic techniques and are mostly beyond the scope of this paper as they do not inherently involve combinatorial techniques. Many of these, including one that we will look at in Section 7.2, are based around the classical *Diffie-Hellman* protocol [30], which we briefly describe for the simple two-party case.

Scheme 7.1 *Let G be a finite multiplicative group of some large prime order q and let g be a generator of G (these parameters are published during the initialisation phase). If U_1 and U_2 wish to establish a key k then:*

1. *U_1 randomly chooses $x_1 \in Z_q^*$ and sends g^{x_1} to U_2;*

2. *U_2 randomly chooses $x_2 \in Z_q^*$ and sends g^{x_2} to U_1;*

3. *U_1 computes $k = (g^{x_2})^{x_1}$ and U_2 computes $k = (g^{x_1})^{x_2}$, both of which are equal to $g^{x_1 x_2}$.*

The security of the Diffie-Hellman protocol relies on the difficulty of taking discrete logarithms (see any standard cryptographic text such as [73]). We will not make any attempt in this paper to review the vast range of extensions and alternatives to Diffie-Hellman that have been proposed for group key agreement, and refer to surveys such as [18] and (more recently and exclusive on key agreement) [31].

7.1 Group key agreement from KPSs and KDSs

In a similar way to our discussion in Section 6.1.2, it is conceptually possible to construct a group key agreement scheme for either a group key predistribution scheme or a group key distribution scheme.

Scheme 7.2 *If we have a $(\mathcal{C}, \mathcal{X})$-KPS then we can realise a $(\mathcal{C}, \mathcal{X})$ key agreement scheme as follows:*

- *Each user stores u_i, as issued in the KPS;*

- *Users $U_i \in A$ establish group key k_A by utilising secure channels amongst themselves that are protected by the group key k_A^* associated with the KPS.*

Precisely how Scheme 7.2 can be manifested very much depends on the mutual trust between users in the scheme. One simple option is that one user U_i could generate k_A and then distribute it to the others encrypted by k_A^*. Another option is that each user $U_i \in A$ generates a component k_A^i, which is then distributed encrypted by k_A^*. Each user in A then decrypts the component and forms $k_A = \sum_{U_i \in A} k_A^i$. Regardless of how this is done, the resulting key agreement scheme will suffer from limitations similar to those of Scheme 6.5 that were noted in Section 6.1.2.

Scheme 7.3 *If we have a $(\mathcal{C}, \mathcal{X})$-KDS then we can realise a $(\mathcal{C}, \mathcal{X})$ key agreement scheme as follows:*

- *During the initialisation phase, each user is provided with data that allows them to fulfill the role of TA in a $(\mathcal{C}, \mathcal{X})$-KDS;*

- *Users $U_i \in A$ establish group key k_A by utilising the group keys k_A^{i*} associated with each of the KDSs.*

Again there are many ways in which Scheme 7.3 could actually manifest itself. Note also that there is no need for the individual KDSs strictly to be $(\mathcal{C}, \mathcal{X})$-KDSs, since it is possible that schemes with smaller communication structures could be cleverly combined. This is precisely what was done in [6], which was later generalised in [15]. This scheme used the broadcast encryption scheme based on a resolvable design discussed in Section 6.1.2 to establish a group key agreement scheme, where each user in A acted as a TA and broadcast an encrypted component key k_A^i, which was then combined to form the group key k_A.

7.2 Key agreement schemes for long term group management

Analogously to the situation discussed in Section 6.1, in dynamic application environments where group membership regularly changes it may be desirable to have schemes that allow long term group keys to be established by key agreement techniques. We have already seen in Section 6.1 that trees underpin a number of group key distribution schemes. There have been several proposals for tree-based group key agreement schemes based on Diffie-Hellman. We will describe the basic set up of just one such scheme, from [42].

Scheme 7.4 *Let H be a subset of m users (chosen from a universe $\mathcal{U}$), G be a finite multiplicative group of some large prime order q and let g be a generator of G. During the initialisation phase these parameters are published and users are issued with information that allows them to communicate with one another using authenticated public broadcast channels. For simplicity, assume that $m = 2^h$. The key establishment phase proceeds as follows:*

1. *Define a (complete) binary tree with m leaves, each associated with a user from H. Label this tree iteratively as in Scheme 6.8 as follows: root by $N_{0,0}$; the left child of $N_{i,j}$ by $N_{i+1,2j}$; and right child of $N_{i,j}$ by $N_{i+1,2j+1}$. Hence the users, which we label $U_{h,0}, \ldots, U_{h,m-1}$, correspond to nodes $N_{h,0}, \ldots, N_{h,m-1}$.*

2. *Each user $U_{h,j}$ generates a secret value $k_{h,j}$ and publicly broadcasts $g^{k_{h,j}}$ to the other members of H.*

3. *Each pair of users $U_{h,j}$ and $U_{h,j+1}$ (for all even $0 \leq j \leq m-1$) compute the Diffie-Hellman key $k_{h-1,j/2}$ using Scheme 7.1, which is then associated with node $N_{h-1,j/2}$. Both $U_{h,j}$ and $U_{h,j+1}$ securely store $k_{h-1,j/2}$ and publicly broadcast $g^{k_{h-1,j/2}}$.*

4. *Each quartet of users $U_{h,j}, U_{h,j+1}, U_{h,j+2}, U_{h,j+3}$ (for all $j \equiv 0 \pmod 4$, $0 \leq j \leq m-1$) compute the Diffie-Hellman key $k_{h-2,j/4}$ using Scheme 7.1, which is then associated with node $N_{h-2,j/4}$. All four users securely store $k_{h-2,j/4}$ and publicly broadcast $g^{k_{h-2,j/4}}$.*

5. *This process is iterated until the last Diffie-Hellman calculation results in $k_{0,0}$, which is adopted as the group key k_H.*

At the end of the above protocol, each user $U_{h,j}$ holds each key that is associated with the nodes on the path from $N_{h,j}$ to the root $N_{0,0}$, as well as $g^{k_{i,j}}$ for each node $N_{i,j}$ in the tree.

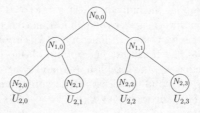

Figure 6: Tree-based Diffie-Hellman node allocation on four nodes

Example 7.5 The underlying tree for $m = 4$ is shown in Figure 6. In this example, $U_{2,j}$ first generates $k_{2,j}$ and broadcasts $g^{k_{2,j}}$ ($1 \leq j \leq 4$). Users $U_{2,0}$ and $U_{2,1}$ conduct a Diffie-Hellman exchange to compute $k_{1,0} = g^{k_{2,0}k_{2,1}}$, and $U_{2,2}$ and $U_{2,3}$ conduct a Diffie-Hellman exchange to compute $k_{1,1} = g^{k_{2,2}k_{2,3}}$. Both $g^{k_{1,0}}$ and $g^{k_{1,1}}$ are then broadcast. Finally the group key $k_{0,0} = g^{k_{1,0}k_{1,1}}$ can be computed by all four users. Each user stores all keys on the path from their leaf node to the root. So, for example, $U_{2,0}$ stores $k_{2,0}$, $k_{1,0}$ and $k_{0,0}$.

As mentioned in Section 6.1, the main advantage of setting up the key agreement tree in Scheme 7.4 is that this structure facilitates relatively efficient user join, user leave, group merge and group partition operations in dynamic environments. We leave the details of these protocols to [42].

7.3 Wireless sensor network schemes

A very interesting class of key establishment schemes that are of combinatorial interest are those designed for application in *wireless sensor networks*. These networks consist of tiny, inexpensive, low-powered *sensors* fitted with wireless transmitters, which can be spatially scattered to form an *ad hoc network*. They are particularly suited to applications in environments where it is difficult to manually establish a communication network, such as during disaster relief operations, seismic data collection, wildlife monitoring or military intelligence gathering. Sensors are distributed around the application environment (perhaps by aeroplane drop) and then attempt to set up a network in order to exchange and return data. What makes wireless sensor networks particularly intriguing is that the actual network topology (defined by a *physical graph*, whose edges represent sensors that are able to communicate with one another at a particular instant in time) is not known prior to deployment and is potentially highly dynamic. Thus we might as well model the physical graph as a random graph.

There are three factors that influence the choice of key establishment techniques for wireless sensor networks. The first is the fact that as there is no network controller after initialisation, there is no entity that can play the role of a TA. This lends itself to either key predistribution or key agreement schemes. However sensors also have very limited storage and computational abilities. This presents a dilemma since:

- As we have already seen in Section 5, the cost of relying solely on key predistribution is often substantial secret storage;

- Relying solely on key agreement involves considerable computational costs and is likely to be hampered by the random nature of the physical graph.

A sensible compromise is thus to predistribute keys "as well as possible", while also permitting a limited amount of communication between sensors to take place during key establishment, effectively making such schemes group key agreement schemes.

7.3.1 A key establishment model for wireless sensor networks
The basic idea behind a *wireless sensor network scheme* (WSN scheme) for communication structure $\mathcal{C}$ is to first establish a $(\mathcal{C}^*, \mathcal{X})$-key predistribution scheme, where both $\mathcal{C}$ and $\mathcal{C}^*$ are defined on the same set of sensors (users). We refer to $\mathcal{C}^*$ as the *network communication structure*. When a set $A \in \mathcal{C}$ of sensors requires a group key k_A:

1. if $A \in \mathcal{C}^*$ then they establish k_A using the KPS;

2. if $A \notin \mathcal{C}^*$ then they use some key agreement protocol to establish a key k_A, potentially using other sensors in the network to assist them.

Note that it is quite reasonable to rely on other sensors to assist in a key agreement process since the random nature of the physical graph necessitates that nodes typically rely on one another for message transmission services. One commonly proposed key agreement technique is to seek a path of secure links in the physical graph that joins the sensors in A, and then get one of the sensors on this path to generate a key and securely relay it.

It is clearly desirable to have C^* matching C as closely as possible. As a result, the effectiveness of a WSN scheme is often measured in terms of the *local connectivity*, which is a notion of the probability that a group of sensors in C either are in C^* or are "close" to being in C^* (where "close" is normally measured in terms of the number of other sensors that need to be involved in establishing k_A if $A \notin C^*$).

There are a wide variety of different approaches to the design of WSN schemes and we refer to [21] for a comprehensive survey. We now review a number of interesting applications of combinatorial structures to the design of WSN schemes. We will mainly restrict our interest here to the case where $C = \{A \subseteq U \mid |A| = t\}$ (and in particular the case $t = 2$), in which case we will refer to t-*wise* (*pairwise*) WSN schemes.

7.3.2 Using a KPS

Any key predistribution schemes that we have already discussed in Section 5 could potentially be adopted as part of a WSN scheme. Thus before considering dedicated designs, it is worth considering existing candidate KPSs. The most appropriate schemes are those with relatively low user storage and fast key computation. This makes schemes such as tree-based key distribution patterns (Section 5.4) attractive candidates. Also of interest are probabilistic schemes such as Scheme 5.8 (the random KPS) and Scheme 5.10, where efficiencies have been gained at the expense of a slightly unpredictable network communication structure. However these KPSs have not all been proposed explicitly for WSNs and, in particular, it is desirable to try to custom-design a KPS that also results in an efficient key agreement phase.

7.3.3 Key ring WSN schemes

We now discuss a class of WSN schemes that are based on key ring predistribution schemes (fundamental Scheme 5.7). Let t be a positive integer and let $U_1, \ldots, U_n$ be a collection of sensors. Let $\mathcal{R} = (\mathcal{I}, \mathcal{B})$ be a key ring, as defined in Section 5.1.2.

Definition 7.6 A $(t, n, \mathcal{R})$-*key ring WSN scheme* (KRWSN) is a WSN scheme arising from a key ring predistribution scheme based on $\mathcal{R}$, where the (network) communication structure is

$$C^* = \{A \subseteq U \mid |A| = t \text{ and } \bigcap_{U_i \in A} u_i \neq \emptyset\}.$$

In other words, a group $U_1, \ldots, U_t$ of t sensors check their public identifier sets $Pub_1, \ldots, Pub_t$ to see if they share any common identifiers. If they do, then they can establish a group key k_A by applying g to the keys k_i that correspond to the identifiers in $\cap_{j=1}^{t} Pub_j$. If not, then they establish the group key by an alternative key agreement mechanism.

The first KRWSN schemes proposed were based on a version of Scheme 5.8, the random key predistribution scheme, where the key ring was defined by $\mathcal{B} = \mathcal{I}^k$ (the collection of all subsets of $\mathcal{I}$ of some fixed size k) [24, 33]. Not surprisingly however, improved schemes can be obtained if the key ring has a more combinatorial structure. In [20], both projective planes and generalized quadrangles were suggested as candidate key rings.

Scheme 7.7 *Let q be a prime power and $\mathcal{R} = (\mathcal{I}, \mathcal{B})$ be a projective plane of order q. For any $n \leq q^2 + q + 1$ we obtain a $(2, n, \mathcal{R})$-KRWSN scheme such that:*

- *Each sensor needs to store $q + 1$ keys;*

- *Every pair of sensors shares precisely one common key.*

Scheme 7.7 is of course also an example of a KDP (see Section 5.3) and thus could have been included in Section 7.3.2. The fact that every pair of sensors share a key means that in this example no key agreement stage is necessary. There is a subtle problem with this scheme however. Many of the applications for wireless sensor networks involve a large number of sensors (potentially tens of thousands). This necessitates a suitable large choice of q in Scheme 7.7, which in turn requires the storage-limited sensor to hold too many keys. In [47] both transversal designs and quadratic curves were considered as candidate key rings and shown to have better properties.

Scheme 7.8 *Let q be a prime and $\mathcal{R} = (\mathcal{I}, \mathcal{B})$ be a transversal design $TD(k, q)$. For any $n \leq q^2$ we obtain a $(2, n, \mathcal{R})$-KRWSN scheme such that:*

- *Each sensor needs to store k keys;*

- *Every pair of sensors share precisely zero or one common key;*

- *The probability that a pair of sensors share a common key is $k/(q + 1)$.*

To see that Scheme 7.8 is an improvement on Scheme 7.7, consider the following example:

Example 7.9 Suppose that we require a WSN with 2400 sensors. If Scheme 7.7 is used then we need to choose $q = 49$ and each node is required to store 50 keys. On the other hand, we could apply Scheme 7.8 with a $TD(30, 49)$ [47], in which case each node only needs to store 30 keys. In this case any pair of nodes share a key with probability 0.6. It is further shown in [47] that if we make certain reasonable assumptions about the valency of the physical graph, the probability that any pair of sensors either share a key, or are connected in the physical graph to a third sensor with whom they both share a common key, is very close to 1.

Example 7.9 motivates the hunt for a more general class of combinatorial structures with similar properties. We say that two sensors U_i and U_j in a WSN can communicate via a *two-hop path* if there exists a third sensor U_k such that both U_i and U_k, and U_j and U_k, share common keys. For a KRWSN scheme this condition equates to requiring that $u_i \cap u_k \neq \emptyset$ and $u_j \cap u_k \neq \emptyset$. It is particularly advantageous if there are several different choices of intermediary node U_k since this increases the chances that one of them is able to act as a direct relay between U_i and U_j in the physical graph. The following type of structure was first proposed in [45].

Definition 7.10 Let $(\mathcal{I}, \mathcal{B})$ be a (v, b, r, k)-configuration. We say that $(\mathcal{I}, \mathcal{B})$ is a (v, b, r, k, μ)-*common intersection design* (CID) if for any distinct pair of blocks $B_i, B_j \in \mathcal{B}$ we have: $|\{B_k \in \mathcal{B} \mid B_i \cap B_k \neq \emptyset \text{ and } B_j \cap B_k \neq \emptyset\}| \geq \mu$.

Clearly common intersection designs make ideal candidates for KRWSN scheme key rings, as well as being of intrinsic combinatorial interest in their own right. We have already seen one example in Scheme 7.8, since a $TD(k, q)$ is an example of a $(qk, q^2, q, k, k(k-1))$-CID. Since we require μ to be as large as possible in a KRWSN scheme, an interesting question is to determine the maximum possible μ when fixing other parameters. Several upper bounds on μ were established in [48] and optimal CIDs were constructed using group-divisible designs, strongly-regular graphs and generalized quadrangles. Further investigation of CIDs is certainly merited.

7.3.4   Graph-based WSN schemes

Given that one of our goals in a WSN scheme is to limit the number of hops between sensors who do not share a common key, another sensible design approach is to base the allocation of keys around a virtual *network graph*, whose vertices are sensors and whose edges join sensors who share a common key (in some sense this is the opposite approach to that taken in Section 7.3.3). We restrict our proposals in this section to pairwise WSN schemes, but the approach could be generalised using hypergraphs. This idea was again first proposed in the literature using random graphs (Scheme 5.8) but we will again see that combinatorial structures provided a more intuitive basis for construction. We will use the following generic scheme.

Scheme 7.11 *A graph-based WSN scheme (GWSN) for a graph $\mathcal{G} = (\mathcal{U}, \mathcal{E})$ is a pairwise WSN scheme based on an underlying node-based KPS where:*

- *Each edge $e \in \mathcal{E}$ is associated with a random key k_e;*

- *$u_i = \{k_e \mid U_i \text{ is adjacent to } e\}$;*

- *$\mathcal{C}^* = \{\{U_i, U_j\} \mid U_i \text{ and } U_j \text{ are joined by an edge } e \in \mathcal{E}\}$.*

One difference between graph-based schemes and KRWSN schemes in general is that all graph-based schemes have full collusion security. In order to exploit this we need to define appropriate graphs on which to base a graph-based WSN scheme. In [46] it was pointed out that (n, r, λ, μ)-strongly regular graphs make ideal candidates, since by definition any pair of sensors in the graph that do not share a key are connected by μ two-hop paths. It was further demonstrated in [46] that careful choices of strongly regular graph have better local connectivity than schemes based on a random graph.

One problem with graph-based WSN schemes is that good connectivity often comes at the expense of requiring a graph where vertices typically have a high degree (and hence sensors have high storage). A clever efficiency improvement that can be applied to certain network graphs was observed in [46]:

Scheme 7.12 *let* $\mathcal{G} = (\mathcal{U}, \mathcal{E})$ *be a network graph that can be decomposed into star-like subgraphs. We can then establish a pairwise WSN scheme based on an underlying node-based KPS where:*

- *Each U_i is associated with an identifier ID_i and a random key k_i;*

- *$u_i = \{k_i \cup \{h(k_j, ID_i) \mid U_i$ is connected to a starlike subgraph centred at $U_j\}\}$, where h is a hash function;*

- *$\mathcal{C}^* = \{\{U_i, U_j\} \mid U_i$ and U_j are joined by an edge $e \in \mathcal{E}\}$.*

The above "trick" allows the sensor storage to be reduced compared to Scheme 7.11, since U_i only needs to store one key k_i for all the edges of the star-like subgraphs centred at U_i (it still needs to store a key for every other edge adjacent to U_i). This saving comes at the cost of reducing the security from unconditional to computational, since security is now dependent on the strength of the hash function.

The last scheme we will look at here is not strictly a graph-based scheme in the notion of Scheme 7.11, but it is based on a complete t-partite network graph. The scheme we describe is a generalisation to t-wise (from pairwise) of a scheme from [46].

Scheme 7.13 *Let $\mathcal{G}_{N_1,...,N_t} = (\mathcal{I}, \mathcal{E})$ denote a complete t-partite graph on n vertices, based on a partition of $\mathcal{I}$ into subsets $N_1, \ldots, N_t$. The t-partite BDVHKY-KPS for $\mathcal{G}_{N_1,...,N_t}$ is defined as follows, where $q \geq n$:*

- *$Pub_i = s_i$, where $s_i \in GF(q)$ and $Pub_i \neq Pub_j$ if $i \neq j$.*

- *The TA (randomly) constructs a secret t-variate polynomial f with coefficients from $GF(q)$,*

$$f(x_1, \ldots, x_t) = \sum_{i_1=0}^{w} \cdots \sum_{i_t=0}^{w} a_{i_1 \ldots i_t} x_1^{i_1} \ldots x_t^{i_t}.$$

- *If $U_i \in N_i$ then $u_i = f(x_1, \ldots, x_{i-1}, s_i, x_{i+1}, \ldots, x_t)$.*

- *$\mathcal{C}^* = \{A \mid A$ contains precisely one member of each $N_i\}$.*

- *For any $A = \{U_{z_1}, \ldots, U_{z_t}\} \in \mathcal{C}^*$, the user U_{z_i} computes $k_A = f(s_{z_1}, \ldots, s_{z_t})$.*

Note that the underlying polynomial in Scheme 7.13 differs from that in Scheme 5.9 by not necessarily being symmetric. The t-partite BDVHKY-KPS provides efficient key agreement since any t sensors A that are not in $\mathcal{C}^*$ must not contain any members of some partition subset N_l. Thus there are at least $|N_l|$ common neighbours of the sensors in A who can potentially act as a two-hop relay to all the sensors in A. Scheme 7.13 also has better resilience than Scheme 5.9, since now $w + 1$ sensors from the same partition subset N_i have to be captured before the scheme is broken.

7.3.5   Hybrid WSN schemes   The last approach that we will briefly mention involves mixing WSN schemes with different properties. The first two ideas each "randomise" in a different way an underlying combinatorial design in order to create schemes that exhibit interesting tradeoffs in comparison to schemes that we have already seen. The third technique combines schemes in a combinatorial way.

- Recall that key ring WSN schemes based on KDPs, such as Scheme 7.7, suffer from the fact that if sensor storage is kept low, the number of possible sensors in the network is restricted. In [20] it was suggested that Scheme 7.7 be combined with a version of the RKPS (Scheme 5.8), essentially "topping up" the number of blocks in a projective plane with some random blocks. This leads to a degradation in the local connectivity but was shown in [20] to offer interesting tradeoffs between local connectivity and sensor storage.

- Recall that Scheme 7.8 was proposed as an alternative to Scheme 7.7 that permitted more sensors at the expense of a loss in connectivity. In [23] it was suggested that this connectivity loss can be avoided by randomly merging blocks of the underlying transversal design, thus creating a key ring with much longer blocks but greater connectivity. This idea thus improves connectivity at the expense of greater sensor storage.

- Recall that Scheme 5.10 was developed from Scheme 5.9 using a randomised product construction to improve resilience at the expense of a loss of connectivity. In [78] a deterministic product construction that combines multiple copies of a key predistribution scheme using a generic set system was studied. Combinatorial properties for desirable set systems were derived and it was shown that special types of 1-designs known as *difference families* made good candidates.

This concept of combining different types of WSN scheme merits further investigation.

7.4   Multisecret sharing schemes

The previous key agreement schemes that we have looked at involve users having to collaborate to construct a group key for practical reasons (such as making key establishment efficient or through restrictions in the connectivity of the network). We now look at a family of key agreement schemes were users are forced to collaborate to construct a group key for *security* reasons. This is most likely to happen in applications where the group keys protect sensitive assets, with no single user being trusted with the sole authority to access them.

Definition 7.14 Let $\mathcal{C} = \{A_1, \ldots, A_m\}$ be a communication structure defined on a set $\mathcal{U}$. An *access structure* for $\mathcal{C}$ is a collection $\Gamma = \{\Gamma_1, \ldots, \Gamma_m\}$ of subsets of $\mathcal{U}$ with the property that:

1. Γ_i consists of subsets of A_i;

2. Γ_i is *monotone* (in other words, if $X \in \Gamma_i$ and $X \subseteq Y \subseteq A_i$ then $Y \in \Gamma_i$).

We will use an access structure on $\mathcal{C}$ to specify the degree of mandatory collaboration between users that is required before a group key can be established. More precisely, we will require the property that the group key k_{A_i} can be established only if users belonging to a set in Γ_i collaborate. This is more clearly specified in the following definition.

Definition 7.15 Let $\mathcal{C} = \{A_1, \ldots, A_m\}$ be a communication structure and $\mathcal{X}$ be an exclusion structure defined on a set $\mathcal{U}$, and let $\Gamma = \{\Gamma_1, \ldots, \Gamma_m\}$ be an access structure for $\mathcal{C}$. A $(\mathcal{C}, \mathcal{X}, \Gamma)$-*multisecret sharing scheme* is a $(\mathcal{C}, \mathcal{X})$-key establishment scheme such that for any set of users B:

1. If $(B \cap A_i) \in \Gamma_i$ then there exists a public function g such that $g(\{u_i \mid U_i \in B\}) = k_{A_i}$. In other words, the users in B can construct k_{A_i} from their collective set of secret values.

2. If $(B \cap A_i) \notin \Gamma_i$ and $B \in \mathcal{X}$ then, even if users in B exchange all their secret values $\{u_i \mid U_i \in B\}$, they will not learn any information about k_{A_i}.

Note that we have deliberately avoided formulating the notion of *not learning any information* and refer to [40] for a combinatorial formalisation and [54] for an information-theoretic formalisation of this concept. We have also avoided a detailed discussion of how the users exchange their values u_i and apply the public function g (typically it is either assumed that users share secret channels or that there exists an entity called a *combiner* that performs this task for them).

Multisecret sharing schemes are generalisations of secret sharing schemes (see Section 4.5), which correspond to the case of a scheme with just one group A_1 (associated with Γ_1) in its communication structure. While bounds on the secret storage have been established for general multisecret sharing schemes under a couple of different threat models [16, 54], we will restrict our attention to the special case of multisecret threshold schemes, defined as follows.

Definition 7.16 A (t, w, λ)-*multisecret threshold scheme* (MTS) is a special class of $(\mathcal{C}, \mathcal{X}, \Gamma)$-*multisecret sharing scheme* where:

1. $\mathcal{C} = \{A \subseteq \mathcal{U} \mid |A| = t\}$;

2. $\mathcal{X} = \{A \subseteq \mathcal{U} \mid |A| \leq w\}$;

3. For each $A_i \in \mathcal{C}$, $\Gamma_i = \{X \subseteq A_i \mid |X| \geq \lambda\}$.

A $(t, w, 1)$-MTS corresponds to a (t, w)-KPS (see Section 5) since in this case there is no requirement for users to collaborate to construct their group keys. In [40] it was shown that for most meaningful choices of w, each user in a (t, w, λ)-MTS needs to be given a secret value u_i that is at least $\binom{w+t-2\lambda+1}{t-\lambda}$ times larger than the size of any key k_{A_i} in the system. This bound is a generalisation of the bound on user storage for KPSs proved in [17] (see Section 5.2). It is thus of particular interest to find MTSs that meet this bound. The following are all "degenerate" cases of optimal MTSs:

- Scheme 5.9 [17] is an optimal $(t, w, 1)$-MTS;

- Optimal (n, w, λ)-MTSs (where $|\mathcal{U}| = n$) correspond to optimal secret sharing schemes (more precisely, if $w = \lambda - 1$ they correspond to *ideal threshold schemes* such as the classical scheme in [69] and for general w they correspond to optimal *ramp schemes*, see [38]);

- An optimal (t, w, t)-MTS is easily constructed by letting u_i be randomly chosen in $GF(q)$ and letting $k_A = \sum_{U_i \in A} u_i$ for any $A \in \mathcal{C}$ [40].

However, the task of constructing optimal MTSs with $1 < \lambda < t$ appears to be intriguingly difficult and to date only two constructions are known. In [41] a family of optimal $(t, n - k + 1, 2)$-MTSs were constructed and in [5] a family of optimal $(3, w, 2)$-MTSs. Both these constructions were based on complex and rather intricate projective geometrical configurations and used a geometrical interpretation of Scheme 5.9 as a building block.

8   Concluding remarks

In this paper we have reviewed a wide variety of applications of combinatorics objects, including designs and graphs, to different types of key establishment scheme. The theory of group key establishment is by no means complete and there are a number of areas where combinatorics can make further contributions to our understanding. Some specific areas where more research would be beneficial include:

- Several combinatorial objects that have direct application to key establishment merit further investigatory work:

  - While a moderate amount of research has been conducted on key distribution patterns (cover free families), there is a great deal of information about these structures to learn. In particular very little is known about KDPs for non-threshold communication structures.

  - Hash-tree key distribution patterns are relatively newly proposed structures and more theoretical work needs to be done concerning both constructions and performance bounds.

  - Cover-based revocation systems have attracted a great deal of interest in the area of broadcast encryption and greater understanding is needed of how efficiently these can be implemented.

  - Common intersection designs provide an interesting solution to the problem of key establishment in wireless sensor networks. More knowledge of how to generate constructions with useful parameters is required.

- There has been some interesting preliminary research conducted on how to take key establishment schemes with nice mathematical structure and convert them into more practical schemes with less inherent structure, but better performance. This can either be through merging schemes, extending schemes or simply using a mathematical scheme as a starting point on which to build a practical solution (Section 7.3.5 describes some work of this type for group key agreement schemes). This area certainly merits further investigation.

- Most of the schemes that we have presented in this review have been discussed in their most basic form. We have not discussed how they can be extended to incorporate all of the extended capabilities mentioned in Section 3.2.5. There remains plenty work to de done in designing schemes with extended capabilities, the most important of which is probably flexibility, in other words the ability to efficiently process dynamic changes to the communication structure over time.

It is hoped that this review has provided convincing evidence that combinatorial mathematics has already made a substantial contribution to the theory of key establishment, and that we can expect it to continue to do so in future cryptographic systems.

References

[1] S.G. Akl and P.D. Taylor. Cryptographic solution to a problem of access control in a hierarchy. *ACM Transactions on Computer Systems*, 1(3):239–248, 1983.

[2] T. Asano. A revocation scheme with minimal storage at receivers. In *Proceedings of the 8th International Conference on the Theory and Application of Cryptology and Information Security*, volume 3108 of *Lecture Notes in Computer Science*, pages 433–450. Springer-Verlag, 2002.

[3] T. Asano. Secure and insecure modifications of the subset difference broadcast encryption scheme. In *Information Security and Privacy: ACISP '04*, volume 3108 of *Lecture Notes in Computer Science*, pages 12–23. Springer-Verlag, 2004.

[4] M.J. Atallah, K.B. Frikken, and M. Blanton. Dynamic and efficient key management for access hierarchies. In *Proceedings of 12th ACM Conference on Computer and Communications Security*, pages 190–202, 2005.

[5] S.G. Barwick and W.-A. Jackson. An optimal multisecret threshold scheme construction. *Designs Codes and Crytography*, 37:367–389, 2005.

[6] A. Beimel and B. Chor. Communication in key distribution schemes. *IEEE Transactions on Information Theory*, 42:19–28, 1996.

[7] S.R. Blackburn. Combinatorial schemes for protecting digital content. In *Surveys in Combinatorics 2003*, pages 43–78. Cambridge University Press, 2003.

[8] B. Blakley. Safeguarding cryptographic keys. In *Proceedings AFIPS 1979 National Computer Conference*, pages 313–317, June 1979.

[9] R. Blom. An optimal class of symmetric key generation systems. In *Eurocrypt '84*, volume 209 of *Lecture Notes in Computer Science*, pages 335–338. Springer-Verlag, 1985.

[10] C. Blundo and P. D'Arco. The key establishment problem. In *FOSAD 2001/2002*, volume 2946 of *Lecture Notes in Computer Science*, pages 44–90. Springer-Verlag, 2004.

[11] C. Blundo and P. D'Arco. Analysis and design of distributed key distribution centers. *Journal of Cryptology*, 18(4):391–414, 2005.

[12] C. Blundo, P. D'Arco, and C. Padrò. A ramp model for distributed key distribution schemes. *Discrete Applied Mathematics*, 128:47–64, 2003.

[13] C. Blundo, P. D'Arco, and A. De Santis. On self-healing key distribution schemes. To appear in IEEE Transactions on Information Theory.

[14] C. Blundo, P. D'Arco, A. De Santis, and M. Listo. Design of self healing key distribution schemes. *Designs Codes and Cryptography*, 32:15–44, 2004.

[15] C. Blundo, L. Frota Mattos, and D.R. Stinson. Generalized beimel-chor schemes for broadcast encryption and interactive key distribution. *Theoretical Computer Science*, 200:313–334, 1998.

[16] C. Blundo, A. De Santis, G. Di Crescenzo, A. Giorgio Gaggia, and U. Vaccaro. Multi-secret sharing schemes. In *Crypto '94*, volume 839 of *Lecture Notes in Computer Science*, pages 150–163. Springer-Verlag, 1994.

[17] C. Blundo, A. De Santis, U. Vaccaro, A. Herzberg, S. Kutten, and M. Yung. Perfectly secure key distribution for dynamic conferences. In *Crypto '92*, volume 740 of *Lecture Notes in Computer Science*, pages 471–486. Springer-Verlag, 1993.

[18] C. Boyd and A. Mathuria. *Protocols for authentication and key establishment*. Springer-Verlag, 2003.

[19] E.F. Brickell and D.M. Davenport. On the classification of ideal secret sharing schemes. *Journal of Cryptology*, 4:123–134, 1991.

[20] S. A. Camtepe and B. Yener. Combinatorial design of key distribution mechanisms for wireless sensor networks. In *ESORICS 2004*, volume 3193 of *Lecture Notes in Computer Science*, pages 293–308. Springer-Verlag, 2004.

[21] S. A. Camtepe and B. Yener. Key distribution mechanisms for wireless sensor networks: a survey. Rensselaer Polytechnic Institute, Computer Science Department, Technical Report TR-05-07, March 2005.

[22] R. Canetti, J. Garay, G. Itkis, D. Micciancio, M. Naor, and B. Pinkas. Multicast security: a taxonomy and some efficient constructions. In *Proceedings of INFOCOM '99*, pages 708–716. IEEE Press, 1999.

[23] D. Chakrabarti, S. Maitra, and B. Roy. A hybrid design of key pre-distribution scheme for wireless sensor networks. In *ICISS 2005*, volume 3803 of *Lecture Notes in Computer Science*, pages 228–238. Springer-Verlag, 2005.

[24] H. Chan, A. Perrig, and D. Song. Random key predistribution schemes for sensor networks. In *IEEE Symposium on Research in Security and Privacy*, pages 197–213, May 2003.

[25] B. Chor, A. Fiat, M. Naor, and B. Pinkas. Traitor tracing. *IEEE Transactions on Information Theory*, 46:893–910, 2000.

[26] C.J. Colbourn, J.H. Dinitz, and D.R. Stinson. Applications of combinatorial designs to communications, cryptography and networking. In *Surveys in Combinatorics 1999*, pages 37–100. Cambridge University Press, 1999.

[27] J. Crampton, K.M. Martin, and P.R. Wild. An explication of key assignment schemes. In preparation, 2006.

[28] J. Crampton, K.M. Martin, and P.R. Wild. Proceedings of 19th computer security foundations workshop. pages 98–111, 2006.

[29] P. D'Arco and D.R. Stinson. On unconditionally secure robust distributed key distribution centers. In *ASIACRYPT 2002*, volume 2501 of *Lecture Notes in Computer Science*, pages 346–363. Springer-Verlag, 2002.

[30] W. Diffie and M. Hellman. New directions in cryptography. *IEEE Transactions on Information Theory*, IT-22(6):644–654, 1976.

[31] R. Dutta and R. Barua. Overview of key agreement protocols. Cryptology ePrint Archive, Report 2005/289, 2005. http://eprint.iacr.org/.

[32] M. Dyer, T. Fenner, and A. Thomason. On key storage in secure networks. *Journal of Cryptography*, 8:189–200, 1995.

[33] L. Eschenauer and V. Gligor. A key management scheme for distributed sensor networks. In *Proceedings of 9th ACM Conference on Computer and Communication Security*, November 2002.

[34] A.L. Ferrara and B. Masucci. An information-theoretic approach to the access control problem. In C. Blundo and C. Laneve, editors, *ICTCS 2003*, volume 2841 of *Lecture Notes in Computer Science*, pages 342–354. Springer-Verlag, 2003.

[35] A. Fiat and M. Naor. Broadcast encryption. In *Advances in Cryptology - CRYPTO'93*, volume 773 of *Lecture Notes in Computer Science*, pages 480–491. Springer-Verlag, 1994.

[36] D. Halevy and A. Shamir. The lsd broadcast encryption scheme. In *Crypto '02*, volume 2442 of *Lecture Notes in Computer Science*, pages 47–60. Springer-Verlag, 2002.

[37] L. Harn and H.Y. Lin. A cryptographic key generation scheme for multilevel data security. *Computers and Security*, 9(6):539–546, 1990.

[38] W.-A. Jackson and K.M. Martin. A combinatorial interpretation of ramp schemes. *Australasian Journal of Combinatorics*, 14:51–60, 1996.

[39] W.-A. Jackson and K.M. Martin. Combinatorial models for perfect secret sharing schemes. *Journal of Combinatorial Mathematics and Combinatorial Computing*, 28:249–265, 1998.

[40] W.-A. Jackson, K.M. Martin, and C.M. O'Keefe. Multisecret threshold schemes. In *Crypto '93*, volume 773 of *Lecture Notes in Computer Science*, pages 126–135. Springer-Verlag, 1994.

[41] W.-A. Jackson, K.M. Martin, and C.M. O'Keefe. A construction for multisecret threshold schemes. *Designs Codes and Crytography*, 9:287–303, 1996.

[42] Y. Kim, A. Perrig, and G. Tsudik. Tree-based group key agreement. *ACM Transactions on Information and System Security*, 7(1):60–96, 2004.

[43] R. Kumar, S. Rajagopalan, and A. Sahai. Coding constructions for blacklisting problems without computational assumptions. In *Advances in Cryptology - CRYPTO'99*, volume 1666 of *Lecture Notes in Computer Science*, pages 609–623. Springer-Verlag, 1999.

[44] K. Kurosawa, K. Okada, and K. Sakano. Security of the center in key distribution schemes. In *Asiacrypt '94*, volume 917 of *Lecture Notes in Computer Science*, pages 333–341. Springer-Verlag, 1995.

[45] J. Lee and D.R. Stinson. A combinatorial approach to key predistribution for distributed sensor networks. In *IEEE Wireless Communications and Networking Conference*, pages 6–11, 2005. CD-ROM, paper PHY53-06, http://www.cacr.math.uwaterloo.ca/dstinson/pubs.html.

[46] J. Lee and D.R. Stinson. Deterministic key predistribution schemes for distributed sensor networks. In *SAC 2004*, volume 3357 of *Lecture Notes in Computer Science*, pages 294–307. Springer-Verlag, 2005.

[47] J. Lee and D.R. Stinson. One the construction of practical key predistribution schemes for distributed sensor networks using combinatorial designs. http://www.cacr.math.uwaterloo.ca/ dstinson/pubs.html, November 2005.

[48] J. Lee and D.R. Stinson. Common intersection designs. *Journal of Combinatorial Designs*, 14:251–269, 2006.

[49] J. Lee and D.R. Stinson. Tree-based key distribution patterns. In *SAC 2005 Proceedings*, volume 3897 of *Lecture Notes in Computer Science*, pages 189–204. Springer-Verlag, 2006.

[50] T. Leighton and S. Micali. Secret-key agreement without public-key cryptography. In *Advances in Cryptology - CRYPTO '93*, volume 773 of *Lecture Notes in Computer Science*, pages 456–479. Springer-Verlag, 1994.

[51] C.-H. Lin. Hierarchical key assignment without public key cryptography. *Computers & Security*, 20(7):612–619, 2001.

[52] D. Liu and P. Ning. Establishing pairwise keys in distributed sensor networks. In *Proceedings of 10th ACM Conference on Computer and Communication Security*, October 2003.

[53] S.J. MacKinnon, P.D. Taylor, H. Meijer, and S.G. Akl. An optimal algorithm for assigning cryptographic keys to control access in a hierarchy. *IEEE Transactions on Computers*, C-34(9):797–802, 1985.

[54] B. Masucci. Sharing multiple secrets: models, schemes and analysis. *Designs Codes and Crytography*, 39:89–111, 2006.

[55] M. J. Mihaljevic. Key management schemes for stateless receivers based on time varying heterogeneous logical key hierarchy. In *Asiacrypt 2003*, volume 2894 of *Lecture Notes in Computer Science*, pages 127–154. Springer-Verlag, 2003.

[56] C. J. Mitchell and F.C. Piper. The cost of reducing key storage requirements in secure networks. *Computers and Security*, 6:339–341, 1987.

[57] C. J. Mitchell and F.C. Piper. Key storage in secure networks. *Discrete Applied Mathematics*, 21:215–228, 1988.

[58] D. Naor, M. Naor, and J. Lotspiech. Revocation and tracing schemes for stateless receivers. In *Advances in Cryptology - CRYPTO '01*, volume 2139 of *Lecture Notes in Computer Science*, pages 41–62. Springer-Verlag, 2001.

[59] M. Naor, B. Pinkas, and O. Reingold. Distributed preudo-random functions and kdcs. In *Advances in Cryptology - Eurocrypt '99*, volume 1592 of *Lecture Notes in Computer Science*, pages 327–346. Springer-Verlag, 1999.

[60] C.M. O'Keefe. A comparison of key distribution patterns constructed from circle geometries. In *Auscrypt '92*, volume 718 of *Lecture Notes in Computer Science*, pages 517–527. Springer-Verlag, 1993.

[61] C.M. O'Keefe. Key distribution patterns using minkowski planes. *Designs Codes and Crytography*, 5:261–267, 1995.

[62] C. Padró, I. Gracia, S.M. Molleví, and P. Morillo. Linear key predistribution schemes. *Designs Codes and Cryptography*, 25:281–298, 2002.

[63] F.C. Piper and S. Murphy. *Cryptography: a very short introduction*. Oxford University Press, 2002.

[64] K.A.S. Quinn. Some constructions for key distribution patterns. *Designs Codes and Cryptography*, 4:177–191, 1994.

[65] K.A.S. Quinn. Bounds for key distribution patterns. *Journal of Cryptology*, 12:227–239, 1999.

[66] M. Ramkumar and N. Memon. An efficient random key pre-distribution scheme for manet security. *IEEE Journal on Selected Areas of Communication*, 2005.

[67] G. Rinaldi. Key distribution patterns using tangent circle structures. *Designs Codes and Crytography*, 31:289–300, 2004.

[68] A. De Santis, A.L. Ferrara, and B. Masucci. Cryptographic key assignment schemes for any access control policy. *Information Processing Letters*, 92:199–2005, 2004.

[69] A. Shamir. How to share a secret. *Communications of the ACM*, 22(11):612–613, 1979.

[70] J. Staddon, S. Miner, M. Franklin, D. Balfanz, M. Malkin, and D. Dean. Self-healing key distribution with revocation. *IEEE Symposium on Security and Privacy*, May 2002.

[71] D.R. Stinson. An explication of secret sharing schemes. *Designs Codes and Cryptography*, 2:357–390, 1992.

[72] D.R. Stinson. On some methods of unconditionally secure key distribution and broadcast encryption. *Designs Codes and Cryptography*, 12:215–243, 1997.

[73] D.R. Stinson. *Cryptography: theory and practice.* Chapman & Hall/CRC, 3rd edition, 2006.

[74] D.R. Stinson and T. Van Trung. Some new results on key distribution patterns and broadcast encryption. *Designs Codes and Cryptography*, 14:261–279, 1998.

[75] D.R. Stinson and R. Wei. An application of ramp schemes to broadcast encryption. *Information Processing Letters*, 69:131–135, 1999.

[76] D.R. Stinson and R. Wei. Generalized cover free families. *Discrete Mathematics*, 279:463–477, 2004.

[77] D.M. Wallner, E.J. Harder, and R.C. Agee. Key management for multicast: issues and architectures. Internet Request for Comments 2627, June, 1999.

[78] R. Wei and J. Wu. Product construction of key distribution schemes for sensor networks. In *SAC 2004*, volume 3357 of *Lecture Notes in Computer Science*, pages 280–293. Springer-Verlag, 2005.

[79] C.K. Wong, M.G. Gouda, and S.S. Lam. Secure group communications using key graphs. *Proceedings of the ACM SIGCOMM '98 conference on Applications, Technologies, Architectures and Protocols for Computer Communication*, pages 68–79, 1998.

[80] Y. Zheng, T. Hardjono, and J. Seberry. New solutions to the problem of access control in a hierarchy. Technical Report 93-2, Department of Computer Science, University of Wollongong, 1993.

[81] S. Zhong. A practical key management scheme for access control in a user hierarchy. *Computers & Security*, 21(8):750–759, 2002.

Keith M. Martin
Information Security Group
Royal Holloway, University of London
Egham Hill, Egham, Surrey TW20 0EX, UK
keith.martin@rhul.ac.uk

Branchwidth of graphic matroids

Frédéric Mazoit and Stéphan Thomassé

Abstract

We prove that the branchwidth of a bridgeless graph is equal to the branch-width of its cycle matroid. Our proof is based on branch-decompositions of hypergraphs. By matroid duality, a direct corollary of this result is that the branchwidth of a bridgeless planar graph is equal to the branchwidth of its planar dual.

1 Introduction.

The notion of branchwidth was introduced by Robertson and Seymour in their seminal paper Graph Minors X [3]. Very roughly speaking, the goal is to decompose a structure S along a tree T in such a way that subsets of S corresponding to disjoint branches of T are pairwise as disjoint as possible. One can define the branchwidth of various structures such as graphs, hypergraphs, matroids, submodular functions... Our goal in this paper is to prove that the definitions of branchwidth for graphs and matroids coincide in the sense that the branchwidth of a bridgeless graph is equal to the branchwidth of its cycle matroid. This answers a question of Thomas [5], also cited in Geelen, Gerards, Robertson and Whittle [1].

Let us now define properly these notions.

Let $H = (V, E)$ be a graph, or a hypergraph, and (E_1, E_2) be a partition of E. The *border* of (E_1, E_2) is the set of vertices which belong to both an edge of E_1 and an edge of E_2. We denote this by $\delta(E_1, E_2)$, or simply by $\delta(E_1)$.

A *branch-decomposition* T of H is a ternary tree T and a bijection from the set of leaves of T into the set of edges of H. In practice, we simply identify the leaves of T with the edges of H. Observe that every edge e of T partitions $T \setminus e$ into two subtrees, and thus corresponds to a bipartition of E, called an *e-separation*. More generally, a T-*separation* is an e-separation for some edge e of T. We will often identify the edge e of T with the e-separation, allowing us to write, for instance, $\delta(e)$ instead of $\delta(E_1, E_2)$, where (E_1, E_2) is the e-separation. Let T be a branch-decomposition of H. The *width* of T, denoted by $\mathrm{w}(T)$, is the maximum value of $|\delta(e)|$ for all edges e of T. The *branchwidth* of H, denoted by $\mathrm{bw}(H)$, is the minimum width of a branch-decomposition of H. A branch-decomposition achieving $\mathrm{bw}(H)$ is *optimal*.

Let us now turn to matroids. Let M be a matroid on ground set E with rank function r. The *width* of every non-trivial partition (E_1, E_2) of E is $\mathrm{w}_m(E_1, E_2) := r(E_1) + r(E_2) - r(E) + 1$. When T is a branch-decomposition of M, i.e. a ternary tree whose leaves are labelled by E, the *width* $\mathrm{w}_m(T)$ of T is the maximum width of a T-separation. Again, the *branchwidth* $\mathrm{bw}_m(M)$ of M is the minimum width of a branch-decomposition of M. One nice fact about branchwidth is that it is invariant under matroid duality (recall that the bases of the dual matroid M^* of M are the complements of the bases of M). Indeed, since $r_{M^*}(U) = |U| + r_M(E \setminus U) - r_M(E)$ for all $U \subseteq E$, $\mathrm{w}_m(E_1, E_2)$ is the same in M and in M^*. Note that since branchwidth is a measure of how complex the matroid is, it is a useful fact that M and M^* have the same branchwidth.

Having defined both the branchwidth of a graph and of a matroid, a very natural question is to compare them when the matroid M is precisely the cycle matroid of a graph G, i.e. the matroid M_G whose ground set is the set of edges of G and whose independent sets are the acyclic subsets of edges. A first observation is that they differ, for instance the branchwidth of the path of length three is 2 whereas the branchwidth of its cycle matroid is 1. The inequality $\mathrm{bw}(M_G) \leq \mathrm{bw}(G)$ always holds, and simply comes from the fact that $\mathrm{w}_m(E_1, E_2) \leq |\delta(E_1, E_2)|$ for every partition of E which has a nonempty border. To see this, define, when $H = (V, E)$ is a hypergraph, a *component* of E to be a minimal - with respect to inclusion - nonempty subset $C \subseteq E$ such that $\delta(C) = \emptyset$. Let F be a subset of E. We denote by $c(F)$ the number of components of the subhypergraph of H spanned by F, i.e. the hypergraph $(V(F), F)$. The hypergraph H is *connected* if $c(E) = 1$ and is moreover *bridgeless* if $c(E \setminus e) = 1$ for all $e \in E$ (since our definition is based on edges, we may have vertices with degree 0 or 1 in a connected bridgeless hypergraph). Observe now that when (E_1, E_2) is a separation of the edges of a graph, we have

$$\mathrm{w}_m(E_1, E_2) = r(E_1) + r(E_2) - r(E) + 1 = n_1 - c(E_1) + n_2 - c(E_2) - n + c(E) + 1,$$

where n_1, n_2, n are the number of vertices respectively spanned by E_1, E_2, E. In particular,

$$\mathrm{w}_m(E_1, E_2) = |\delta(E_1, E_2)| + c(E) + 1 - c(E_1) - c(E_2) \leq |\delta(E_1, E_2)|,$$

since $c(E) + 1 - c(E_1) - c(E_2) \leq 0$ when $\delta(E_1, E_2)$ is not empty.

Let us define a new branchwidth, the *matroid branchwidth* $\mathrm{bw}_m(H)$ of a hypergraph H in which the separations (E_1, E_2) are evaluated with the function $\mathrm{w}_m(E_1, E_2) = |\delta(E_1, E_2)| + 1 + c(E) - c(E_1) - c(E_2)$ instead of the function $|\delta(E_1, E_2)|$. We also write $\mathrm{w}_m(E_1)$ instead of $\mathrm{w}_m(E_1, E_2)$. In particular, when G is a graph, we have $\mathrm{bw}_m(G) = \mathrm{bw}_m(M_G)$.

The main result of this paper, Theorem 1, is that when H is connected and bridgeless, there exists a branch-decomposition $\mathcal{T}$ of H achieving $\mathrm{bw}_m(H)$ such that every $\mathcal{T}$-separation (E_1, E_2) is such that $c(E_1) = c(E_2) = 1$. Thus we have $\mathrm{w}(\mathcal{T}) = \mathrm{w}_m(\mathcal{T})$, and since $\mathrm{bw}_m(H) \leq \mathrm{bw}(H)$ and $\mathcal{T}$ is optimal, we have $\mathrm{bw}_m(H) = \mathrm{bw}(H)$. This implies in particular that the branchwidth of a bridgeless graph is equal to the branchwidth of its cycle matroid. Moreover, the case $\mathrm{bw}(G) > \mathrm{bw}_m(M_G)$ happens if and only if the graph G has a bridge, $\mathrm{bw}_m(M_G) = 1$ and $\mathrm{bw}(G) = 2$. In other words, if G is a tree that is not a star.

Another consequence of our result concerns planar graphs. The key-fact here is that planar duality corresponds to matroid duality, i.e. when G is planar and G^* is the planar dual of G, we have $(M_G)^* = M_{G^*}$. Therefore, when G is a planar bridgeless graph, we derive:

$$\mathrm{bw}(G) = \mathrm{bw}_m(M_G) = \mathrm{bw}_m((M_G)^*) = \mathrm{bw}_m(M_{G^*}) = \mathrm{bw}(G^*).$$

Which is a new proof of the fact that for bridgeless graphs, the branchwidth is invariant under taking planar duality. The first proof of this result was a direct corollary of a result of Seymour and Thomas in [4].

The paper is organized as follows. In Sections 2 and 3, we analyze the properties of a possible minimal counterexample H to our main theorem. We get more and more structure, step by step. At the end of Section 3, the hypergraph H is very constrained, tripartite, triangle-free etc, but no further simple step follows. The contradiction is achieved via a particular separation of H. The existence of such a separation relies on a (technical) partition lemma on multigraphs, the proof of which is postponed to Section 4.

Unless stated otherwise, we always assume that $\mathcal{T}$ is a branch-decomposition of a hypergraph $H = (V, E)$. Also, when speaking about width, branchwidth, etc, we implicitly mean the matroid one.

2 Faithful branch-decompositions.

Let (E_1, E_2) be a $\mathcal{T}$-separation. The decomposition $\mathcal{T}$ is *faithful* to E_1 if for every component C of E_1, the partition $(C, E \setminus C)$ is a $\mathcal{T}$-separation. The *border graph* $G_{\mathcal{T}}$ has vertex set V and contains as edges all pairs xy for which there exists a $\mathcal{T}$-separation e such that $\{x, y\} \subseteq \delta(e)$. A branch-decomposition $\mathcal{T}'$ is *tighter* than $\mathcal{T}$ if $\mathrm{w}_m(\mathcal{T}') < \mathrm{w}_m(\mathcal{T})$ or if $\mathrm{w}_m(\mathcal{T}) = \mathrm{w}_m(\mathcal{T}')$ and $G_{\mathcal{T}'}$ is a subgraph of $G_{\mathcal{T}}$. Moreover, $\mathcal{T}'$ is *strictly tighter* than $\mathcal{T}$ if $\mathcal{T}'$ is tighter than $\mathcal{T}$, and $\mathcal{T}$ is not tighter than $\mathcal{T}'$. Finally, $\mathcal{T}$ is *tight* if no $\mathcal{T}'$ is strictly tighter than $\mathcal{T}$.

Lemma 1 *Let (E_1, E_2) be a partition of E. For any union E_1' of connected components of E_1 and E_2, we both have $\delta(E_1') \subseteq \delta(E_1)$ and $\mathrm{w}_m(E_1') \leq \mathrm{w}_m(E_1)$.*

Proof. Clearly, $\delta(E_1') \subseteq \delta(E_1)$. Moreover, every vertex of $\delta(E_1)$ belongs to one component of E_1 and one component of E_2. Therefore, if C is a component of E_1' which is the union of k components of E_1 and E_2, there are at least $k - 1$ vertices of $C \setminus \delta(C)$ which belong to $\delta(E_1)$, In all, the weight of the separation increased by $k - 1$ since we merged k components into one, but it also decreased by at least $k - 1$ since we lost at least that many vertices on the border. Since this is the case for every component of E_1' and of $E \setminus E_1'$, we have $\mathrm{w}_m(E_1') \leq \mathrm{w}_m(E_1)$. $\blacksquare$

Lemma 2 *Let (E_1, E_2) be an e-separation of $\mathcal{T}$. Let $\mathcal{T}_1$ be the subtree of $\mathcal{T} \setminus e$ with set of leaves E_1. If $\mathcal{T}$ is not faithful to E_1, we can rearrange $\mathcal{T}_1$ in $\mathcal{T}$ to form a tighter branch-decomposition $\mathcal{T}'$ of H which is faithful to E_1.*

Proof. Fix the vertex $e \cap \mathcal{T}_1$ as a root of $\mathcal{T}_1$. Our goal is to change the binary rooted tree $\mathcal{T}_1$ into another binary rooted tree $\mathcal{T}_1'$. For every connected component C of E_1, consider the subtree $\mathcal{T}_C$ of $\mathcal{T}_1$ which contains the root of $\mathcal{T}_1$ and has set of leaves C. Observe that $\mathcal{T}_C$ is not necessarily binary since $\mathcal{T}_C$ may contain paths having internal vertices with only one descendant. We simply replace these paths by edges to obtain our rooted tree $\mathcal{T}_C'$. Now, consider any rooted binary tree BT with $c(E_1)$ leaves and identify these leaves with the roots of $\mathcal{T}_C'$, for all components C of E_1. This rooted binary tree is our $\mathcal{T}_1'$. We denote by $\mathcal{T}'$ the branch-decomposition we obtain from $\mathcal{T}$ by replacing $\mathcal{T}_1$ by $\mathcal{T}_1'$. Roughly speaking, we merged all subtrees of $\mathcal{T}_1$ induced by the components of E_1 together with $\mathcal{T} \setminus \mathcal{T}_1$ to form $\mathcal{T}'$. Let us prove that $\mathcal{T}'$ is tighter than $\mathcal{T}$. For this, consider an edge f' of $\mathcal{T}'$. If $f' \notin \mathcal{T}_1'$, the f'-separations of $\mathcal{T}$ and $\mathcal{T}'$ are the same. If $f' \in BT$, by Lemma 1, we have $w_m(f') \leq w_m(e)$ and $\delta(f') \subseteq \delta(e)$. In both cases $\mathcal{T}'$ is tighter than $\mathcal{T}$. So the only case we have to consider is when f' is an edge of some tree $\mathcal{T}'_C$, where C is a component of E_1. Recall that f' corresponds to a path P of $\mathcal{T}_C$. Let f be any edge of P. Let $(F, E \setminus F)$ be the f-separation of $\mathcal{T}$, where $F \subseteq E_1$. Therefore, the f'-separation of $\mathcal{T}'$ is $\big(F \cap C, E \setminus (F \cap C)\big)$. Since F is a subset of E_1, the connected components of F are subsets of the connected components of E_1. Thus $F \cap C$ is a union of connected components of F. By Lemma 1, we have $\delta(f') \subseteq \delta(f)$ and $w_m(f') \leq w_m(f)$.

We have proved that $w(\mathcal{T}') \leq w(\mathcal{T})$ and that $G_{\mathcal{T}'}$ is a subgraph of $G_{\mathcal{T}}$, thus $\mathcal{T}'$ is tighter than $\mathcal{T}$.   ∎

3 Connected branch-decompositions.

Let $F \subseteq E$ be a set of edges such that $c(F) = 1$. The hypergraph on vertex set V and edge set $(E \setminus F) \cup \{V(F)\}$ is denoted by $H * F$. In other words, $H * F$ is obtained by merging the edges of F into one edge. A partition (E_1, E_2) of E is *connected* if $c(E_1) = c(E_2) = 1$. A branch-decomposition $\mathcal{T}$ is *connected* if every $\mathcal{T}$-separation is connected.

Lemma 3 *If $\mathcal{T}$ is a tight branch-decomposition of a connected hypergraph H, every $\mathcal{T}$-separation (E_1, E_2) is such that E_1 or E_2 is connected.*

Proof. Suppose for contradiction that there exists a $\mathcal{T}$-separation (E_1, E_2) such that neither E_1 nor E_2 is connected. By Lemma 2, we can assume that $\mathcal{T}$ is faithful to E_1 and to E_2. Let $\mathcal{C}_1$ and $\mathcal{C}_2$ be respectively the sets of components of E_1 and E_2. Consider the graph on set of vertices $\mathcal{C}_1 \cup \mathcal{C}_2$ where $C_1 C_2$ is an edge whenever $C_1 \in \mathcal{C}_1$ and $C_2 \in \mathcal{C}_2$ have nonempty intersection. This graph is connected since H is connected and is not a star since both E_1 and E_2 are not connected. Thus, it has a vertex-partition into two connected subgraphs, each having at least two vertices. This vertex-partition corresponds to a partition $(\mathcal{C}'_1, \mathcal{C}'_2)$ of $\mathcal{C}_1 \cup \mathcal{C}_2$.

Consider any rooted binary tree BT with $|\mathcal{C}'_1|$ leaves. Since every $C \in \mathcal{C}'_1$ is an element of $\mathcal{C}_1 \cup \mathcal{C}_2$ and $\mathcal{T}$ is faithful to E_1 and to E_2, $(C, E \setminus C)$ is an e-separation of $\mathcal{T}$. We denote by $\mathcal{T}_C$ the tree of $\mathcal{T} \setminus e$ with set of leaves C. Root $\mathcal{T}_C$ with the vertex $e \cap \mathcal{T}_C$ in order to get a binary rooted tree. Now identify the leaves of BT with the roots of $\mathcal{T}_C$, for $C \in \mathcal{C}'_1$. This rooted tree is our $\mathcal{T}'_1$. We construct similarly $\mathcal{T}'_2$. Adding an edge between the roots of $\mathcal{T}'_1$ and $\mathcal{T}'_2$ gives the branch-decomposition $\mathcal{T}'$ of H. By Lemma 1, $w_m(\mathcal{T}') \leq w_m(\mathcal{T})$ and $G_{\mathcal{T}'}$ is a subgraph of $G_{\mathcal{T}}$. Let us now show that $G_{\mathcal{T}'}$ is a strict subgraph of $G_{\mathcal{T}}$. Indeed, since $\mathcal{C}'_1$ is connected and has at least two elements, it contains $C_1 \in \mathcal{C}_1$ and $C_2 \in \mathcal{C}_2$ such that $V(C_1) \cap V(C_2)$ is nonempty. By construction, every vertex x of $V(C_1) \cap V(C_2)$ is such that $x \notin \delta(\mathcal{C}'_1)$ and $x \in \delta(\mathcal{C}_1)$. Similarly, there is a vertex y spanned by $\mathcal{C}'_2$ such that $y \notin \delta(\mathcal{C}'_2)$ and $y \in \delta(\mathcal{C}_2)$. Thus xy is an edge of $G_{\mathcal{T}}$. The selected x is not in the union of the members of $\mathcal{C}'_2$ and the selected y is not in the union of the members of $\mathcal{C}'_1$, so x and y are not in the vertex-boundary of any separation displayed by $G'_{\mathcal{T}}$. Consequently xy is not an edge of $G_{\mathcal{T}'}$, contradicting the fact that $\mathcal{T}$ is tight.   ∎

Theorem 1 *For every branch-decomposition $\mathcal{T}$ of a connected hypergraph H, there exists a tighter branch-decomposition $\mathcal{T}'$ such that for every $\mathcal{T}'$-separation (E_1, E_2) with $c(E_1) > 1$, E_1 consists of components of $H \setminus e$, for some $e \in E_2$. In particular, if H is bridgeless, it has an optimal connected branch-decomposition.*

Proof. Let us prove the theorem by induction on $|V| + |E|$. The statement is obvious if $|E| \leq 3$, so we assume now that H has at least four edges. Call a branch-decomposition *achieved* if it satisfies the conclusion of Theorem 1. If $\mathcal{T}$ is not tight, we can replace it by a tight branch-decomposition tighter than $\mathcal{T}$. So we may assume that $\mathcal{T}$ is tight.

If there is an edge $e \in E$ such that $H \setminus e$ is not connected, we can assume by Lemma 2 that $\mathcal{T}$ is faithful to $E \setminus e$. Let E_1 be a connected component of $E \setminus e$. Let $\mathcal{T}_1$ be the branch-decomposition induced by $\mathcal{T}$ on $E_1 \cup e$. Let $\mathcal{T}_2$ be the branch-decomposition induced by $\mathcal{T}$ on $E \setminus E_1$. Observe that both $E_1 \cup e$ and $E \setminus E_1$ are connected, so by the induction hypothesis, there exist two achieved branch-decompositions $\mathcal{T}_1'$ and $\mathcal{T}_2'$, respectively tighter than $\mathcal{T}_1$ and $\mathcal{T}_2$. Identify the leaf e of the tree $\mathcal{T}_1'$ with the leaf e of $\mathcal{T}_2'$ and attach a leaf labelled by e to this identified vertex. Let $\mathcal{T}'$ denote this branch-decomposition of H. Observe that $\mathcal{T}'$ is tighter than $\mathcal{T}$. Moreover, since both $\mathcal{T}_1'$ and $\mathcal{T}_2'$ are achieved, $\mathcal{T}'$ is also achieved.

So we assume now that H is bridgeless. We can also assume that all the vertices of H have degree at least two, since we can simply delete the vertices of H with degree 0 or 1, and apply induction. The key-observation is that if there is a connected $\mathcal{T}$-separation (E_1, E_2) with $|E_1| \geq 2$ and $|E_2| \geq 2$, we can apply the induction hypothesis on $H * E_1$ and $H * E_2$ and merge the two branch-decompositions to obtain an optimal connected branch-decomposition of H. Therefore, we assume that every $\mathcal{T}$-separation (E_1, E_2) with $|E_1| \geq 2$ and $|E_2| \geq 2$ is such that E_1 or E_2 is not connected. We now orient the edges of $\mathcal{T}$. If (E_1, E_2) is an e-separation such that E_2 is connected and $|E_2| > 1$, we orient e from E_1 to E_2. Since H is bridgeless, every edge of $\mathcal{T}$ incident to a leaf is oriented from the leaf. By Lemma 3, every edge has at least one orientation. And by the key-observation, every edge of $\mathcal{T}$ has exactly one orientation.

This orientation of $\mathcal{T}$ has no circuit, thus there is a vertex $t \in \mathcal{T}$ with outdegree zero. Since every leaf has outdegree one, t has indegree three. Let us denote by A, B, C the sets of leaves of the three trees of $\mathcal{T} \setminus t$. Observe that by construction, $A \cup B$, $A \cup C$ and $B \cup C$ are connected. By Lemma 2, we can assume moreover that $\mathcal{T}$ is faithful to A, B and C. We claim that A is a disjoint union of edges, i.e. the connected components of A are edges of H. To see this, assume for a contradiction that a component C_A of A is not an edge of H. Since $\mathcal{T}$ is faithful to A, $(C_A, E \setminus C_A)$ is a $\mathcal{T}$-separation. But this is simply impossible since $B \cup C$ being connected, $E \setminus C_A$ is also connected, contrary to the fact that every edge of $\mathcal{T}$ has a unique orientation. So the hypergraph H consists of three sets of disjoint edges A, B, C. Call this partition the *canonical partition* of $\mathcal{T}$. Call $(A, E \setminus A)$, $(B, E \setminus B)$ and $(C, E \setminus C)$ the *main $\mathcal{T}$-separations*. Note that the width of every other $\mathcal{T}$-separation is at most $\mathrm{bw}_m(H)$. Since every vertex of H belongs to two or three edges, it is spanned by at least two of the sets $\delta(A), \delta(B), \delta(C)$. In particular $G_\mathcal{T}$ is the complete graph on V, and thus every optimal branch-decomposition of H is tighter than $\mathcal{T}$. Therefore, every optimal branch-decomposition of H has a canonical partition, otherwise we can conclude by induction. Set $\delta_{AB} := |\delta(A) \cap \delta(B)|$, $\delta_{AC} := |\delta(A) \cap \delta(C)|$, $\delta_{BC} := |\delta(B) \cap \delta(C)|$ and $\delta_{ABC} := |\delta(A) \cap \delta(B) \cap \delta(C)|$. We now prove some properties of H.

1. Two of the sets A, B, C have at least two edges. Indeed, assume for a contradiction that $A = \{a\}$ and $B = \{b\}$. Since $|E| \geq 4$, there are at least two edges in C. Let $c \in C$. Observe that c intersects both a and b since $A \cup C$ and $B \cup C$ are connected. Assume without loss of generality that $|a \cap c| \geq |b \cap c|$. Now form a new branch-decomposition T' by *moving* c to A, i.e. T' has a separation $(A \cup c, B \cup (C \setminus c))$ and then four branches with respective leaves $A, c, B, (C \setminus c)$. We have

   $$\mathrm{w}_m(A \cup c, B \cup (C \setminus c)) \leq |\delta(A \cup c)| = |\delta(A)| + |b \cap c| - |a \cap c| \leq |\delta(A)| = \mathrm{w}_m(A).$$

   In particular T' is tighter than T, and since the T'-separation $(A \cup c, B \cup (C \setminus c))$ is connected and both of its branches have at least two vertices, we can apply induction to conclude.

2. Each set A, B, C has at least two edges. Indeed, assume for a contradiction that A consists of a single edge a. Observe that since $A \cup B$ and $A \cup C$ are connected, a intersects every edge of H. Let b be an edge of B. Call $|b \cap \delta(C)| - |b \cap a|$ the *excess* of an edge b of B. Let us prove that the excess of b is positive. Indeed, if $|b \cap \delta(C)| \leq |b \cap a|$, we can as previously move b to A in order to form a tighter branch-decomposition T'. If moreover $(B \cup C) \setminus b$ is connected, we are done since we now have a connected separation $(A \cup b, (B \cup C) \setminus b)$, on which we can apply induction. If $(B \cup C) \setminus b$ is not connected, the canonical partition of T' must be $A, b, (B \cup C) \setminus b$ since the union of each two of these branches is connected. Since A consists of a single edge, we conclude as in Fact 1. Similarly, the *excess* $|c \cap \delta(B)| - |c \cap a|$ of an edge $c \in C$ is positive. Let s be the minimum excess of an edge e_s of $B \cup C$. Observe that $s \geq 1$ and that every $b \in B$ satisfies $|b \cap \delta(C)| \geq |b \cap a| + s$. Thus, summing for all edges of B, we obtain $\delta_{BC} \geq \delta_{AB} + s|B|$. Similarly, $\delta_{BC} \geq \delta_{AC} + s|C|$. Note also that $\mathrm{bw}_m(H) \geq \mathrm{w}_m(C) = \delta_{BC} + \delta_{AC} - \delta_{ABC} - |C| + 1$ and $\mathrm{bw}_m(H) \geq \mathrm{w}_m(B) = \delta_{BC} + \delta_{AB} - \delta_{ABC} - |B| + 1$. In all

   $$2\,\mathrm{bw}_m(H) \geq 2\delta_{BC} - 2\delta_{ABC} + \delta_{AC} - |C| + \delta_{AB} - |B| + 2.$$

   Then $2\,\mathrm{bw}_m(H) \geq \delta_{AB} + s|B| + \delta_{AC} + s|C| - 2\delta_{ABC} + \delta_{AC} - |C| + \delta_{AB} - |B| + 2$. Finally, $\mathrm{bw}_m(H) \geq \delta_{AC} + \delta_{AB} - \delta_{ABC} + 1 + ((s-1)|C| + (s-1)|B|)/2$. Since $|\delta(A)| = \delta_{AC} + \delta_{AB} - \delta_{ABC}$, we have $\mathrm{bw}_m(H) \geq |\delta(A)| + s$. But then we can move e_s to A to conclude since $|\delta(A \cup e_s)| \leq \mathrm{bw}_m(H)$.

3. Observe that canonical partitions A, B, C now satisfy that A, B and C are disconnected.

4. We have $\mathrm{bw}_m(H) = \mathrm{w}_m(A)$. If not, pick two edges a, a' of A and merge them together. The hypergraph we obtain is still connected and bridgeless, and the branch-decomposition still has the same width. Apply induction to get an achieved branch-decomposition. Then replace the merged edge by the two original edges. This branch-decomposition T' is optimal but does not have a disconnected canonical partition. Thus we have a contradiction to Fact 3. Similarly, $\mathrm{bw}_m(H) = \mathrm{w}_m(B) = \mathrm{w}_m(C)$.

5. We have $\mathrm{bw}_m(H) \geq \beta + 1$, where β is the maximum size of an edge of H. Observe that H has no edge of size one. Indeed if such an edge e belongs to, say, A, it is also included in another edge, say in B. But then moving e to B would give a canonical partition which does not consist of disjoint edges, a contradiction. So the size of an edge of H is at least two. In particular every separation which is not a main one has width at most $\mathrm{bw}_m(H) - 1$. Since the separation $(a, E \setminus a)$, where a is an edge of H of size β, is not a main separation, it follows that $\mathrm{bw}_m(H) \geq \beta + 1$.

6. We have $\delta_{ABC} = 0$. Indeed, suppose for a contradiction that there exists a vertex z in $\delta(A) \cap \delta(B) \cap \delta(C)$. Consider the hypergraph H_z obtained from H by removing the vertex z from all its edges. Observe that H_z is connected since z is incident to three edges and H is bridgeless. The branch-decomposition T induces a branch-decomposition T_z of H_z having width at most $\mathrm{w}_m(T) - 1$. We apply induction on T_z to obtain an achieved branch-decomposition T_z' of H_z. Now reinsert the vertex z into the edges of H_z and call T' the branch-decomposition obtained from T_z'. Let us show that T' is optimal. Observe that if a T_z'-separation (E_1, E_2) is connected, adding z will raise by at most one its width in T'. Moreover if a T_z'-separation (E_1, E_2) is not connected, say $c(E_2) > 1$, adding z can raise by at most two its width in T' (either by merging three components of E_2 into one, or by merging two and increasing the border by one). Since T_z' is achieved, E_2 is a set of components of $E \setminus e$ for some edge e of H_z. But then in T_z', we have $\mathrm{w}_m(E_1, E_2) \leq |\delta(E_2)| - 3 + 2 \leq |e| - 1 \leq \beta - 1 \leq \mathrm{bw}_m(H) - 2$, and thus $\mathrm{w}_m(E_1, E_2) \leq \mathrm{bw}_m(H)$ in T'. Therefore T' is optimal. Moreover every T'-separation (E_1, E_2) is connected. Indeed, if (E_1, E_2) is connected in T_z', we are done. If E_1 is not connected in T_z', E_1 consists of components of $H_z \setminus e$, for some edge e of H_z. But since H is bridgeless, every component of E_1 in H must contain z, otherwise they would be components of $H \setminus e$. Consequently E_1 is connected in H.

7. Every edge of H is incident to at least four other edges. Indeed, assume for a contradiction that an edge a of A is incident to only one edge b of B and at most two edges of C (the case where a is only incident to edges of C is obvious, we just move a to C and thus decrease the border of A). Moving a to B increases $\mathrm{w}_m(B)$ by $|a \cap \delta(C)| - |a \cap b|$ and does not increase $\mathrm{w}_m(A)$ and $\mathrm{w}_m(C)$. Therefore, if $|a \cap \delta(C)| \leq |a \cap b|$, we can move a to B, and this new branch-decomposition T' is strictly tighter than T since the vertices of $a \cap b$ are no longer joined to $(\delta(A) \setminus a) \cap \delta(C)$ in the graph $G_{T'}$. Thus $|a \cap \delta(C)| \geq |a \cap b| + 1$. Moreover, moving a to C increases $\mathrm{w}_m(C)$ by at most $|a \cap b| - |a \cap \delta(C)| + 1$, since at most two components of C can merge. So $|a \cap b| + 1 > |a \cap \delta(C)|$, a contradiction.

8. Fact 7 implies in particular that the size of any edge of H is at least four. It follows that $\mathrm{w}_m(e) \leq \mathrm{bw}_m(H) - 3$ whenever e is not one of the main T-separations. Therefore $\beta \leq \mathrm{bw}_m(H) - 3$.

9. The hypergraph H is triangle-free. Indeed, suppose that there exist three edges $a \in A$, $b \in B$ and $c \in C$ and three vertices $x \in a \cap b$, $y \in b \cap c$ and

$z \in c \cap a$. Let H/xyz be the hypergraph obtained by contracting x, y, z to a single vertex v. The branch-decomposition $\mathcal{T}$ induces a branch-decomposition $\mathcal{T}/xyz$ of H/xyz. Note that H/xyz is still connected and bridgeless, and that $w_m(\mathcal{T}/xyz) = w_m(\mathcal{T}) - 1$ since we decrease by one the border of every main separation. By induction, we can find an achieved branch-decomposition $\mathcal{T}'$ of H/xyz which is tighter than $\mathcal{T}/xyz$. We claim that $\mathcal{T}'$ is also an achieved branch-decomposition of H. Consider for this a $\mathcal{T}'$-separation (E_1, E_2) of E. If a, b, c belong to the same part, say E_1, the width of (E_1, E_2) is the same in H/xyz and in H. If a, b belong to one part and c to the other, the width of (E_1, E_2) is one less in H/xyz than in H. Thus $bw_m(H) \leq bw_m(H/xyz) + 1$, and in particular $\mathcal{T}'$ is optimal. Finally, since (E_1, E_2) is connected in H/xyz, it is also connected in H. Thus, $\mathcal{T}'$ is achieved.

Now we are ready to finish the proof. Note that $bw_m(H) = (w_m(A) + w_m(B) + w_m(C))/3 = (2|V| - |E|)/3 + 1$. Consider the line multigraph $L(H)$ of H, i.e. the multigraph on vertex set $A \cup B \cup C$ and edge set V such that $v \in V$ is the edge which joins the two edges e, f of H such that $v \in e$ and $v \in f$. The multigraph $L(H)$ satisfies the hypothesis of Lemma 4 (proved in the next section), thus it admits a vertex-partition as in the conclusion of Lemma 4. This corresponds to a partition of $A \cup B \cup C$ into two subsets $E_1 := A_1 \cup B_1 \cup C_1$ and $E_2 := A_2 \cup B_2 \cup C_2$ such that $|\delta(E_1, E_2)| \leq (2|V| - |E|)/3 + 1$ and both E_1 and E_2 have at least $\lfloor |E|/2 \rfloor - 1$ internal vertices. In particular, the separation (E_1, E_2) has width at most $bw_m(H)$. Let us show that one of $w_m(A_1 \cup B_1)$, $w_m(B_1 \cup C_1)$, and $w_m(C_1 \cup A_1)$ is also at most $bw_m(H)$. For this, observe that the set $\delta(A_1 \cup B_1) \cup \delta(B_1 \cup C_1) \cup \delta(C_1 \cup A_1)$ covers every vertex of V which is not an internal vertex of E_2 twice. Thus

$$|\delta(A_1 \cup B_1)| + |\delta(B_1 \cup C_1)| + |\delta(C_1 \cup A_1)| \leq 2|V| - 2\lfloor |E|/2 \rfloor + 2 \leq 2|V| - |E| + 3.$$

Without loss of generality, we can assume that $\delta(A_1 \cup B_1) \leq (2|V| - |E|)/3 + 1 = bw_m(H)$, and thus we split E_1 into two branches $A_1 \cup B_1$ and C_1. We similarly split E_2 to obtain an optimal branch-decomposition $\mathcal{T}'$ of H. Observe that in the graph $G_{\mathcal{T}'}$, there is no edge between the internal vertices of E_1 and E_2. This contradicts the fact that $\mathcal{T}$ is tight.   ∎

4   The partition Lemma.

Let G be a multigraph on vertex set V and X, Y be two subsets of V. We denote by $e(X, Y)$ the number of edges of G between X and Y. We also denote by $e(X)$ the number of edges in X. The *degree* of a vertex x in a subset Y of G is $d_Y(x) := e(x, Y)$. When $Y = V$, we simply write $d(x)$. The *underlying degree* of x in Y is the number of neighbours of x in Y, i.e. we forget the multiplicity of edges. A graph is *2-connected* if it is connected and the removal of any vertex leaves it connected.

Lemma 4 *Let G be a 2-connected triangle-free multigraph on $n \geq 5$ vertices and m edges. Assume that its minimum underlying degree is at least four and that its maximum degree is at most $(2m - n)/3 + 1$. There exists a partition (X, Y) of the* *vertex set of G such that* $e(X) \geq \lfloor n/2 \rfloor - 1$, $e(Y) \geq \lfloor n/2 \rfloor - 1$ *and* $e(X, Y) \leq (2m - n)/3 + 1$.

Proof. Call a partition *good* if it satisfies the conclusion of Lemma 4. Assume first that there are vertices x, y such that $e(x, y) \geq \lfloor n/2 \rfloor - 1$. The minimum degree in $V \setminus \{x, y\}$ is at least two, so $e(V \setminus \{x, y\})$ is at least $n - 2$ and hence at least $\lfloor n/2 \rfloor - 1$. Thus, if the partition $(V \setminus \{x, y\}, \{x, y\})$ is not good, we necessarily have $d(x) + d(y) - 2e(x, y) > (2m - n)/3 + 1$. By the maximum degree hypothesis, both $d(x)$ and $d(y)$ are greater than $2e(x, y)$. Since G is triangle-free, there exists a partition (X, Y) where $(N(x) \cup x) \setminus y \subseteq X$ and $(N(y) \cup y) \setminus x \subseteq Y$. Observe that $e(X) \geq d(x) - e(x, y) > e(x, y) \geq \lfloor n/2 \rfloor - 1$. Similarly $e(Y) \geq \lfloor n/2 \rfloor - 1$. We then have:

$$e(X, Y) \leq m - (d(x) + d(y) - 2e(x, y)) < m - (2m - n)/3 - 1 = (m + n)/3 - 1.$$

Moreover, since $m \geq 2n$ by the minimum degree four hypothesis, we have $e(X, Y) \leq (2m - n)/3 + 1$ and finally (X, Y) is a good partition. We assume from now on that the multiplicity of an edge is less than $\lfloor n/2 \rfloor - 1$.

Let $a + b = n$, where $a \leq b$. A partition (X, Y) of V is an *a-partition* if $|X| \leq a$, $e(X) \geq a - 1$, $e(Y) \geq b - 1$, $e(X, Y) \leq (2m - n)/3 + 1$, and the additional requirement that X contains a vertex of G with maximum degree.

Note that there exists a 1-partition, just consider for this $X := \{x\}$, where x has maximum degree in G (the minimum degree in Y is at least three, ensuring that $e(Y) \geq n - 2$). We consider now an a-partition (X, Y) with maximum a. If $a \geq b - 1$, this partition is good and we are done. So we assume that $a < b - 1$. In particular $e(X) = a - 1$.

The key-observation is that there exists at most one vertex y of Y such that $e(Y \setminus y) < b - 2$. Indeed, if there is a vertex of Y with degree one in Y, we simply move it to X, and we obtain an $(a + 1)$-partition ($e(X)$ increases, $e(Y)$ decreases by one, and $e(X, Y)$ decreases). Thus the minimum degree in Y is at least two, and hence $e(Y) \geq |Y|$. Moreover, if there is a vertex z of Y with degree two in Y, we can still move it to X: indeed $e(X)$ increases, $e(Y \setminus z) \geq |Y| - 2$ and $e(X, Y)$ does not increase. So the minimum degree in Y is at least three (but the minimum underlying degree may be one). This implies that $e(Y) \geq 3|Y|/2$. Let $Y := \{y_1, \ldots, y_{|Y|}\}$ where the vertices are indexed in order of increasing degree in Y. For every $i \neq |Y|$, we have $e(Y) \geq (3(|Y| - 2) + d_Y(y_i) + d_Y(y_{|Y|}))/2$. Furthermore,

$$e(Y \setminus y_i) \geq (3(|Y| - 2) + d_Y(y_i) + d_Y(y_{|Y|}))/2 - d_Y(y_i) \geq 3(|Y| - 2)/2 \geq |Y| - 2 \geq b - 2.$$

We now discuss the two different cases depending on whether or not there exists $y \in Y$ such that $e(Y \setminus y) < b - 2$. In the following, the *excess* of a vertex $y \in Y$ is $exc(y) := d_Y(y) - d_X(y)$.

- Assume that $e(Y \setminus y) \geq b - 2$ for every $y \in Y$. We denote by Y' the (nonempty) set of vertices of Y with at least one neighbour in X. We let $Y'' := Y \setminus Y'$, by definition every vertex of Y'' has underlying degree at least four in Y. Note that we can move a vertex of Y' to X if it does not have positive excess. Denote by c the minimum excess of a vertex of Y'. We have $c > 0$. The sum of the degrees of the vertices of Y' is at least $2e(X, Y) + c|Y'|$. Now, summing the degrees of all the vertices of Y, we get $2e(Y) + e(X,Y) \geq 4|Y''| + 2e(X,Y) + c|Y'|$, and hence:

$$2e(Y) \geq e(X,Y) + 4|Y''| + c|Y'|. \tag{4.1}$$

Let $y \in Y'$ satisfy $exc(y) = c$. Since the partition $(X \cup y, Y \setminus y)$ is not an $(a+1)$-partition, we have $e(X,Y) + c > (2m - n)/3 + 1$. Since $m = e(X,Y) + e(X) + e(Y)$, this implies

$$e(X,Y) + 3c > 2e(X) + 2e(Y) - n + 3. \tag{4.2}$$

Equations (4.1) and (4.2) give:

$$3c > 2e(X) + 4|Y''| + c|Y'| - n + 3. \tag{4.3}$$

Since $e(X) \geq a - 1 \geq n - |Y| - 1$, we get $3c > n - 2|Y| + 4|Y''| + c|Y'| + 1$. From $|Y| = |Y'| + |Y''|$, we get $3c > n + 2|Y| + (c - 4)|Y'| + 1$, and finally $n + 2|Y| < (c - 4)(3 - |Y'|) + 11$. If $c = 4$, we get $n + 2|Y| \leq 10$, which is impossible since $n \geq 5$ and $|Y| > n/2$. If $c = 3$, we get $n + 2|Y| - |Y'| \leq 7$, again impossible. If $c = 2$, we get $n + 2|Y| - 2|Y'| \leq 4$, again impossible. If $c = 1$, we get $n + 2|Y| - 3|Y'| \leq 1$, which can only hold if $|Y| = |Y'| = n - 1$. Thus, X consists of a single vertex, completely joined to Y, contrary to the fact that G is triangle-free and has minimum underlying degree 4. Finally $c > 4$, and consequently $|Y'| < 3$. Observe that $|Y'| > 1$ since G is 2-connected. Thus $|Y'| = 2$. Let y_1, y_2 be the vertices of Y', indexed in such a way that $e(y_1, X) + e(y_2, Y'') \geq e(y_2, X) + e(y_1, Y'')$. Let $X_1 := X \cup y_1$ and $Y_1 := Y \setminus y_1$. Since $y_1 \in Y'$, we have $e(X_1) \geq a$. Moreover $e(Y_1) \geq b - 2$. We claim that $e(y_1, y_2) \leq e(Y'')$. Indeed, since G has minimum underlying degree four, the minimum degree in $Y \setminus \{y_1, y_2\}$ is at least two. So

$$e(Y'') = e(Y \setminus \{y_1, y_2\}) \geq |Y| - 2 \geq \lfloor n/2 \rfloor - 1 \geq e(y_1, y_2).$$

Thus

$$e(X_1, Y_1) = e(y_1, y_2) + e(y_1, Y'') + e(y_2, X) \leq e(Y'') + e(y_2, Y'') + e(y_1, X).$$

In particular $e(X_1, Y_1) \leq e(X_1) + e(Y_1)$, or equivalently $e(X_1, Y_1) \leq m/2$. Since $m \geq 2n$, we have $e(X_1, Y_1) \leq (2m - n)/3 + 1$. So the partition (X_1, Y_1) is good.

- Now assume that there exists a vertex $y \in Y$ such that $e(Y \setminus y) \leq b - 3 \leq |Y| - 3$. We denote by Y' the set of vertices of $Y \setminus y$ with at least one neighbour in X. Set $Y'' := Y \setminus (Y' \cup y)$. Observe that since every vertex of Y'' has underlying degree four in Y, we have $e(Y \setminus y) \geq 3|Y''|/2$. Thus, $|Y''| \leq (2|Y| - 6)/3$. Since $|Y| > 3$, we have $|Y''| < |Y| - 3$, and finally $|Y'| \geq 3$. Denote by c the minimum excess of a vertex of Y', again $c > 0$. Summing the degrees of the vertices of Y gives $2e(Y) \geq e(X,Y) + 4|Y''| + c|Y'| + exc(y)$. Equation (4.2) still holds, so

$$exc(y) < 3c + n - 3 - 2e(X) - 4|Y''| - c|Y'| \leq 3c - 1 - e(X) - 3|Y''| - (c-1)|Y'|$$

since $e(X) + |Y''| + |Y'| \geq n - 2$. Therefore $exc(y) < -e(X) - 3|Y''| - (c - 1)(|Y'| - 3) + 2$. Since $|Y'| \geq 3$, $|Y''| \geq 0$, and $c \geq 1$, we have $exc(y) \leq 1 - e(X)$. Recall that the minimum degree in Y is at least three, hence summing the degrees in Y of the vertices of $Y \setminus y$ gives $3(|Y| - 1) \leq 2e(Y \setminus y) + d_Y(y) \leq 2b - 6 + d_Y(y)$. Finally, $d_Y(y) \geq |Y| + 3$ and by the fact that $exc(y) \leq 1 - e(X)$, we have $d_X(y) \geq |Y| + e(X) + 2$. In all, we have $d(y) \geq 2|Y| + e(X) + 5$. Recall that X contains a vertex x with maximum degree in G. In particular both x and y have degree at least $2|Y| + e(X) + 5$. Observe that $d_X(x)$ is at most $e(X)$, and consequently $d_Y(x)$ is at least $2|Y| + 5$. Now the end of the proof is straightforward, it suffices to switch x and y to obtain the good partition $(X_1, Y_1) := ((X \cup y) \setminus x, (Y \cup x) \setminus y)$. We now need to consider the value of $e(x, y)$. Indeed if $e(x, y)$ is at most $e(X)$, we have:

1. $e(Y_1) \geq d_{Y_1}(x) \geq 2|Y| + 5 - e(x, y) \geq 2|Y| - e(X) \geq |Y| \geq n/2$.

2. $e(X_1) \geq d_{X_1}(y) \geq |Y| + e(X) + 2 - e(x, y) \geq n/2$.

3. Finally, since the excess of y is at most $1 - e(X)$, we have $d_{X_1}(y) + e(x, y) = d_X(y) \geq d_Y(y) + e(X) - 1$, hence $d_{X_1}(y) \geq d_Y(y) - 1$. Moreover $d_{Y_1}(x) \geq 2|Y| + 5 - e(X) \geq e(X) + 5 \geq d_X(x) + 5$. Thus, $e(X_1, Y_1) = e(X, Y) + d_Y(y) - d_{X_1}(y) + d_X(x) - d_{Y_1}(x) \leq e(X, Y) - 4$. Therefore $e(X_1, Y_1) \leq (2m - n)/3 + 1$, since (X, Y) is an a-partition.

To conclude, we just have to show that $e(x, y)$ is at most $e(X)$. Assume for a contradiction that $e(x, y) \geq a$. We consider the partition into $X_2 := \{x, y\}$ and $Y_2 := V \setminus \{x, y\}$. Observe that the minimum underlying degree in Y_2 is at least two. Thus $e(Y_2) \geq n - 2 \geq b - 2$. By the maximality of a, (X_2, Y_2) is not an $(a + 1)$-partition, therefore $e(X_2, Y_2) > (2m - n)/3 + 1$, hence

$$d(x) + d(y) - 2e(x, y) > (2m - n)/3 + 1. \qquad (4.4)$$

We now claim that any partition (X_3, Y_3) such that $(x \cup N(x)) \setminus y \subseteq X_3$ and $(y \cup N(y)) \setminus x \subseteq Y_3$ is good. Indeed, we have $e(X_3) \geq d(x) - e(x, y) \geq 2|Y| + e(X) + 5 - n/2 \geq n/2$. Similarly $e(Y_3) \geq n/2$. So, if (X_3, Y_3) is not good, we must have $e(X_3, Y_3) > (2m - n)/3 + 1$. Therefore $m - (d(x) + d(y) - 2e(x, y)) > (2m - n)/3 + 1$, and by Equation (4.4), we have $m > 2(2m - n)/3 + 2$. Finally $m < 2n - 6$ which is impossible since the minimum degree in G is at least four. ∎

An independent proof of the equality of branchwidth of cycle matroids and graphs was also given by Hicks and McMurray [2]. Their method is based on matroid tangles.

References

[1] J.F. Geelen, A.M.H. Gerards, N. Robertson and G.P. Whittle, On the excluded minors for the matroids of branch-width k, *J. Combin. Theory Ser. B* **88** (2003), 261–265.

[2] I.V. Hicks and N.B. McMurray, The branchwidth of graphs and their cycle matroids, *preprint*.

[3] N. Robertson and P. Seymour, Graph minors. X. Obstructions to tree-decomposition, *J. Combin. Theory Ser. B* **52** (1991), 153–190.

[4] P. Seymour and R. Thomas, Call routing and the ratcatcher, *Combinatorica* **14** (1994), 217–241.

[5] R. Thomas, Tree-decompositions of graphs, *Lecture notes 1996*.

Frédéric Mazoit
LIF - CMI
39 rue Joliot-Curie
13453 Marseille Cedex 13
France
Frederic.Mazoit@lif.univ-mrs.fr

Stéphan Thomassé
LIRMM-Université Montpellier II
161 rue Ada
34392 Montpellier Cedex 5
France
thomasse@lirmm.fr